UNIVERSITY OF BATH • SCIENCE 16-19

Project Director: J. J. Thompson, CBE

ENVIRONMENTAL SCIENCE

KEVIN BYRNE

Nelson

Thomas Nelson and Sons Ltd
Nelson House, Mayfield Road
Walton-on-Thames, Surrey
KT12 5PL UK

Thomas Nelson Australia
102 Dodds Street
South Melbourne
Victoria 3205 Australia

Nelson Canada
1120 Birchmount Road
Scarborough, Ontario
M1K 5G4 Canada

First published by Thomas Nelson and Sons Ltd 1997

I(T)P Thomas Nelson is an International Thomson Publishing Company.

I(T)P is used under licence.

ISBN 0-17-448243-4

NPN 9 8 7 6 5 4 3 2

Printed in China

Acknowledgements

The author and publishers wish to thank the following who have kindly given permission for the use of copyright material:

Blackie (Arthur Rothstein) *p 199 bottom*; Keith Hiscock *p 263*; Hutchison *p 1*; Images Colour Library *pp 200 centre, 226, 235*; Oxford Scientific Films *pp 39, 200 left, 261 top*; Planet Earth Pictures *pp 57 left and right, 76, 199 top, 248, 261 bottom, 266*; Rex Features *pp 15, 44, 66, 73, 143, 150, 172, 186, 218, 269*; Science Photo Library: *pp 36* (David Parker), *38* (Alfred Pasieka), *60 and 79* (Adam Hart-Davis), *179* (NASA), *259* (Martin Dohrn); Spectrum Colour Library *p 135*; Superstock *p 214*; Tony Stone *pp 93, 161, 169, 176, 200 right, 201.*

Picture research by Image Select International Ltd.

Cambridge University Press *Figure 1.10, page 10*, from C. J. Barrow, *Land degradation*, 1994, and *diagram on page 159 (question 12.7)* from J. F. Allen *Energy Resources for a Changing World*, 1992; English Nature stimulus material and diagrams in *Question 10.5, page 126* adapted from *Data support for Education*; World Wide Fund for Nature UK, Data Support for Education Service, *diagram in question 13.4, page 186*; Addison Wesley Longman *Table 8.1, page 84* from J. Tivy *Agricultural Ecology*, 1990 and *Figure 10.7, page 117* from A. Wellburn, *Air Pollution and Climate Change*, reprinted by permission of Addison Wesley Longman; Hodder and Stoughton Educational *Figure 7.4, page 77* and *Figure 7.11, page 81* from P. Freeland, *Habitat and the Environment*, 1991; *The Guardian* for an *extract from 'Green shoots of recovery', page 258*, and *'National Trust gets tough', page 267 The Guardian*; *The Independent* extracts from *'Raiders of the lost bark' 'On yer bike'* and *'RSPB want new marine protection'*; HarperCollins Publishers Ltd. *Figure 2.10, page 27, Figure 2.12, page 29* and *Figure 4.9, page 53*, from D. Money, *Climate and Environmental Systems*, 1992, for *Figure 2.8, page 25* and *Figure 2.9, page 26* from B. Knapp *Systematic Geography*, 1992, for *Figure 2.17, page 31* from Pollard, Hooper and Moore *Hedges*, 1974 and for *Figure 7.10, page 80* and *Figure 8.6, page 89* from Packham and Harding, *Ecology of Woodland Processes*, Edward Arnold Publishers, 1982;
The Controller of Her Majesty's Stationery Office *Figure 2.15, page 19; Table 3.1, page 29; Figure 5.3, page 55; Figure 5.20, page 68; Figure 5.21, page 68; Figure 5.25, page 70; Figure 15.12, page 218, from The UK Environment*, 1992. Crown copyright is reproduced with the permission of the Controller of Her Majesty's Stationery Office; Blackwell Scientific *Figure 1.6, page 5* and *Figure 1.9, page 7* from J. Andrews et al. *An Introduction to Environmental Chemistry*, 1991; The Forestry Industry Committee of Great Britain for *Figure 13.5, page 178* and *Figure 13.11, page 181*, from *The Forestry Industry Yearbook, 1993;* The Environment Press *for Figure 9.12, page 106;*
The Centre for Alternative Technology for *Figure 12.8* and *Figure 12.9, page 164* from *Where the Wind Blows, 1993.*

The authors and publishers are grateful to the following examination organisations for permission to reproduce examination questions:

The Associated Examining Board
Edexcel Foundation, London Examinations
Northern Examinations and Assessment Board
Southern Examining Group

Answers are the sole responsibility of the author and have not been provided or approved by the organisations. The organisations accept no responsibility whatsoever for the accuracy or method of working in the answers given.

Every effort has been made to trace all the copyright holders, but if any have been overlooked the publishers will be pleased to make the necessary arrangement at the first opportunity.

Contents

Theme 2: Human impact

The Project: an introduction

The University of Bath Science 16–19 Project grew out of reappraisal of how far sixth form science had travelled during a period of unprecedented curriculum reform, and an attempt to evaluate future development. Changes were occurring both within the constitution of 16–19 syllabuses themselves and as a result of external pressures from 16+ and below: syllabus redefinition (starting with the common cores); the introduction of AS-level and its academic recognition; the originally optimistic outcome to the Higginson enquiry; new emphasis on skills and processes; and the balance of continuous and final assessment at GCSE level.

This activity offered fertile ground for the School of Education at the University of Bath to join forces with a team of science teachers, drawn from a wide spectrum of education experience, to create a flexible curriculum model and then develop resources to fit it. The group addressed the task of satisfying those requirements:

- the new syllabus and examination demands of A- and AS-level courses;
- the provision of new materials suitable for both the core and options parts of the syllabuses;
- the striking of an appropriate balance of opportunities for students to acquire knowledge and understanding, to develop skills and concepts, and to appreciate the applications and implications of science;
- the encouragement of a degree of independent learning through highly interactive texts;
- the satisfaction of the needs of a wide ability range of students at this level.

Some of these objectives were easier to achieve than others. Relationships to still evolving syllabuses demand the most rigorous analysis and a sense of vision – and optimism – regarding their eventual destination. Original assumptions about AS-level, for example, as a distinct though complementary sibling to A-level, needed to be revised.

The Project, though, always regarded itself as more than a provider of materials, important as this is, and concerned itself equally with the process of provision – how material can best be written and shaped to meet the requirements of the educational market-place. This aim found expression in two principal forms: the idea of secondment at the University and the extensive trialling of early material in schools and colleges.

Most authors enjoyed a period of secondment from teaching, which allowed them not only to reflect and write more strategically (and, particularly so, in a supportive academic environment) but, equally, to engage with each other in wrestling with the issues in question.

The Project saw in the trialling a crucial test for the acceptance of its ideas and their execution. Over one hundred institutions and one thousand students participated, and responses were invited from teachers and pupils alike. The reactions generally confirmed the soundness of the model and allowed for more scrupulous textual housekeeping, as details of confusion, ambiguity or plain misunderstanding were revised and reordered.

The test of all teaching must be in the quality of the learning, and the proof of these resources will be in the understanding and ease of accessibility that they generate. The Project, ultimately, is both a collection of materials and a message of faith in the science curriculum of the future.

J. J. Thompson

How to use this book

Earth has existed for 4.65 billion years. Humans appeared 350 000 years ago. If we condense all of earth's history into one hour, then humans appeared approximately four and a half minutes ago. In those 'few minutes', humans have had a major effect on the environmental systems upon which we and millions of other plant and animal species depend. We frequently hear in the media that humans are changing climates, destroying habitats, eroding the soil and polluting waters. Many people think that if we carry on like this, the prospects for future life on earth look grim. This concern is leading, in turn, to increased environmental awareness and an understanding that we must look after our planet in order to preserve the living organisms that live on it.

A very exciting thing about environmental science is learning about how components of the environment are inter-connected. Drinking a can of cola can be argued to have global effects; tin and aluminium mines cause air and water pollution, soil erosion and habitat destruction. Hundreds of gallons of water, which may be in short supply, are used in the manufacture of every can. Paints, additives and wasted heat pollute water supplies, harming plants and animals. Energy is used at every stage, from the initial mining to the distribution of the cans to the shop – all of this increases the chances of air and water pollution and contributes to global climatic change. Just by considering the example of a canned drink, we touch upon every chapter of this book. It is precisely because everything is connected, that each one of us can do so much to help change things for the better. We hope that this book will help you to understand some of the connections.

Before we can appreciate what we are doing wrong, we need to understand the key scientific and ecological principles upon which life on earth is based. These are covered in Theme 1 of this book. Once we have grasped these basic facts and principles, we can begin to appreciate why so many of our activities could be harmful and we can look at ways in which these effects can be reduced. Human impact on the environment is the subject of Theme 2.

You may wish to turn straight to the issues covered in Theme 2. These chapters assume a knowledge of the basic scientific facts and principles but, wherever possible, cross-references between chapters are given to help you find relevant earlier information if you need it. Using the index will help you find the appropriate pages within the chapter. Alternatively, you may prefer to work your way through the basic principles before exploring the connections to human activity in Theme 2.

I would like to thank Richard Plant, Cath Brown, Jim Sharpe, Margaret Dobson and Ros Brownsword for contributing to the material, and particularly Jim for great assistance in editing. Finally, I would like to thank my students at Abbey, who acted as willing but critical guinea pigs for much of the material contained here.

Learning objectives

These are given at the beginning of each chapter, and they outline what you should gain from the chapter. They are statements of attainment and often link closely to statements in a course syllabus. Learning objectives can help you to make notes for revision as well as check your progress.

Questions

In-text questions occur at points when you should consolidate what you have just learned, or prepare for what is to follow by thinking along the lines required by the question. Some of these are actual examination questions. Some questions can be answered from the material covered in the previous section, while others may require additional thought or information. Suggested answers to most in-text questions are given in Appendix D.

Analysis

Where you are asked to think about a particular idea, or to develop an idea further, you will find text and questions presented together as an analysis box. Sometimes these will require you to refer to other resources. The answers to these questions are given in Appendix C.

Summary assignments

Each chapter ends with a summary assignment. These help to check your understanding of the learning objectives, and so allow you to check your own progress. Some of these are examination questions. The answers are given in Appendix E.

Case studies

We have included a number of case studies at appropriate points in the text. These draw on the topic that you have just been studying, and are intended to show you how an understanding of environmental science can be applied to solving a variety of environmental problems.

Glossary

An extensive glossary defining key terms used in *Environmental Science* is provided in Appendix A.

Investigations

Suggestions for practical work associated with each chapter are summarised in Appendix B.

Other resources

At some points, we have assumed that you will have access to other books, leaflets, CD-ROMs and videos. You should always look out for recent articles in this field in periodicals, as well as in newspapers and magazines.

Theme **1**

ECOLOGICAL PRINCIPLES

Environmental science involves the study of how living things interact with each other and with their environment.

To help us understand how people can live sustainably on planet earth, we will look at the dynamics of our environment, including energy flow, food sources and soil studies. In Theme 1 we will put together knowledge from several scientific disciplines such as biology, chemistry, geology and hydrology, to explain basic ecological principles.

Chapter 1

THE EARTH'S ATMOSPHERE

There are nine major planets in our solar system and there may be billions of star and planet systems but, as far as we know, ours is the only planet that supports life. For that we can largely thank the sun which, at 150 million kilometres away, provides just the right amount of light and heat for life to exist (Figure 1.1). A better understanding of the conditions which allow life to exist on earth should help us to take better care of our planet.

LEARNING OBJECTIVES

After completing the work in this chapter you will be able to:

1. outline the importance of the sun for life on earth
2. describe the forms of radiation which the sun emits
3. explain the biological and environmental importance of these different forms of radiation
4. describe and explain the effect of the atmosphere on incoming and outgoing radiation
5. describe the harmful consequences of human alteration of the composition of the atmosphere.

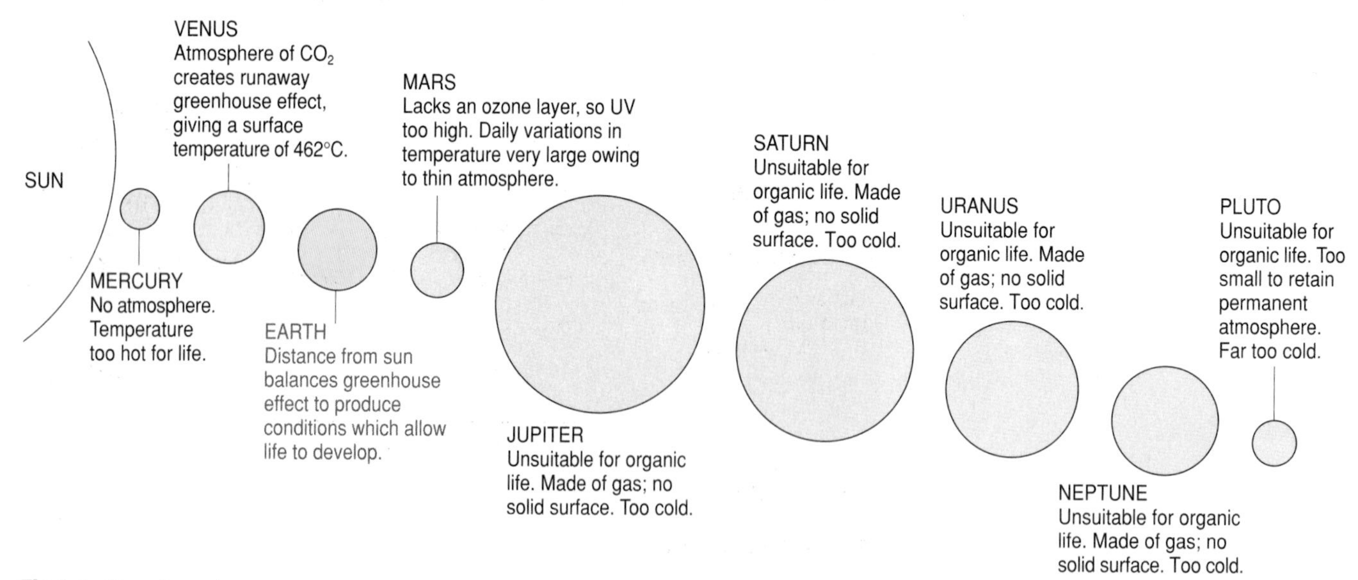

Fig 1.1 The solar system.

The main organisms that can use the sun's energy directly are green plants, cyanophytes and specialised bacteria. Green plants convert solar energy into chemical energy (Chapter 5) through the process of photosynthesis, and animals gain their energy by eating plants or by eating other animals which have fed on plants (Chapter 6). Besides providing the basis of most food chains, the sun's energy drives the water and nutrient cycles and is the basis of many of the energy sources upon which humans depend for life, including the fossil fuels (coal, oil, gas, peat and shale), wood, wind and waves (Figure 1.2).

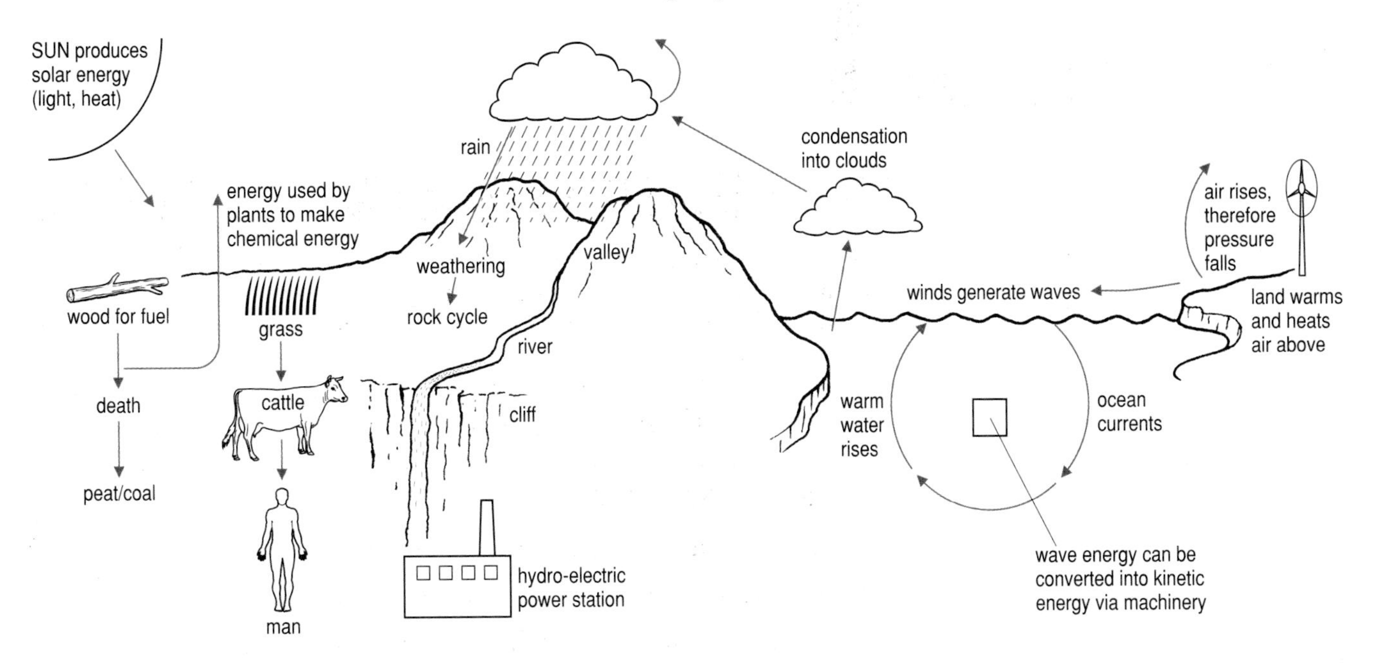

Fig 1.2 The sun powers most of the energy cycles on earth.

1.1 THE SUN AS A SOURCE OF ENERGY

The sun is an average yellow star that has been radiating energy for nearly five billion years. It consists mainly of hydrogen (70%) and helium (28%). The sun converts 508 million tonnes of hydrogen atoms into helium atoms every second. This joining together of hydrogen atoms to form helium is known as **nuclear fusion** (Figure 1.3). During the fusion process, which occurs at 15 million °C, a huge amount of energy is released (3.7×10^{26} J s^{-1}) and it is some of this energy that heats and lights the earth.

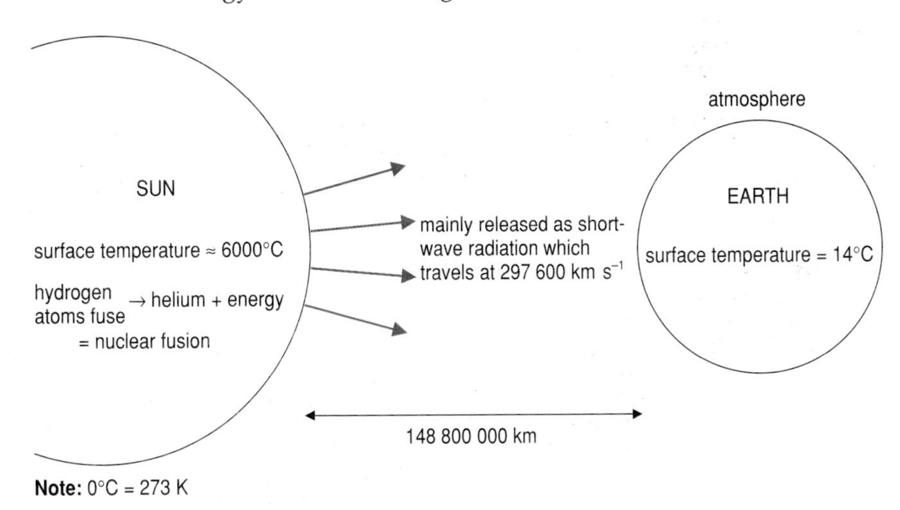

Note: 0°C = 273 K

Fig 1.3 Nuclear fusion.

This energy is in the form of electromagnetic radiation (emr) which travels as waves and can be described in terms of its wavelength and frequency. The wavelengths of different electromagnetic radiation vary enormously. X-rays, for example, have a wavelength of 1×10^{-9} metres, which is a thousand millionth of a metre, while radiowaves may have a wavelength of a million metres. **Frequency**, which is inversely proportional to wavelength, refers to the number of waves in a second. The energy of a particular form of electromagnetic radiation is proportional to its frequency. The different types of electromagnetic radiation which are emitted from the sun can be shown in an **electromagnetic spectrum** (Figure 1.4).

Fig 1.4 Electromagnetic radiation.

Frequency decreases but wavelength increases from left to right, that is, from gamma radiation to radio waves.

It is not just the sun that emits radiation; everything and everybody emits radiation continuously. However, different forms of radiation are not emitted in equal amounts and the relative proportion of the wavelengths emitted is determined by the temperature of the object. The hotter the object, the greater the proportion of short-wave radiation emitted. The sun, for example, which is extremely hot, emits a relatively high proportion of ultra-violet and visible light, whereas most 'hot' objects encountered on earth emit mainly infra-red with some visible light (Figure 1.5). When an electric fire is turned on, heat can be felt before any visible changes have occurred. This is due to infra-red radiation. However, as the temperature of the bars increases, they emit more visible light and begin to glow.

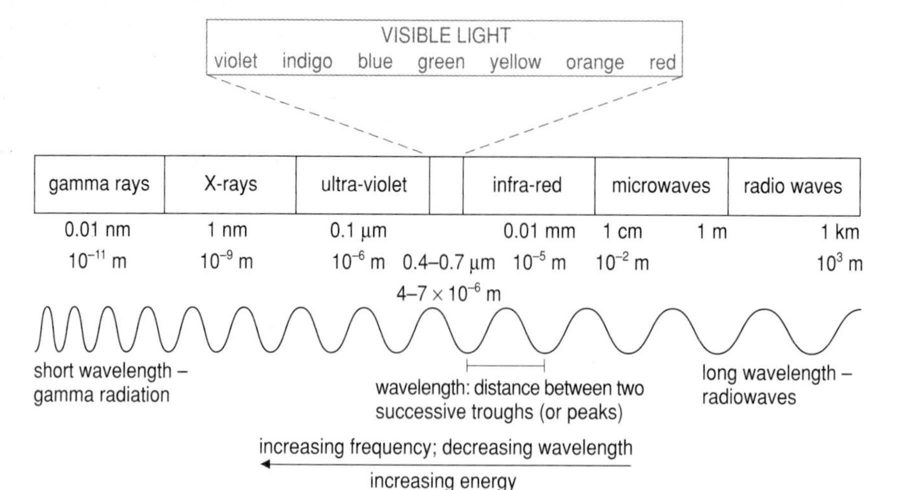

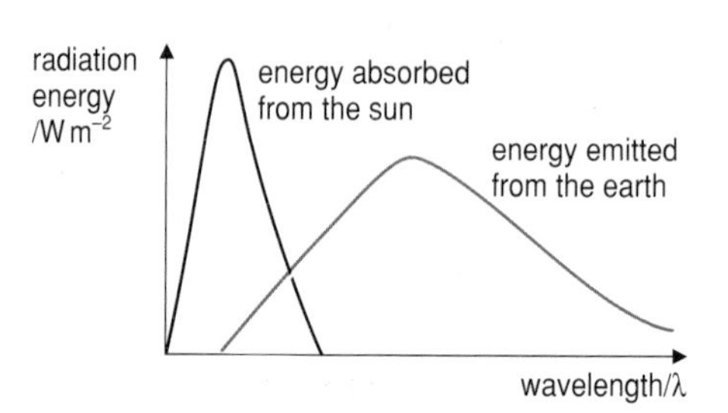

Fig 1.5 Absorption and emission of energy from the earth.

All of these forms of electromagnetic radiation hurtle through the near-vacuum of space at 300 000 km s^{-1} and, after eight minutes, they enter the upper part of the earth's atmosphere (Figure 1.3).

1.2 THE STRUCTURE AND COMPOSITION OF THE ATMOSPHERE

The atmosphere is a relatively thin layer of gases and suspended particles surrounding the earth. It is like the skin around an apple, and just as the skin protects the apple, so our atmosphere protects and makes possible all life on earth. The lower atmosphere, up to an altitude of about 80 km, is mostly made up of nitrogen and oxygen, with ten other gases present in much smaller quantities (Table 1.1).

Table 1.1 The composition of the lower atmosphere

Gas	% by volume	Parts per million (ppm)	Formula	Biological and environmental importance
nitrogen	78.08	780 840.0	N_2	biologically inert, thus atmospheric concentration remains stable; needed to make proteins and DNA
oxygen	20.95	209 460.0	O_2	needed for aerobic respiration
argon	0.93	9 340.0	Ar	inert
carbon dioxide	0.035	350.0	CO_2	major contributor to the 'greenhouse effect' but essential for photosynthesis
neon	0.001 8	18.0	Ne	inert
helium	0.000 52	5.2	He	inert
methane	0.000 14	1.4	CH_4	major contributor to the 'greenhouse effect' and acid rain
krypton	0.000 10	1.0	Kr	inert
nitrous oxide	0.000 05	0.5	N_2O	major contributor to the 'greenhouse effect'
hydrogen	0.000 05	0.5	H_2	none
xenon	0.000 009	0.09	Xe	inert
ozone	0.000 007	0.07	O_3	stratospheric ozone filters out UV radiation, tropospheric ozone is a health hazard

Above 80 km, the proportion of heavier gases decreases rapidly, and above 150 km, the main atmospheric gas is oxygen in its atomic (O) rather than its molecular (O_2) state. This is a result of molecular oxygen being broken down by ultra-violet radiation in a process called photodissociation, an important process which we shall examine later.

In studying the atmosphere, it is convenient to subdivide it into a number of layers based upon the relationship between altitude and change of temperature (Figure 1.6). This is an important diagram and one which you should study carefully.

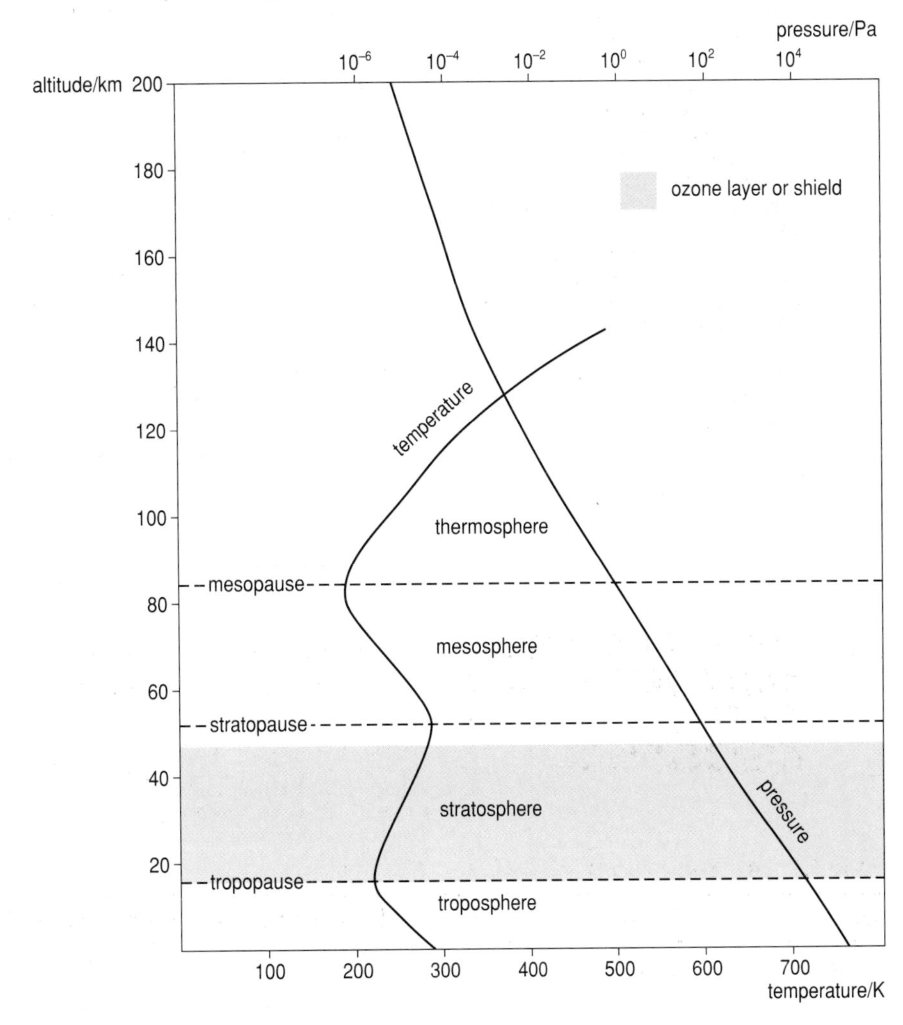

Fig 1.6 Vertical structure of the atmosphere.

The atmosphere can be divided into layers known as the troposphere, the stratosphere, the mesosphere and the thermosphere. Imagine that you are standing on the ground and are about to be whisked up through the atmosphere...

- As you move up through the troposphere it will get colder. This is because the troposphere is heated by the radiation emitted from the ground (long-wave radiation). The higher you go, that is, the further you get from the source of heat, the cooler it will be. By the time you get to the top of the troposphere (12 000 metres up) you are at –60°C. The troposphere is the region of the atmosphere that contains the weather. Above it, the stratosphere is continuously calm and sunny. Within the troposphere both temperature and pressure decrease with height.
- As you move up through the stratosphere the temperature will increase because this layer contains a thin band of ozone (the ozone layer) which,

along with water vapour, absorbs ultra-violet radiation from the sun. The stratosphere is heated from above, so the higher you go, the hotter it will be. The fact that it is warmer than the troposphere means that the stratosphere is relatively stable. The heavier, colder air below it has no tendency to rise and, in effect, the stratosphere acts as a 'lid' on the turbulent troposphere.

- Within the mesosphere temperature decreases with altitude, as in the troposphere. This is because there are no gases, particles or water vapour to absorb ultra-violet radiation. Within the mesosphere temperatures drop to their lowest value.
- Higher up still, into the thermosphere, temperatures increase once more.

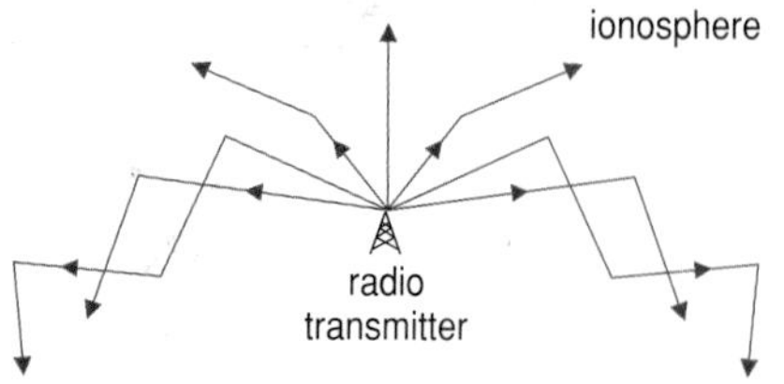

Fig 1.7 The range of radio transmissions is greatly increased by radio waves being reflected by the ionosphere.

The upper mesosphere and thermosphere (above 80 km) are called the ionosphere, since this layer contains many **ions** (electrically charged particles), which can absorb ultra-violet radiation. These ions are also important since they give rise to the northern and southern lights and because they are capable of reflecting radio signals which can then bounce back and forth between the earth's surface and the ionosphere, eventually allowing radio signals to travel all the way around the planet (Figure 1.7). There is no clearly defined boundary which marks the end of the atmosphere; gases are continuously lost from the atmosphere into space, which is made up mainly of helium and hydrogen.

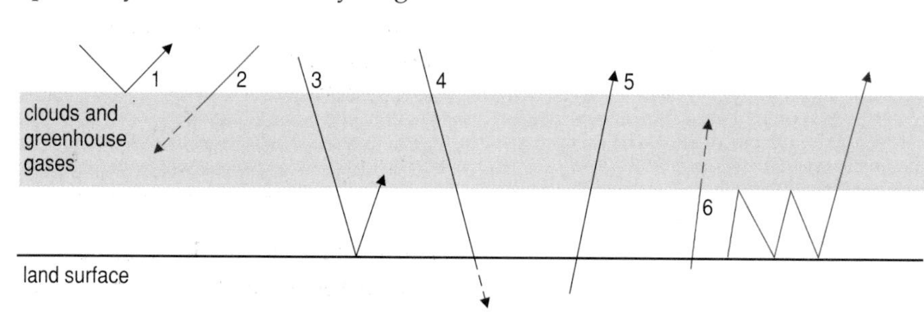

1 Some of the incoming radiation is reflected back into space from clouds and the atmosphere.
2 Some of the incoming radiation is absorbed by the gases in the atmosphere. This warms the atmosphere.
3 Some of the incoming radiation is reflected back from the earth's surface.
4 Some of the incoming radiation is absorbed by the earth's surface.
5 As the earth's surface and atmosphere are warmed by the absorption of incoming radiation, they begin to reradiate heat, some of which escapes to space.
6 Some of the outgoing radiation emitted from the earth's surface is absorbed, scattered or reflected by the clouds and gases in the atmosphere, delaying the release of energy. This trapping in of the heat in the atmosphere is called the greenhouse effect, and the gases which are responsible are called greenhouse gases.

Fig 1.8 The effect of the atmosphere on incoming and outgoing radiation.

1.3 HUMAN EFFECTS ON THE ATMOSPHERE

The enhanced greenhouse effect

Many people regard the greenhouse effect as the most serious environmental threat to our present way of life on earth. The greenhouse effect is the natural 'trapping in' of heat by gases such as carbon dioxide and methane in the atmosphere. By allowing short-wave radiation in, but trapping some of the long-wave radiation which is trying to get out, these gases act in a similar way to the glass in a greenhouse, hence the term greenhouse effect.

Without the greenhouse effect the average temperature on earth would be about –17°C and life on earth as we know it would be impossible. The problem is that over the last few hundred years human activities have resulted in rising concentrations of many of the greenhouse gases. This has led to an increased or 'enhanced' greenhouse effect which, in turn, has led to an increased average global temperature. We now know that the mean global temperature has risen by 0.3–0.6°C over the last one hundred years. Some scientists have estimated that if we do nothing to reduce greenhouse gas emissions, the average global temperature will rise by 0.3°C every decade. However, temperatures will not necessarily rise everywhere. Some areas of the world may become cooler, and we simply do not know what overall effect this will have on regional and world climate. Possible effects will be discussed but first we need to look at why the concentrations of the greenhouse gases are rising.

The most important greenhouse gases are shown in Table 1.2.

Table 1.2 The greenhouse gases

Greenhouse gas	Atmospheric concentration / ppmv	Rate of increase /% per annum	Global warming potential	Lifetime /years
carbon dioxide	356	0.5	1	120
methane	1.72	0.6 – 0.75	21	10.5
nitrous oxide	0.31	0.2 – 0.3	206	132
CFC-11	0.000 255	4	3400	55
CFC-12	0.000 453	4	7100	116
carbon monoxide	small and very variable	small and very variable	small and very variable	months
NMHC (non-methane hydrocarbons)	small and very variable	small and very variable	small and very variable	days/ months
NO_x (NO_2, NO)	small and very variable	small and very variable	very variable	days

Water vapour is, in fact, the most effective greenhouse gas but its concentration in the atmosphere is not directly affected by human activities.

Carbon dioxide (CO_2)

Anthropogenic (human) release of carbon dioxide contributes most to the enhanced greenhouse effect. All fossil fuels, such as coal and oil, contain carbon; burning them releases carbon dioxide. As world demand for energy increases, so more fossil fuels are being burnt, hence CO_2 levels are rising (Chapter 12).

All green plants photosynthesise. They remove carbon dioxide from the atmosphere and use sunlight energy to convert it into carbohydrates. Thus they provide the food upon which all other living organisms directly or indirectly depend (Chapter 5). Humans are destroying huge areas of natural vegetation such as the tropical rainforests, hence global photosynthesis is declining, leaving more carbon dioxide in the atmosphere. This problem is exacerbated if we burn destroyed vegetation, as combustion releases more CO_2.

By analysing air bubbles trapped in ice, scientists have deduced how atmospheric carbon dioxide concentrations have changed over the last few thousand years. There have been many fluctuations, with low concentrations corresponding with Ice Ages and high concentrations corresponding with warmer interglacial periods. However, it is important to note that most of the carbon in the biosphere is not in the atmosphere as carbon dioxide, but is locked up in vegetation, in soil and in the oceans (Figure 1.9). These are called **carbon sinks** and their relative importance is not clear. In other words, we simply do not know for sure where all the carbon is.

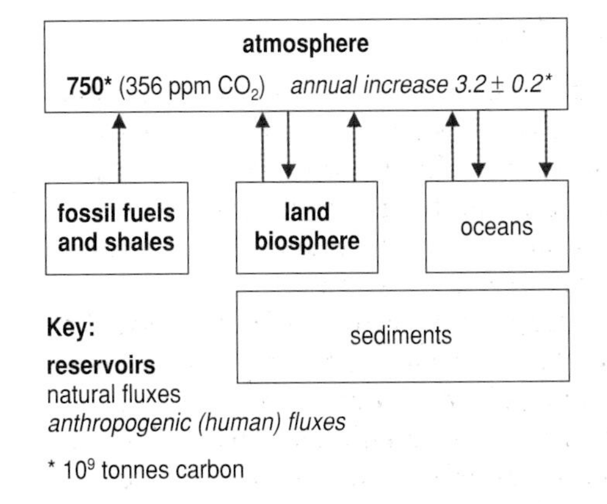

Fig 1.9 The global carbon cycle.

Chlorofluorocarbons (CFCs)

CFCs are used as coolants in refrigerators, as propellants in aerosols and as expanders in expanded foam products. Their increasing concentration in the atmosphere over the last two decades reflects their increased use. The Montreal Protocol has led to a huge decline in the production of CFCs but, unfortunately, those CFCs that have already been released into the atmosphere will stay there for 40 to 80 years. They have a long residence time.

Methane (CH_4)

Methane is produced by anaerobic bacteria which live in marshes, landfill sites, rice paddies and in the guts of ruminants (for example, cattle, sheep, camels). A cow may release a phenomenal 200 litres of methane a day! Atmospheric methane concentrations have risen as the ruminant population and the area given over to landfill have significantly increased. Leaking gas pipes and coal mines also release methane.

Nitrous oxide (N_2O)

Nitrous oxide is released during fossil fuel combustion and from denitrifying bacteria which act on nitrates and nitrites. Increasing use of nitrate fertilisers and increasing cultivation of soils may result in increasing concentrations of nitrous oxide.

Tropospheric ozone (O_3)

Tropospheric ozone (O_3) is produced through a complex series of reactions involving pollutants from car exhausts, such as nitrous oxide, and hydrocarbons, which react with sunlight. Thus more vehicle-use means a greater greenhouse effect.

ANALYSIS

The case of the missing CO_2

The concentration of CO_2 in the atmosphere has risen from a fairly constant 280 ppm in the pre-industrial period before 1700, to 355 ppm (parts per million) today. Scientists have long known that this is a consequence of the increased burning of fossil fuels which all contain carbon, but they have been puzzled, as the known rates of fossil fuel combustion should have led to much higher CO_2 concentrations in the atmosphere, even when taking into account increased absorption by the sea.

For every year since 1890 scientists have been able to calculate and compare the amount of CO_2 released by fossil fuels and the amount measured in the atmosphere. These figures were used to calculate how much should have been absorbed by the oceans each year. Between 1890 and 1940 the oceans absorbed all of the CO_2 produced by fossil fuels that did not remain in the atmosphere. However, between 1940 and the early 1980s, some 35 billion tonnes of carbon went missing. It did not stay in the atmosphere nor was it absorbed by the sea. Where did the missing carbon go? Some scientists now believe that a great unacknowledged carbon sink is the forests of the northern hemisphere, representing a quarter of the world's forests. Certainly, there is good evidence that the missing carbon went north of the equator. While 95% of CO_2 from fossil fuels is released in the northern hemisphere, the difference in atmospheric CO_2 levels between the northern and southern hemispheres is only 3 ppm. This is a very small difference and not one which can be explained by atmospheric mixing. Again, the carbon sink could account for the missing carbon.

To help verify this, plant ecologists measured the widths of annual rings of hundreds of trees in the boreal forests and so obtained a record of over a century's growth. This data confirmed that trees in the boreal forests have been growing faster than usual in the last one hundred years. The extra growth may account for the missing carbon.

There are two reasons why the trees may have been growing faster. The extra CO_2 from fossil-fuel combustion might have stimulated faster growth – low levels of CO_2 often act as a limiting factor in photosynthesis. However, to build tissues, the trees would have needed other nutrients, particularly nitrogen and phosphorus. Ironically, pollution from fossil fuels may have provided the boreal forests with these extra nutrients. Secondly, higher temperatures in the north may have caused the forest to grow faster.

This is not the whole story, however. The tree-ring data showed that the boreal forests have acted alternately as carbon sinks and sources. Before 1890, mainly because of forest fires and tree felling, the boreal forests were a net source of CO_2. Deforestation released more CO_2 mainly because debris from the forest floor rots more quickly when the trees are cleared. After 1920, a steep increase in the tree growth turned forests from a source into a sink.

In the late 1970s the boreal forests again seemed to have become a net source of CO_2. Massive deforestation, forest fires and the increased activity of pests have been blamed. Forest biologists believe that forest fires and pest activity have increased because of warmer temperatures. Thus the increase in temperature and decrease in the volume of wood in northern forests may be reinforcing each other in a vicious circle. Increasing levels of atmospheric CO_2 could be causing an increase in temperature that, in turn, helps to destroy trees. This releases more CO_2, causing even higher temperatures, and so on; in other words, creating a cycle of positive feedback.

Source: adapted from D. Mackenzie, 'Where has all the carbon gone?' *New Scientist* January 1994 (No. 1907) © *New Scientist*

Using the information in the passage and your own knowledge, answer the following questions.

1. What is the precise cause of the long-term increase in CO_2 levels?

2. Explain the term 'carbon sink'.

3. What evidence is given in the passage that the missing carbon dioxide ended up in the northern hemisphere?

4. How can tree-ring data be used to provide evidence of changing CO_2 concentrations?

5. What does the passage suggest may be limiting factors to tree growth?

6. Why is it 'ironic' that pollution from fossil fuels may have caused tree growth to increase?

7. What is meant by the phrase '... before 1890, the boreal forests were a net source of CO_2'?

8. Suggest why deforestation increases the rate of organic matter decomposition.

9. Why may rising temperature and rising CO_2 concentrations be an example of positive feedback?

Consequences of the enhanced greenhouse effect

Since climate determines the type of vegetation that develops in any area, global warming will result in a change in the regional distribution of vegetation, especially towards the poles where the greatest increase in temperature is expected. For example, the boreal forests of Asia, Scandinavia and Canada may push further northwards, replacing the tundra, while the southern edge of the forests might also move northwards because of increasing competition from deciduous tree species and grasses which find the conditions more favourable. This would obviously have dramatic effects on the wildlife that inhabit these areas.

In the northern latitudes, growing seasons may be extended, increasing the potential for more food to be produced, but over some areas increasing temperatures may lead to longer dry seasons and droughts. This could have serious implications for world food production. The grain belts of the USA and the former USSR may both become much less productive, leading to economic and political implications.

Sea-levels will rise due to glaciers melting and the thermal expansion of water. This may cause serious flooding of low-lying areas, coastal erosion and salination of soils, making them much less fertile. No one is sure by how much the sea-level will rise around Britain, but a frequently quoted estimate for the year 2050 appears to be 0.3 to 0.6 metres. The most vulnerable areas in Britain are in the south-east, where the land is actually sinking relative to the sea. Any rise in sea-level might thus pose a serious threat to property, agriculture, coastal and inland ecosystems, and amenities and recreation.

CASE STUDY

Life in the UK greenhouse

Assuming a global average increase of 1.4°C, government scientists have predicted that:

- average summer temperatures will rise by 1.5°C; winter temperatures by 2°C
- species will move northwards and to higher elevations
- some sensitive species may become extinct, but invasion and migration may mean an increase in the total number of species
- habitats such as estuaries and peatlands may be threatened
- mean sea-level will rise by 20 cm, threatening habitats and property in the UK on the coasts of East Anglia, Lancashire, Yorkshire, Lincolnshire, Essex and Sussex
- soils will dry, altering the types of crops and trees that can be grown in particular areas
- shrinkage of soils will lead to structural damage to buildings
- underground water supplies will shrink, particularly in the south and east.

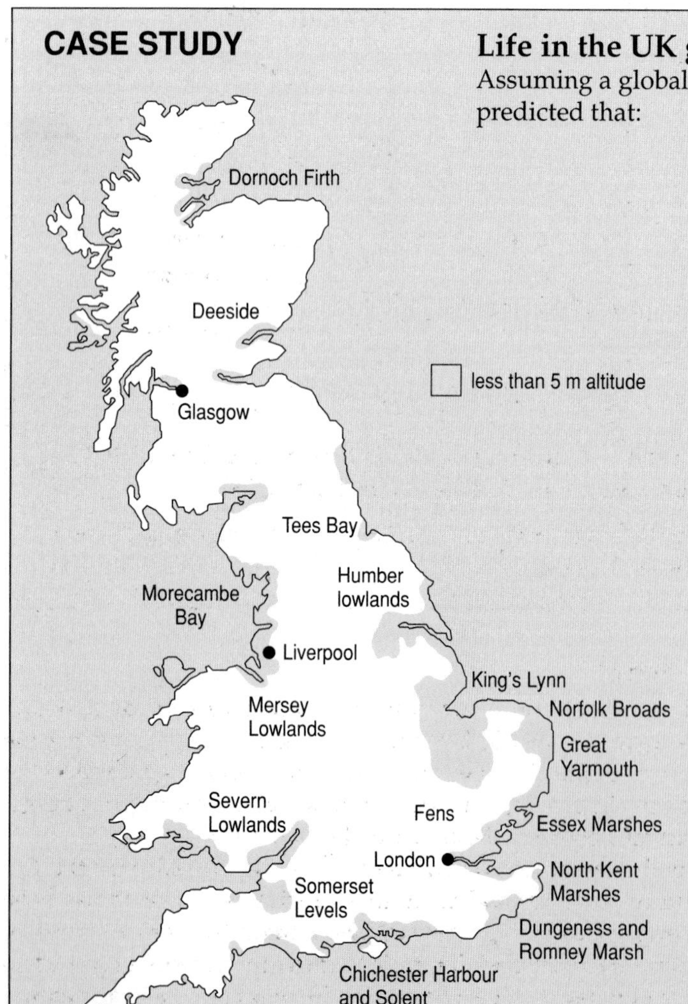

Fig 1.10 The greenhouse effect and land degradation.

Negative and positive feedback

If a system begins to move out of equilibrium, for example if global temperatures start to rise, some processes are initiated which tend to pull the temperature back down to the equilibrium value again. Other mechanisms may begin which actually accelerate the move away from equilibrium, that is, speed up the temperature increase. The first type of mechanism, which tries to move the system back to equilibrium, is called **negative feedback**, while mechanisms which accelerate the move away from equilibrium are called **positive feedback**.

In terms of the greenhouse effect, one possible negative feedback may be that with increasing temperature there will be an increase in evaporation, which will lead to an increase in cloud cover through condensation of the water vapour. Such clouds may reflect incoming radiation, thus reducing the amount which actually reaches the earth's surface, causing a cooling effect.

Positive feedback may also be involved. Higher temperatures will cause greater evaporation, which will lead to a wetter atmosphere. As mentioned earlier, water vapour is a powerful greenhouse gas and will cause temperatures to rise further, which will lead to greater evaporation. We do not know, however, which type of feedback will have the greatest effect. It has proved extremely difficult to build negative and positive feedback components into climate change models, and this is one reason for the uncertainty about possible changes.

QUESTIONS

The graph shows predicted contributions to rising sea-level in the next century if we do nothing to reduce greenhouse gas emissions.

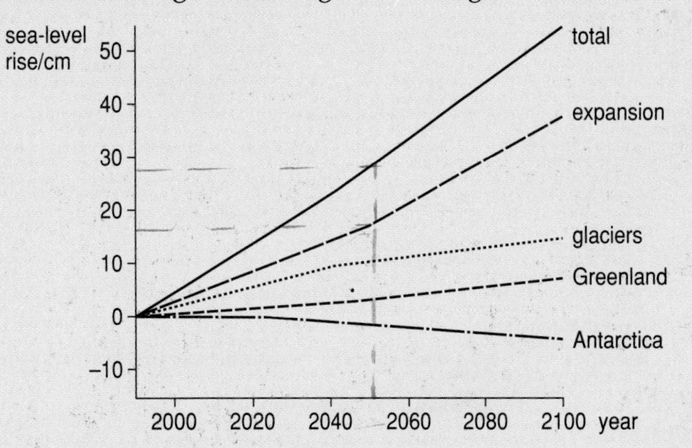

1.1 Calculate the percentage of total sea-level rise which is predicted to occur as a consequence of thermal expansion in 2050.

1.2 Suggest why the effect of global warming in Antarctica might lead to a **decrease** in sea-level.

What can be done?

On a world scale we should reduce emissions of carbon dioxide by reducing our consumption of fossil fuels. This can be achieved by energy conservation measures, improvements in the efficiency of energy generation or transmission, or by increasing our use of renewable sources of energy. Afforestation and reforestation should exceed deforestation.

In the UK, total carbon dioxide emissions have decreased over the last twenty years and the only individual source which has increased has been transport. However, coal-fired power stations remain the major source of carbon dioxide, and industry and domestic homes are the two biggest users of the electricity generated. The use of gas in domestic central heating systems is another large source of emissions. Some of the key strategies for reducing the enhanced greenhouse effect are listed below:

- reduce total energy demand
- increase energy efficiency of power stations
- increase fuel efficiency of vehicles
- raise energy efficiency standards in new or existing buildings
- provide bigger grants for home insulation measures
- increase the total contribution which renewables make to the total energy supply (Chapter 12)
- introduce carbon taxes – effectively, make those fuels which contain most carbon more expensive
- introduce carbon and nitrogen filters in power stations (Chapter 10)
- phase out CFCs and HCFCs as soon as possible
- collect all methane which leaks from landfill sites (Chapter 16)
- reduce deforestation
- encourage afforestation (Chapter 13)
- encourage soil management techniques which reduce soil disturbance (Chapter 14).

A global problem

The enhanced greenhouse effect is a global pollution problem; the decisions which we all make about energy use in our homes or about our personal transport may affect someone's life in Bangladesh, Egypt or East Anglia. The problem has been caused by wealthy countries of the west where total and per capita energy consumption is much higher than in the developing world. The development of a country, the building of roads, cities and industry, requires huge amounts of energy, and in the past this has been based almost entirely upon carbon-emitting fossil fuels. Developing countries, such as India, are understandably keen to catch up with the west, and pressure for such development is increased by rapidly expanding populations. If developing countries choose fossil fuels as their major source of energy, it seems inevitable that global warming will accelerate. The disparity between energy use in the developed and developing worlds and the choices which both face will be discussed in Chapter 12.

QUESTIONS

Some of the greenhouse gases have a much greater potential warming effect than others and can therefore be regarded as a greater environmental problem. To compare greenhouse gases, the concept of Global Warming Potential (GWP) is used. For convenience, the GWP of CO_2 is given as 1. The GWP of some of the other greenhouse gases is given below.

	GWP
carbon dioxide	1
methane	21
CFC-11	3400
nitrous oxide	206
HCFC (possible CFC substitute)	9940

1.3 What factors will affect the GWP of a greenhouse gas?

1.4 What effect might each of the following have on global warming:
(a) government subsidies for public transport
(b) the promotion of incineration of waste rather than dumping in landfill.

1.5 Figure 1.11 shows some of the sources and sinks of methane in the atmosphere. Figure 1.12 compares changes in atmospheric methane levels with changing human population.

	Annual release $t \times 10^6\ yr^{-1}$	Range
Sources		
wetlands (bogs, swamps, tundra, etc.)	115	100–200
termites	20	10–50
ocean	10	5–20
freshwater	5	1–25
CH_4 hydrate	5	0–5
Anthropogenic		
coal mining, natural gas and petroleum industry	100	70–120
rice paddies	60	20–150
enteric fermentation	80	65–100
animal wastes	25	20–30
domestic sewage treatment	25	–
landfills	30	20–70
biomass burning	40	20–80
Sinks		
atmospheric (tropospheric and stratospheric) removal	470	420–520
removal by soils	30	15–45
atmospheric increase	32	28–37

Fig 1.11 Sources and sinks of methane in the atmosphere.

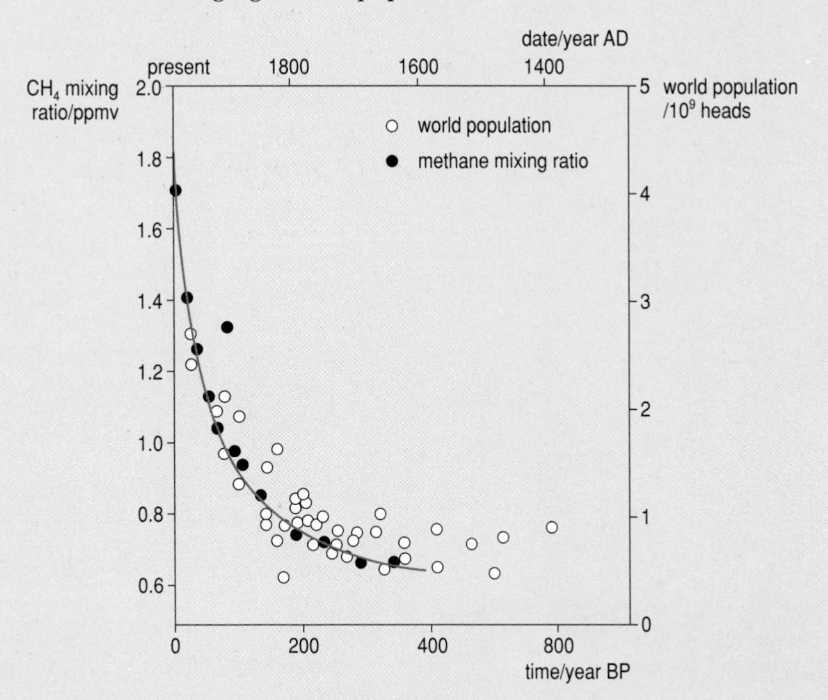

Fig 1.12 Changes in atmospheric methane levels.

Explain possible connections between human population growth and rising concentrations of methane.

CASE STUDY

The 'hole' in the ozone layer

There are two ozone layers, one that is anthropogenic in the troposphere and another, much larger and natural, in the stratosphere (Figure 1.14). When we talk of damage to the ozone layer we are referring to the layer of ozone in the stratosphere. Stratospheric ozone is formed as follows:

- UV radiation splits an oxygen molecule (O_2) into two oxygen atoms (O).

  $O_2 \longrightarrow O + O$

- An oxygen atom then combines with an oxygen molecule to make ozone (O_3).

  $O + O_2 \longrightarrow O_3$

Some stratospheric ozone descends into the troposphere, increasing the concentration there. Stratospheric ozone is continually made and destroyed, and there is usually a balance between the rate of production and the rate of destruction. Humans have upset this natural balance by releasing chemicals which speed up the destruction of ozone (Table 1.3).

The stratospheric ozone layer is essential to life on earth because it absorbs much of the harmful incoming ultra-violet radiation, thus helping to regulate atmospheric temperature and preventing serious biological harm. For this reason, it is sometimes referred to as the ozone shield. Over-exposure to UVB radiation (280–320 nm) can cause:

- mutations in DNA
- increased sunburn and skin cancer
- increased radiation blindness and cataracts
- damage to the human immune system
- decreased photosynthesis and therefore decreased crop growth rates
- disruption of insects which use UV to see or navigate.

Although the chemicals that destroy the ozone layer are also important greenhouse gases, the overall picture is more complicated. Destruction of ozone in the stratosphere actually leads to a cooling of that layer which, in turn, has a cooling effect on the troposphere. Oddly enough then, some of the global warming effect of CFCs is offset by their effect on the ozone layer.

Table 1.3 Anthropogenic ozone destroyers

Chemical	Use	Comments
halogenated hydrocarbons, e.g. chlorofluorocarbons (CFCs)	aerosol propellants, foam expanders, refrigerants	releases chlorine and bromine, which destroy ozone; rapidly being phased out
hydrochlorofluorocarbons (HCFCs)	as replacements for the above	have a shorter lifetime but act more quickly than CFCs
methyl bromide	pesticide	releases bromine, which is forty times more effective at destroying ozone than chlorine is
halons	fire extinguishants	releases bromine
tetrachloroethane	solvent in medicines, pesticides and paints	releases chlorine
1,1,1 trichloroethane	solvent	releases chlorine
methyl chloroform	cleaning solvent	releases chlorine
NO, NO_2, N_2O (NO_x)	bacterial breakdown of nitrates and nitrites in soil and from exhaust gases of supersonic aircraft	N_2O oxidised to NO, which reacts with so reducing ozone levels in the atmosphere

The mechanism by which ozone is destroyed by CFCs is summarised in the equations below:

- $\underset{\text{(a CFC)}}{CFCl_3} + \text{uv (190–220 nm)} \longrightarrow CFCl_2 + Cl$
- $Cl + O_3 \longrightarrow O_2 + ClO$
- $ClO + O \longrightarrow O_2 + Cl$

The chlorine then goes on to destroy more ozone.

CASE STUDY

Dying to get a suntan

Perhaps the most serious fashion victims of our times are sun-worshippers, who are literally risking death in search of a tan. The familiar bronze tan is caused by a dark brown pigment called melanin, produced by specialist skin cells called melanocytes. Exposure to ultra-violet light, particularly one type of ultra-violet called UVB, triggers the melanocytes to produce extra melanin as a way of protecting the sensitive skin cells below. This is the basis of the suntan.

Such a tan does not give complete protection against UVB. Firstly, it takes several days from the initial exposure to UVB for a tan to develop, during which time serious damage can occur to the skin, and secondly, even the darkest tan will not stop all UVB radiation from reaching sensitive skin cells.

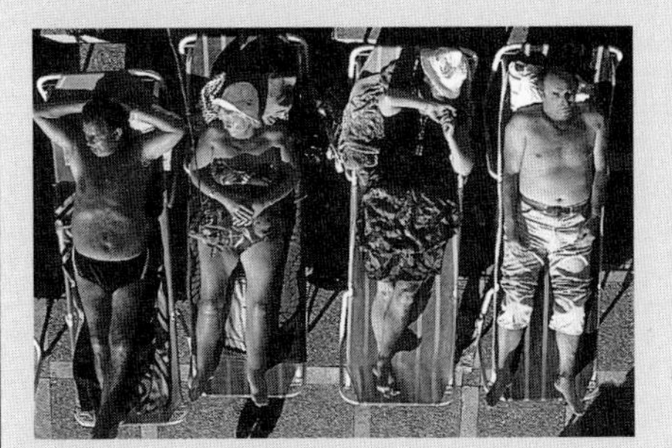

Fig 1.13 Suntan lotion gives variable protection but less than a sunblock.

Excessive exposure to UVB radiation can lead to skin cancer and it is sensible therefore to take serious precautions to try to prevent this. It is very easy to underestimate the amount of UVB to which we are exposed; UVB penetrates clouds, reflects off sand, penetrates through one metre of seawater and is transmitted through wet T-shirts. Sitting under the beach umbrella and taking the occasional dip, even on a cloudy day would not therefore be sensible without protection.

Exposure to the sun should be brief until a semi-protective tan has developed, and even then a sunblock which filters out the UVB and the UVA is recommended.

CASE STUDY

The Antarctic ozone 'hole'

Since the 1950s scientists have measured the thickness of the ozone layer at Halley Bay in the Antarctic. Usually the ozone layer thins during the southern spring (September and October) and recovers in November, but over the last 30 years the minimum thickness has been decreasing and recovery has been taking longer. Why does the ozone hole appear in the first place?

During the long, dark, Antarctic winter, temperatures in the stratosphere may fall to as low as –80°C and all available water turns to ice. This ice provides a surface upon which harmless chlorine compounds are converted to ozone-destroying forms of chlorine. The spring sun in September then activates these dangerous forms which begin to destroy ozone, hence an ozone hole appears in spring. In November, warmer and ozone-rich air from lower latitudes usually enters the Antarctic stratosphere and fills the hole. In recent years the hole has been getting bigger and bigger and taking much longer to fill.

Furthermore, the protective ozone layer is becoming thinner over many other latitudes, and in 1992 scientists discovered high concentrations of chlorine monoxide over the Arctic. Although the Arctic hole is much smaller than the hole over the Antarctic, it is likely to extend outward over the densely populated areas of North America and Europe, thus posing a serious threat to human health.

What can be done?

In 1987, 31 countries signed an agreement – The Montreal Protocol – to protect the ozone layer by agreeing to cut CFC production by 50% by the end of the century. However, many scientists argued that this would not be enough, and in 1989 over 80 countries agreed to stop production of CFCs, halogens and CCl_4 by the year 2000. A fund was established in 1991 to help industry in developing countries switch away from CFCs to ozone-friendly technologies.

However, developing countries have repeatedly complained that the funds are insufficient, hence reduction of CFC-use has been disappointingly slow.

Despite continuing problems in developing countries, the Montreal Protocol has been very successful in reducing CFC-output (Figure 1.15 a and b). In 1992, scientists in Germany perfected the use of hydrocarbons as replacements for CFCs as refrigerants. However, the development of fridges which actually used these alternatives was extremely slow until the international environmental campaigners, Greenpeace, commissioned greenfreeze fridges from a single factory in east Germany. Today, virtually all of the east German domestic market use greenfreeze technology, and the first UK greenfreeze factory opened in February 1996.

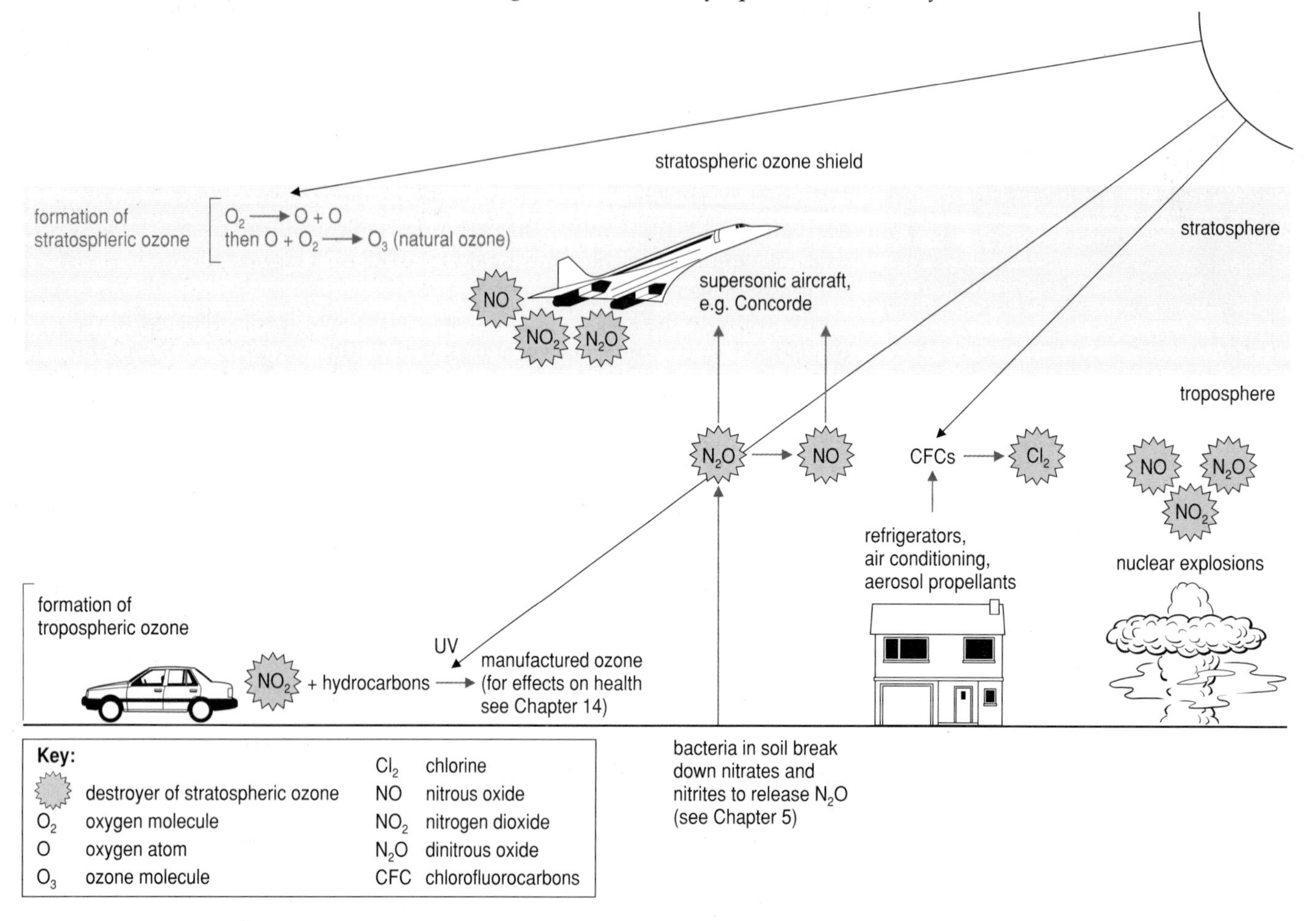

Fig 1.14 Formation and destruction of the atmospheric ozone shield.

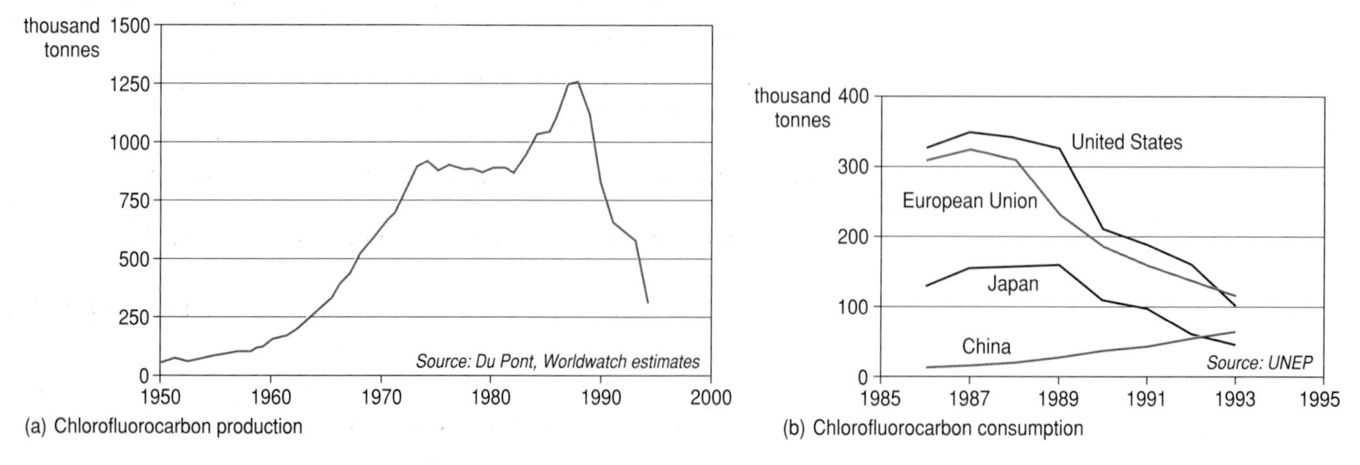

Fig 1.15 Chlorofluorocarbon production and consumption.

SUMMARY ASSIGNMENT

1. Explain the significance of the stratosphere for life on earth.
2. Summarise the effects of air pollutants on the passage of solar radiation through the atmosphere.
3. **(a)** The relationship between altitude, temperature and ozone concentration is shown in graphs A and B.

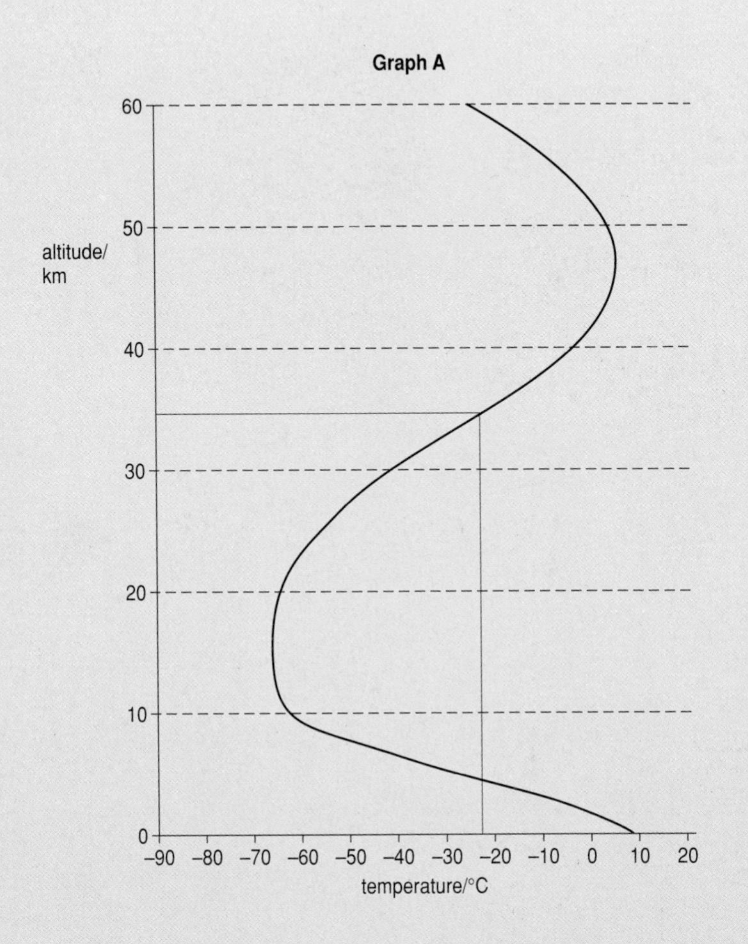

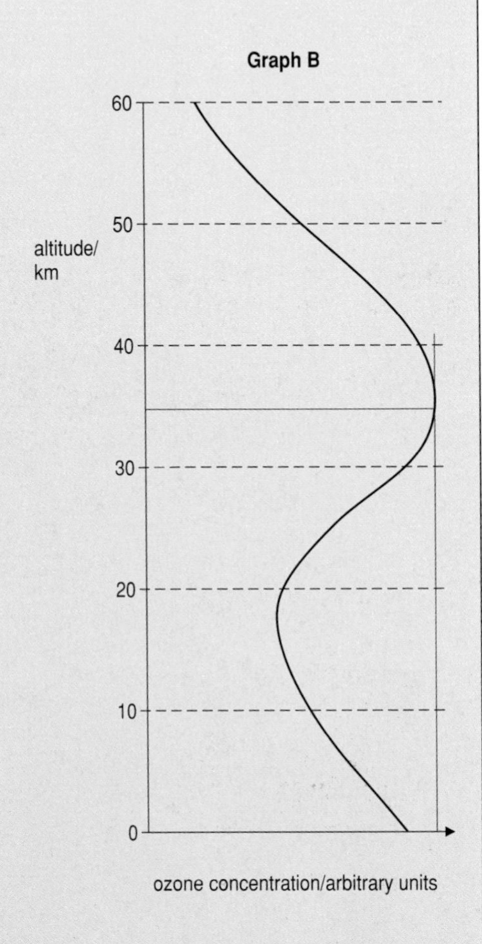

Using the information in the graphs and your knowledge:

(i) describe the relationship between atmospheric temperature and altitude.

(ii) state at what temperature and altitude ozone is most abundant in the atmosphere.

(b) Describe, briefly, the formation of ozone in the stratosphere.

(c) Explain why the ozone layer is important to life on earth.

AEB June 1993

4. The flow diagram shows some of the possible consequences of global warming.

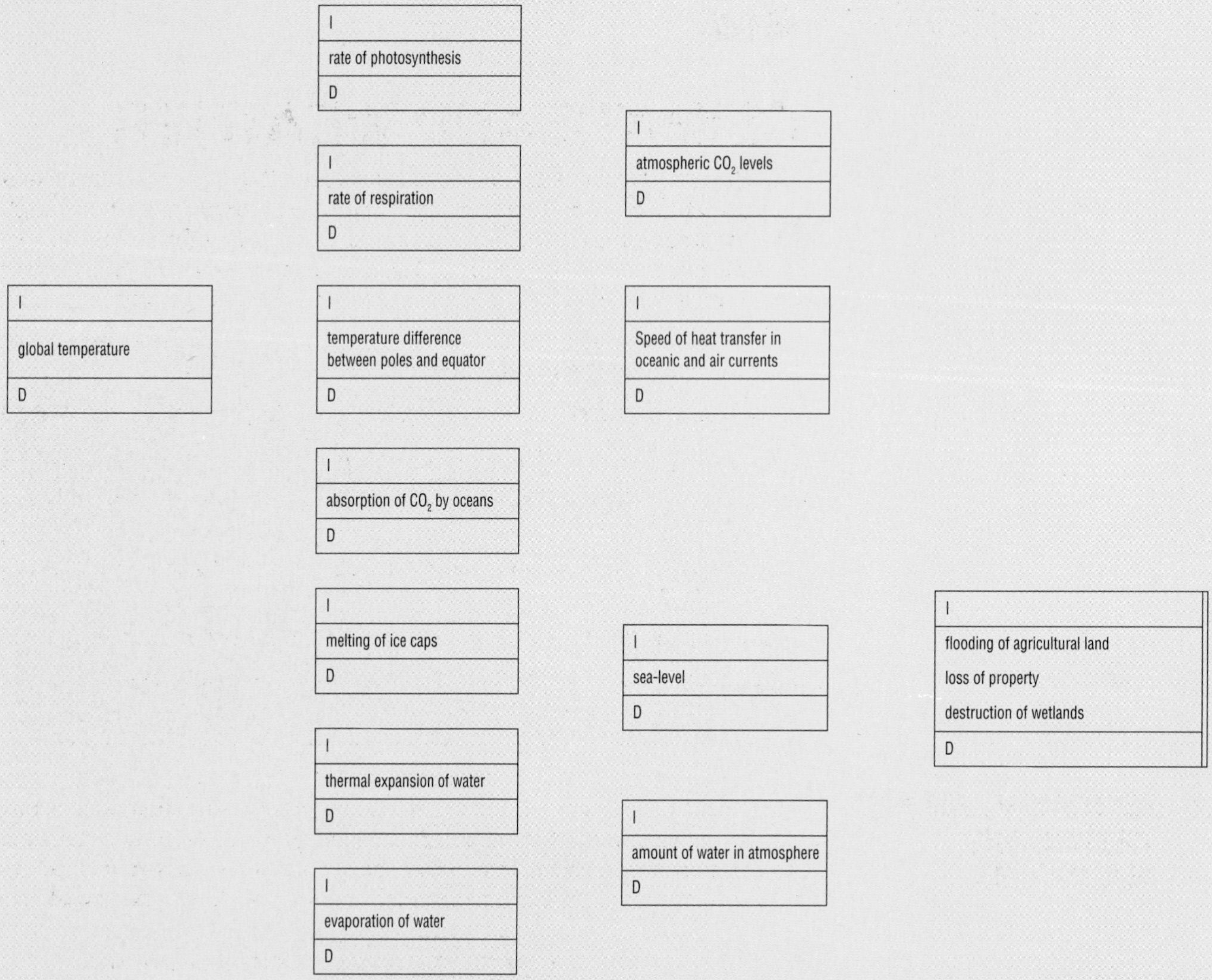

I
rate of photosynthesis
D

I
atmospheric CO_2 levels
D

I
rate of respiration
D

I
global temperature
D

I
temperature difference between poles and equator
D

I
Speed of heat transfer in oceanic and air currents
D

I
absorption of CO_2 by oceans
D

I
melting of ice caps
D

I
sea-level
D

I
flooding of agricultural land
loss of property
destruction of wetlands
D

I
thermal expansion of water
D

I
amount of water in atmosphere
D

I
evaporation of water
D

(a) Complete the diagram by completing or drawing in as many connections as you can. For example, if you think that an increase in global temperature would decrease the rate of photosynthesis, complete the arrow as follows:

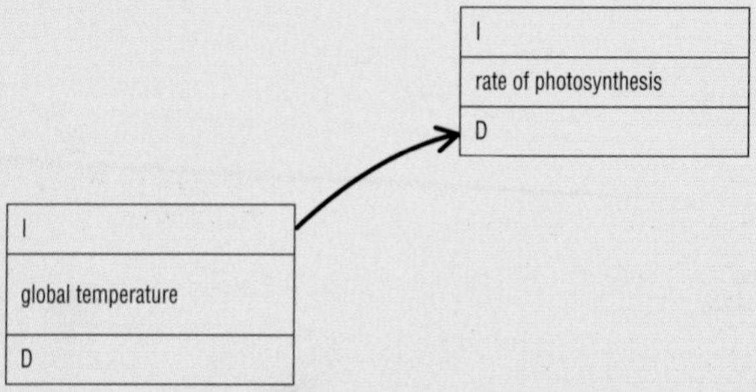

I
rate of photosynthesis
D

I
global temperature
D

(b) Identify as many examples of **positive** and **negative** feedback as you can.

Chapter 2

CLIMATE AND WEATHER

The climate plays a major role in weathering (Chapter 3), soil formation (Chapter 4), the water and mineral cycles (Chapter 8) and, through photosynthesis, the growth of all green plants, including those that humans depend upon for food (Chapter 14). The climate affects almost every environmental system and almost every aspect of human society.

LEARNING OBJECTIVES

After completing the work in this chapter you will be able to:

1. outline the fate of incoming solar radiation
2. explain why temperature varies, both from place to place and with time
3. explain how these temperature variations affect the movement of air and water
4. describe the processes which interact to form the UK climate
5. describe the microclimatic effects of topography, land use, vegetation and urban areas.

2.1 INCOMING AND OUTGOING RADIATION

Insolation is the term used to describe the amount and duration of incoming solar radiation. This radiation provides the energy for life on earth (Chapter 6) and also powers the climatic systems. The sun emits solar radiation, much of which is short-wave, and the cooler earth emits mainly long-wave radiation.

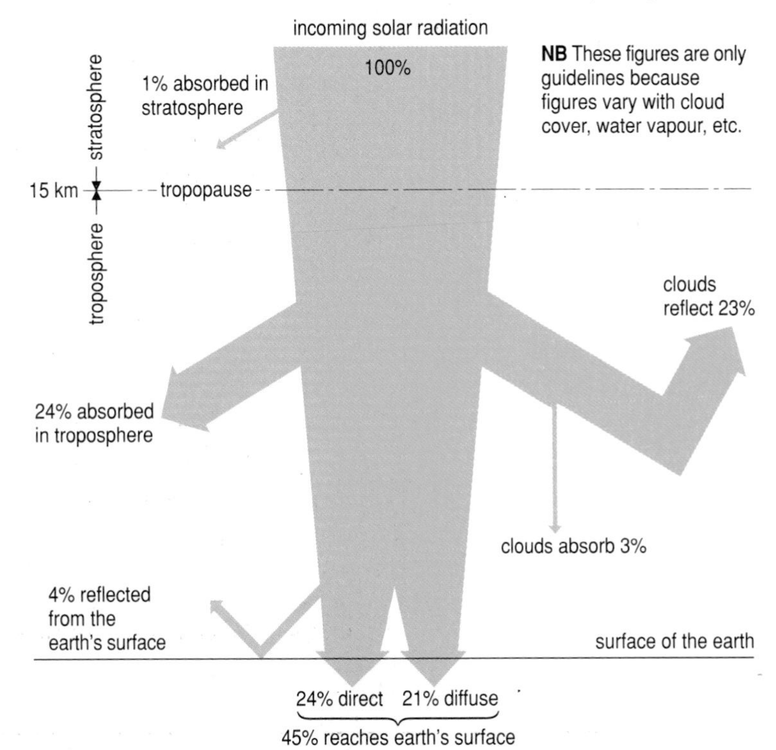

Fig 2.1 The fate of incoming solar radiation.
Note that when radiation enters the atmosphere it may:
- pass directly to the earth's surface
- become absorbed by the constituents of the atmosphere
- be reflected or scattered by constituents in the atmosphere (diffuse radiation).

Incoming radiation is reflected from the clouds and the atmospheric components (Chapter 1), and finally by the surface itself.

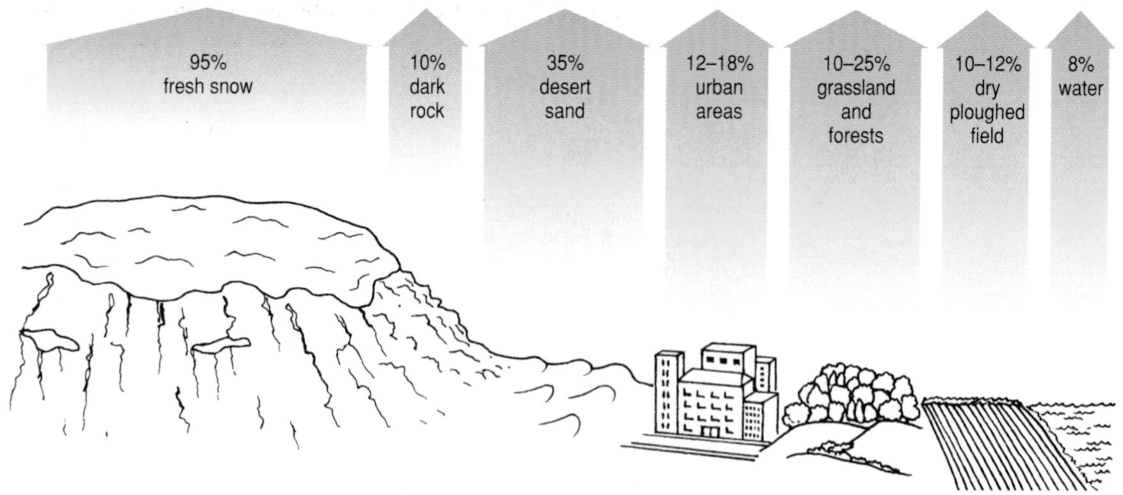

Fig 2.2 Ground surface albedo values.

The albedo effect can be defined as the reflection and scattering of light from a surface. The amount of albedo is the proportion of solar radiation which is reflected by a particular body. The highest surface albedo is from fresh snow, which reflects 95% of the incident radiation, whereas dark objects, such as tarmac, absorb more radiation than they reflect.

A thick cloud layer may reflect up to 80% of incoming solar radiation, but will also cause heat to be retained close to the earth's surface by reflecting back the outgoing radiation (Chapter 1). This is the reason why cloudy summer evenings are warmer than those with clear skies.

Imagine astronauts repairing the outside of their spacecraft in orbit. As they move from shadow into full sunlight, the temperature of their spacesuits will increase by several hundred degrees centigrade. However, at the earth's surface the temperature range is limited to only a few degrees because the atmosphere retains heat (Chapter 1). This temperature balance is extremely important to life on earth, which can only operate within certain limits. For example, photosynthesis in many temperate plants will stop when the temperature is below 0°C or above 40°C.

It is possible to calculate an overall **heat budget** for the earth. The main input is from the sun, with a small addition from the earth itself. The balance between incoming and outgoing radiation at any one point on the earth's surface is, however, not straightforward. The reasons for and the effects of this imbalance will now be discussed.

2.2 SPATIAL, DIURNAL AND SEASONAL TEMPERATURE VARIATIONS

The radiation which reaches the earth's surface will vary both from place to place (spatially) and in time (temporally). **Spatial temperature variations** depend on latitude. Latitude is calculated by measuring the angular distance either side of the equator, so that the equator itself is at 0° and the poles at 90°, with the distance halfway between the pole and the equator being at 45° (Figure 2.3). Having travelled such a great distance, incoming solar radiation, although still diverging away from the sun, is basically travelling in straight lines. The amount of radiation is therefore the same throughout line A on Figure 2.3. The inverse square law states that as distance from the sun increases, the intensity of radiation decreases. This can be expressed in the equation as:

$$\text{Intensity} = \frac{1}{d^2} \qquad \text{where d = distance}$$

However, because the surface of the earth is curved, incoming radiation at high latitudes has to pass through a greater depth of atmosphere and is spread out over a greater surface area than incoming radiation at lower latitudes. As a result, there is a huge difference in the heat budget of different latitudes, and this effect also means that light intensity, which is an important factor in photosynthesis (Chapter 5), is on average greater near the equator. From Table 2.1 it can be seen that there is a positive heat balance in the tropics; the amount of radiation absorbed exceeds radiation lost. In high latitudes, there is a negative heat balance, where heat re-radiated is greater than heat absorbed. This temperature difference provides the power for ocean currents and the global climatic systems, which transfer heat from areas with a positive heat balance to areas with a negative heat balance.

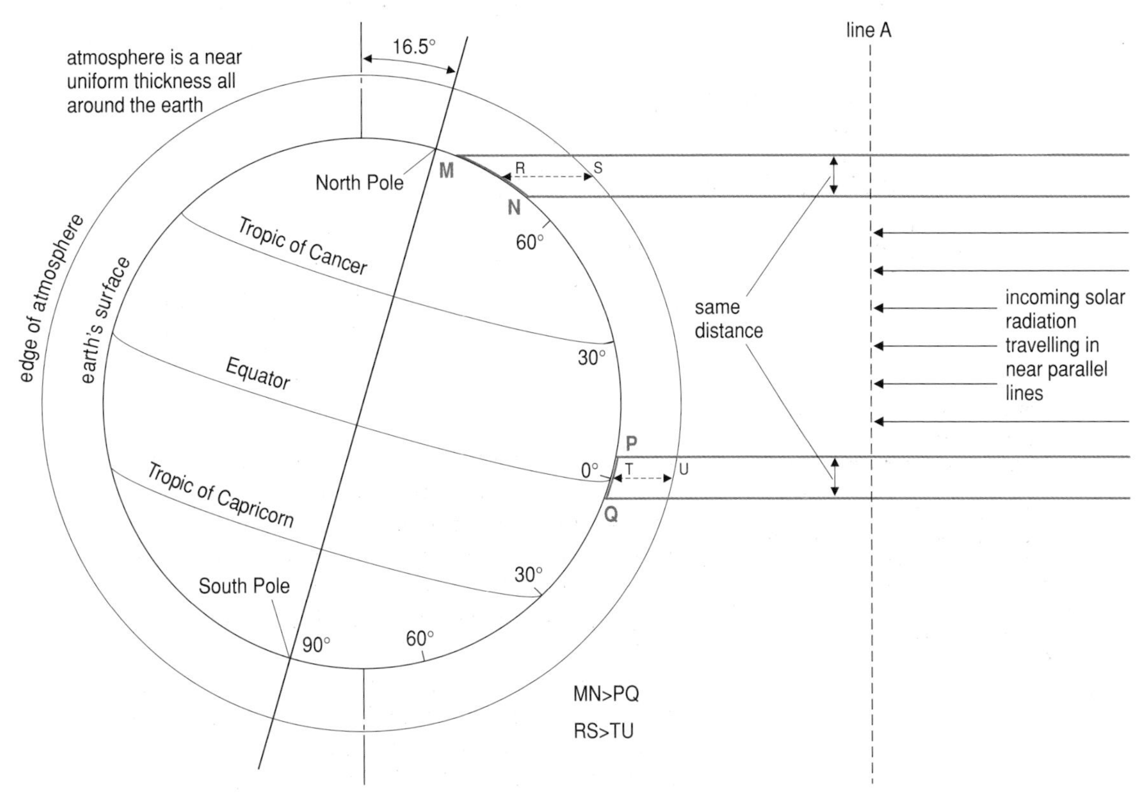

Fig 2.3 The effect of latitude on incoming solar radiation.
Radiation has to travel through a greater depth of atmosphere (RS as compared to TV) in high latitudes. When it reaches the surface the radiation is also spread out over a greater area (MN as compared to PQ) than in lower latitudes.

Table 2.1 Annual radiation budgets in the northern hemisphere

	Latitude	Radiation absorbed /W m^{-2}	Radiation re-radiated /W m^{-2}	Overall radiation balance /W m^{-2}
Positive heat balance	0° (equator)	327	189	48
	30°	207	198	9
Negative heat balance	60°	135	190	–55
	90° (North Pole)	98	176	–78

Note: These are average figures. As will be seen later in the chapter, the exact radiation balance will depend on the small-scale physical features present.

Temperatures may also vary at different times of the day (diurnal), at different times of the year (seasonal) and in geological time (glacial-interglacial cycles – Chapter 3).

Diurnal variations occur because the earth rotates. Thus a point on the earth's surface will point towards or away from the sun, depending upon the stage of rotation.

Seasonal variations result from the earth's orbit around the sun. From Figure 2.4 it can be seen that the earth pivots on an axis that is not at right-angles to the orbital plain but at an angle between 21.5° and 24.5°. Throughout the orbit, the distance from the sun to the equator remains constant, whereas the distance from the sun to the poles varies. Consequently, there is greater temperature variation and seasonality at the poles than at the equator. Hot 'summer' conditions prevail over the hemisphere which is closest to the sun and cold 'winter' conditions over the hemisphere which is furthest away.

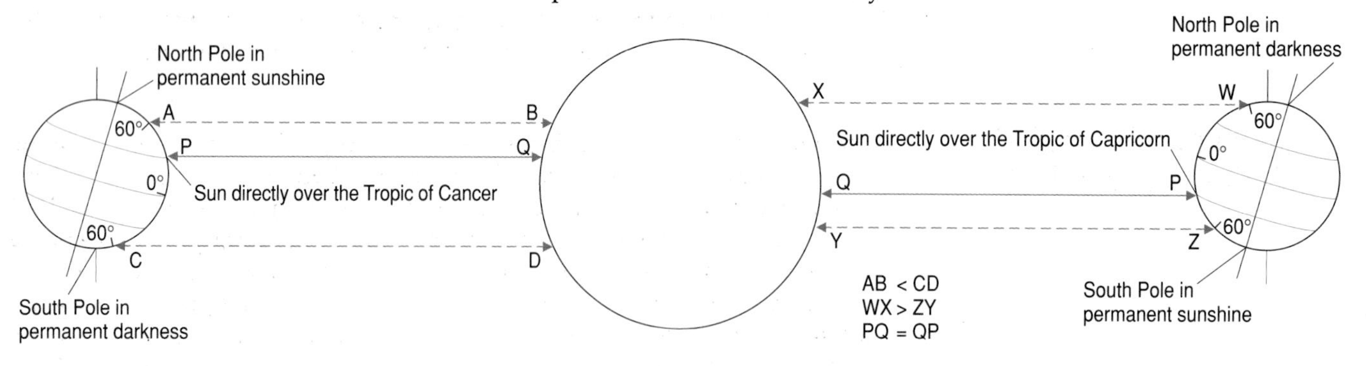

Fig 2.4 The earth's orbit and the seasons.

2.1 (a) What important facts can you note from Table 2.1?

(b) Draw a diagram to show why, during winter in the northern hemisphere, insolation energy per unit area is greater at 60°S than at 60°N.

The temperature of two points on the same latitude is not necessarily the same. Later in the chapter we will discuss how temperature will also vary with the prevailing wind, altitude, ocean currents, and with maritime or continental climates.

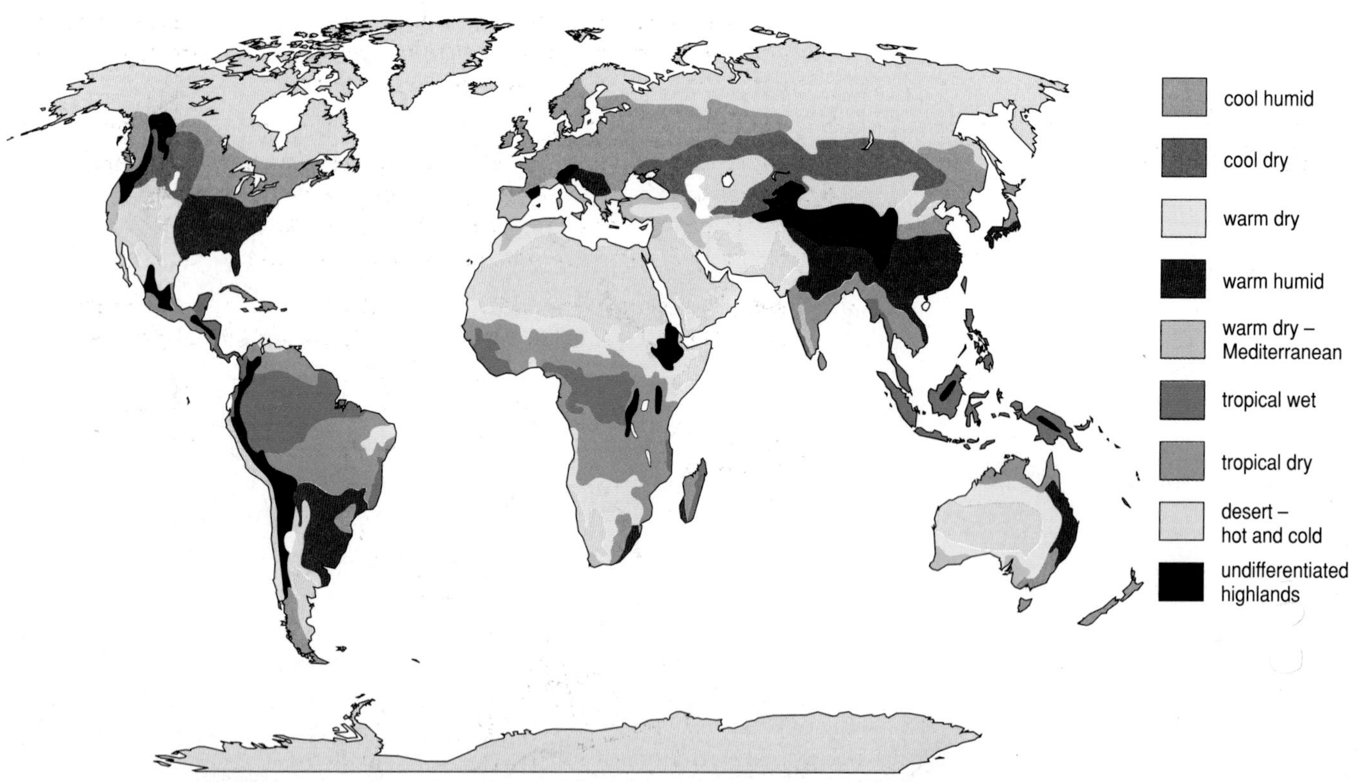

The world can be mapped into agroclimatic zones on the basis of temperature and rainfall. Certain crops will only grow in certain climatic zones.

Fig 2.5 Agroclimatic zones.
This is a very simplified diagram. It will be explained later in Chapter 2, Section 2.5 and in Chapter 4, that the actual ecosystems present and the crops grown will vary depending on much smaller-scale factors.

The climate has a major impact on agriculture because only certain crops will grow in a particular climatic zone (Chapter 14). Thus, for the purpose of agriculture, the world can be classified into several agroclimatic zones on the basis of temperature and rainfall (Figure 2.5). These zones are a useful indication of how climate has a major impact on human activities.

2.3 ATMOSPHERIC CIRCULATION

The wind is a mass of moving air resulting from differences in atmospheric pressure. Air pressure is largely determined by temperature. As a body of air is warmed, it will rise relative to its surroundings (this is called **convection**) and as it rises it will expand, resulting in a reduced density which causes a lowering of pressure. Conversely, pressure will increase as cold air sinks.

When warm air rises, it leaves behind an area of low pressure (called a depression or cyclone), and air will be sucked into this area, forming winds. The winds always blow from areas of high pressure (which form anticyclones) to areas of low pressure.

Air pressure is measured in millibars (mb) by a barometer. On a weather map (synoptic chart), pressure is shown by isobars, lines which connect points of equal pressure. The closer the isobars, the steeper the pressure gradient and therefore the higher the wind velocity. Look out for this effect the next time you watch the weather report on television.

Clouds can form through convection, as warm air rises or as air is forced to rise, for example over mountain ranges. As a body of air rises, it will expand and cool due to the loss of radiant heat. The rate at which temperature decreases with height is called the **lapse rate**. As the temperature drops, water vapour within the air will condense and form clouds. The temperature below which further cooling results in condensation is called the **dew point**. Cloud formation only occurs when particles, such as dust, act as nuclei for condensation, and is most efficient around **hygroscopic nuclei**, such as salt crystals which have an affinity for water. Humans have flown over clouds and dropped salt crystals onto it in an attempt to cause rain. This is called salting clouds or cloud seeding.

QUESTIONS

2.2 Saturated air is that which can contain no more water vapour. The lapse rate of unsaturated air is approximately 1°C with every 100 metres. If the dew point is reached at 8°C, and the temperature at the surface is 20°C, calculate the height at which clouds form.

2.3 (a) The diagram represents the volume changes occurring in a mass of warm, moist air as it moves upwards through the atmosphere.

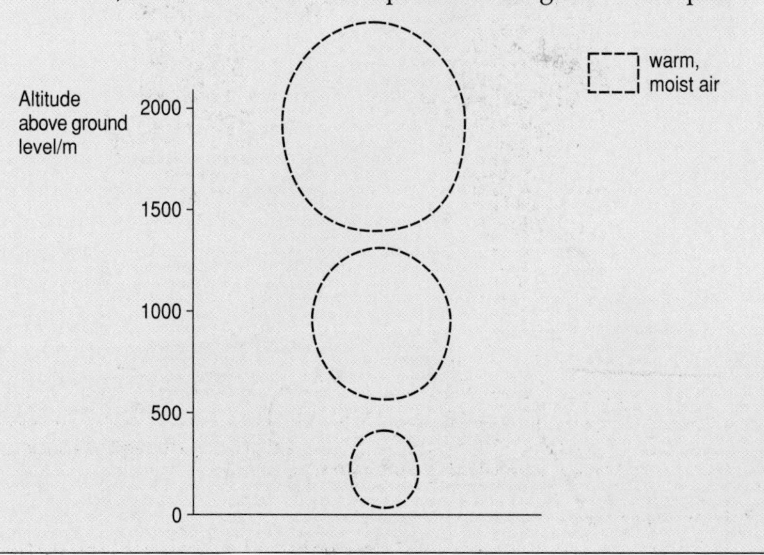

(i) By reference to the diagram, explain why the volume of the mass of air changes with altitude.
(ii) Describe what must happen before the moisture in the rising air forms cloud.

(b) The table shows changes in air temperature with altitude. Complete the table, giving the temperature of the air which you predict at 2000 metres.

Altitude/m	Temperature/°C
ground level	22
1000	12
2000	

(c) Explain how changes in the amount of cloud cover may cause variations in the diurnal temperature range from day to day at any one place.

AEB June 1993

In 1856 William Ferrel suggested that a global convection system existed, which consisted of three cells (Figure 2.6). In reality, Ferrel's model is complicated by smaller-scale factors, but it is a useful basis for understanding climatic events.

- The Polar cell develops as cold air sinks over the poles, resulting in increasing density and pressure. The air moves along the surface towards lower latitudes as an easterly wind.
- The Hadley cell develops as hot air rises over equatorial regions. This air moves at high altitude, until it sinks over sub-tropical areas (around 30° latitude). Some of the air moves back at surface level towards the equator as the easterly trade winds. The rest moves westwards towards the pole. However, the winds move at an angle and not directly towards the pole because they are deflected by the **Coriolis force**. The Coriolis force is caused by the rotation of the earth, and causes a body of air, or water, to be deflected in the opposite direction to rotation. In the northern hemisphere the coriolis force deflects the wind and ocean currents from east to west, and in the southern hemisphere from west to east.
- The Ferrel cell is a zone of variable weather conditions, which form in response to events within the previous two cells.

This model can be used to explain why a belt of deserts forms in the sub-tropics around the world. Figure 2.6 shows that at 30° latitude, air is sinking, and not rising to form clouds. Without clouds, there can be no rain.

An **air mass** is a large body of air in which the characteristics of any one horizontal level are uniform. The types of air mass that affect the UK are shown in Figure 2.7. It can be noted that a polar air mass is cold and moves southwards, whereas a tropical air mass is warm and moves northwards. A maritime air mass has come from the ocean, whereas a continental air mass has by definition travelled over land. Thus in the UK an air mass which comes from the south-west (a tropical maritime – mT) is moist and warm because it has originated over the southern Atlantic Ocean. An air mass which comes from the north-west (a polar maritime – mP) has travelled across the northern Atlantic Ocean and will therefore be moist but will have a low temperature.

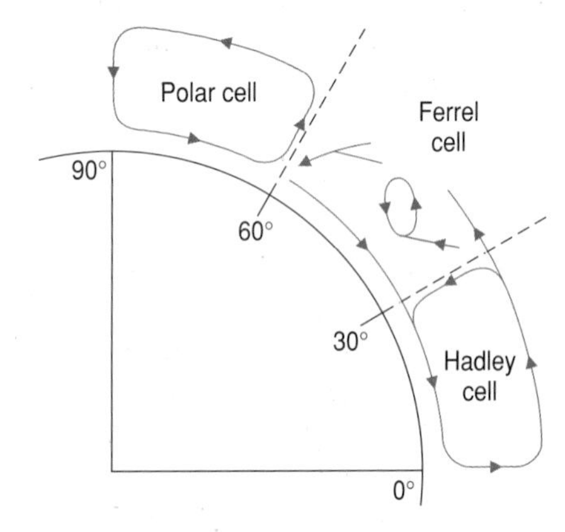

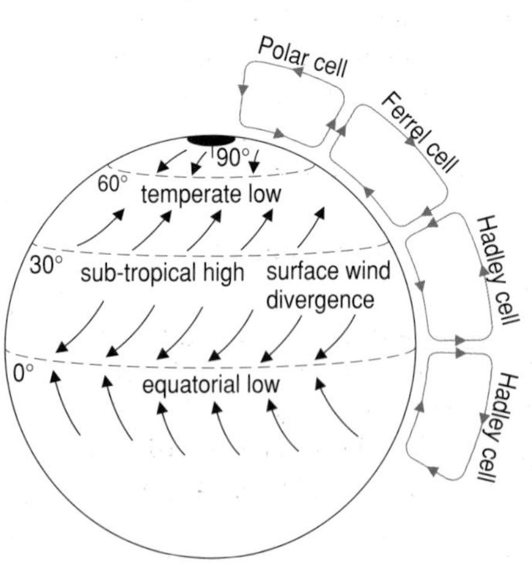

Fig 2.6 Global-scale wind movements.

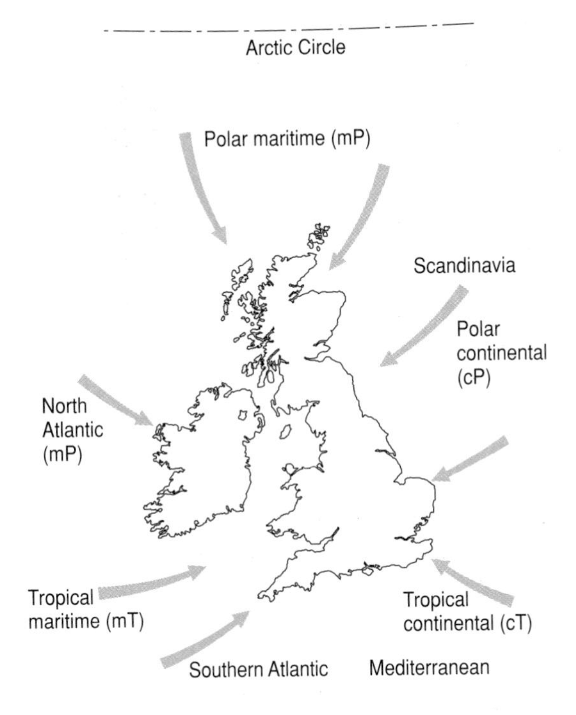

Fig 2.7 Air masses that affect the UK.

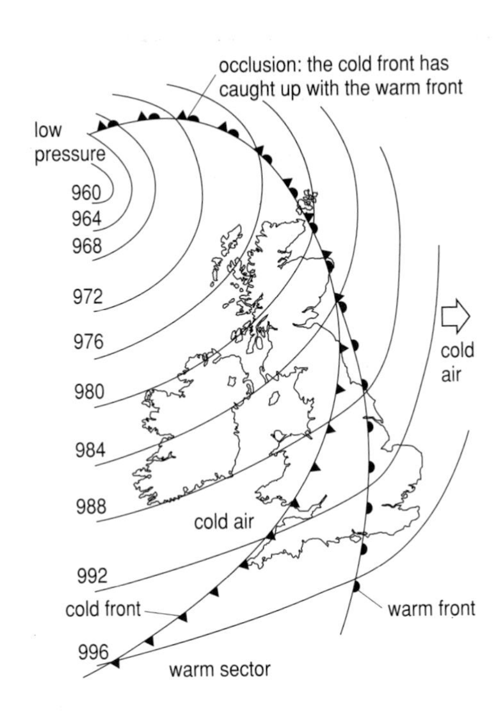

Fig 2.8 Depression.

When two air masses meet, they will remain separate because they have different characteristics. The junction between them is called a **front**.

In mid-latitudes a front exists between cold air in the north and warm air in the south. When cold air advances into the warm air along this front, the latter is forced to rise above the cold, forming an area of low pressure. Surface winds begin to spiral inwards towards the centre of the low pressure. This system may grow to cover an area over 1000 km across, and is called a depression (Figure 2.8).

CASE STUDY

UK weather

There is a subtle difference between climate and weather. The study of climate (climatology) describes atmospheric events over a large area and over a long time (for example the colder temperatures in the northern hemisphere during the last Ice Age). However, the study of weather (meteorology) describes events in a smaller area and over a shorter time (for example what the weather was like last weekend).

The UK weather is dominated by about 100 **depressions** a year, each of which stays over the country for around three days. Depressions can bring high winds (due to a steep pressure gradient) and precipitation. There are two phases of precipitation associated with a passing depression:

- ahead of the warm front, where clouds form as warm air pushes itself above a cold air mass
- behind the cold front, where clouds form as warm air is forced to rise above the advancing cold air mass.

The weather experienced under an **anticyclone** lasts for up to a week and depends on the time of year. Anticyclones bring calm conditions and, as air is sinking, the sky is cloudless and there is little or no precipitation. In summer clear skies result in hot days and cold nights, but in winter temperatures may drop very low as insolation is greatly reduced.

2.4 THE EFFECT OF CONTINENTS, OCEAN CURRENTS AND MOUNTAINS

Continents and oceans

The climate is influenced by the physical differences between oceans and continents. Land that is close to an ocean is said to have a **maritime climate**, and land that is far away from an ocean will be subject to the effects of **continentality**. Britain is such a small land mass that it has a largely maritime climate, although the weather is also sometimes affected by the proximity to continental Europe. Continentality has several different effects:

- Land masses lose and gain heat more rapidly than oceans because they have a lower specific heat capacity than water. Therefore, during the summer, continental interiors that are far away from oceans are subject to a much greater temperature variation than land near oceans. In other words, islands and coasts are kept warm in winter, and mild in summer by the water that surrounds them.
- Air that travels over a large land mass will lose much of its moisture by the time it reaches the continental interior. Precipitation in such areas therefore tends to be irregular. Continental interiors, such as the large mass of central Asia, are therefore subject to much drier climates than coastal regions.

Ocean currents also affect the climate of the surrounding areas. Figure 2.9 shows the main ocean currents throughout the world, which move in response to the position of continents and the coriolis force. An ocean current that moves from high latitudes brings cold water, whereas a current that moves from low latitudes brings with it warm water. Figure 2.10 shows mean temperatures at sea-level in January and in June.

The northern United Kingdom (59° latitude) is subject to a much milder climate than Siberia, which is at approximately the same latitude. For example, Verkhoyansk (67° latitude) has experienced –68°C, which is in sharp contrast to the minimum temperature of John o' Groats. The UK has a more favourable climate because of the Gulf Stream and because we do not suffer from the effects of continentality.

QUESTION

2.4 Table 2.2 shows the climatic characteristics of the UK and of Kamchatka.

(a) Describe in detail why the climates are different (for example temperature, amount and type of rainfall).

(b) Describe what effects the climate will have on agricultural activity in the two areas.

Table 2.2 Climatic characteristics of the UK and of Kamchatka

Climatic characteristic	UK	Kamchatka (eastern Russia)
latitude	similar	similar
mean winter temperature /degrees Celsius	0–8	–16 to –24 inland and –8 to –16 on the coast
mean summer temperature	similar	similar
total annual precipitation	similar	similar
main months of precipitation	all year	summer

Note: Assume that all other factors are similar.

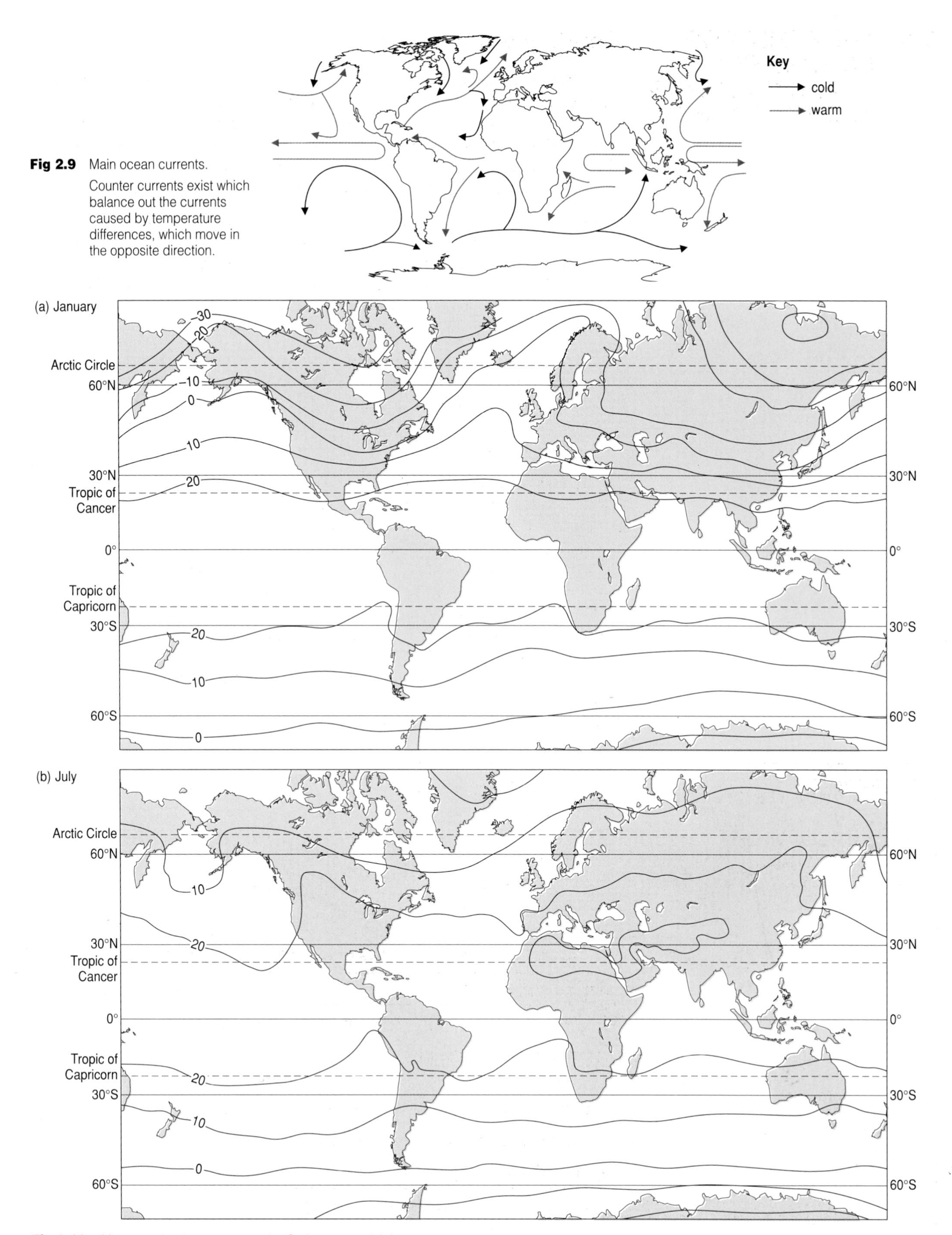

Fig 2.9 Main ocean currents.
Counter currents exist which balance out the currents caused by temperature differences, which move in the opposite direction.

Fig 2.10 Mean sea-level temperatures in °C, January and July.
It can be seen that the temperature in the Atlantic Ocean has been influenced by the ocean current, which is called the Gulf Stream, because it brings warm water towards Western Europe from the Gulf of Mexico.

Mountains

It may seem that because mountains are closer to the sun than sea-level, the temperature should be higher. However, this is not the case because the sun is so far away from the earth that the distance up a mountain makes very little difference to its temperature. Mountain tops are colder because the atmosphere is thinner than at sea-level. Less energy is therefore being absorbed and re-radiated, and the pressure is lower, so air is expanding and losing radiant heat. Mountains receive less reflected energy from the earth's surface simply because they are further away.

Mountain ranges force air to rise. Remember that when air rises and the dew point is reached, clouds develop and precipitation may occur. This is known as **orographic rain** (Figure 2.11).

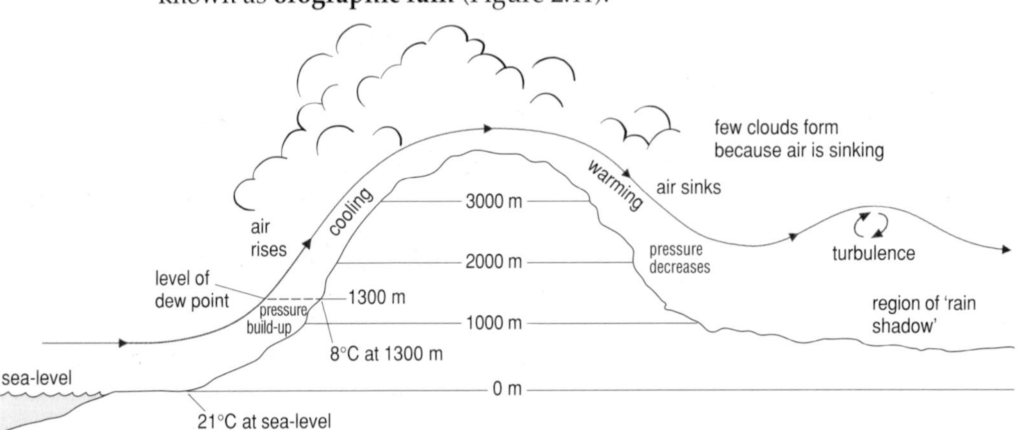

Fig 2.11 The formation of orographic rain.

On the leeward side of a mountain range there is a **rain shadow**, where there is less precipitation because air is falling and so clouds do not develop. The falling air warms and can therefore hold more water. The majority of water within the air has already fallen on the mountains.

In the UK there is much more precipitation in the west. This is because in the UK the predominant wind direction is from the south-west, and the mountain ranges are generally situated in the west (for example Snowdonia, the Lake District and the Brecon Beacons). The air approaching the UK has a high moisture content because it has travelled over the Atlantic Ocean. Further orographic rain occurs over the central Pennine range, and thus the low-lying eastern UK is in a rain shadow.

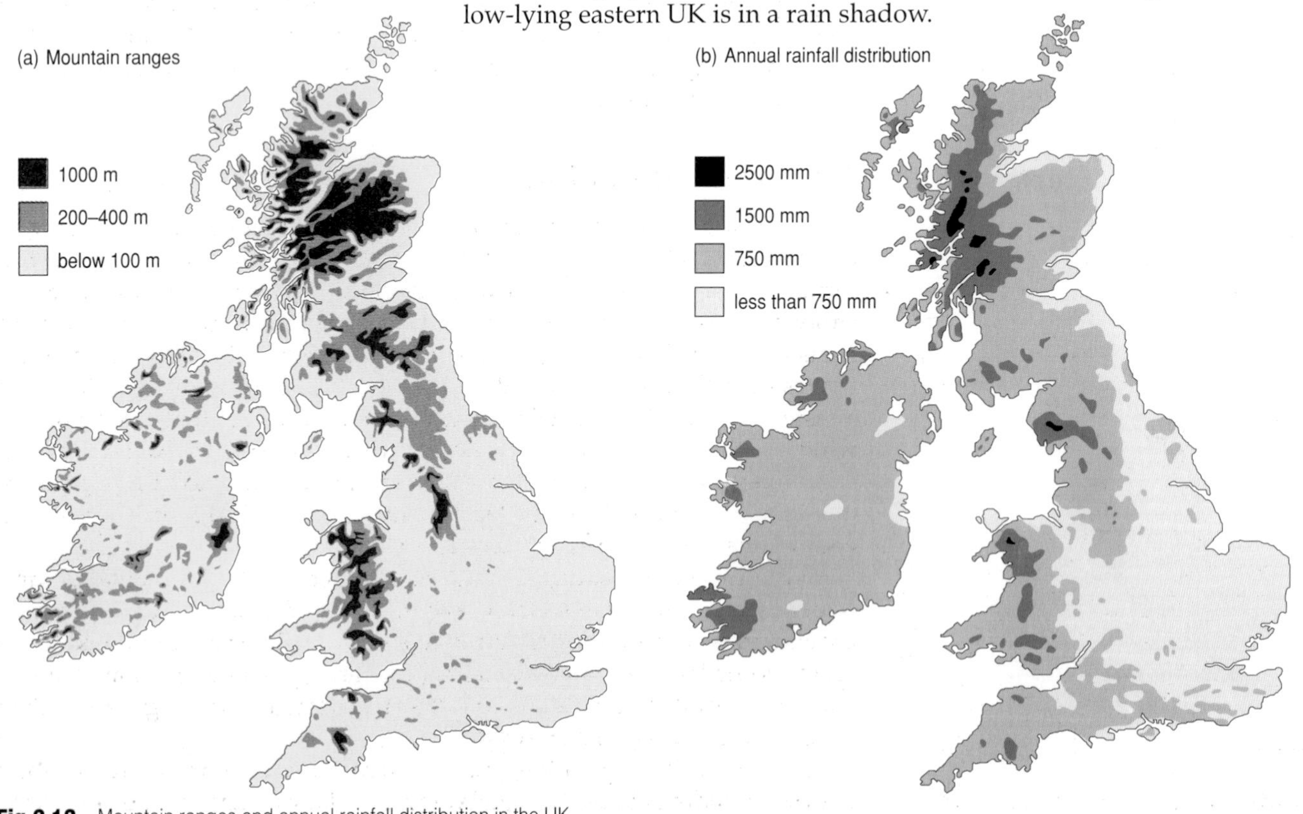

Fig 2.12 Mountain ranges and annual rainfall distribution in the UK.

QUESTION

2.5 The climate of Britain is described as temperate.
  (a) By reference to temperature and precipitation, define a temperate climate.
  (b) Britain is surrounded by shallow seas. What effect does this have on the British climate?

AEB June 1990

2.5 SMALL-SCALE CLIMATES

On global, continental and regional scales, the climate is largely determined by factors such as atmospheric circulations and ocean currents. However, on a microscale, the climate of any point on the earth's surface is heavily modified by localised, site-specific factors, such as topography, land use and vegetation.

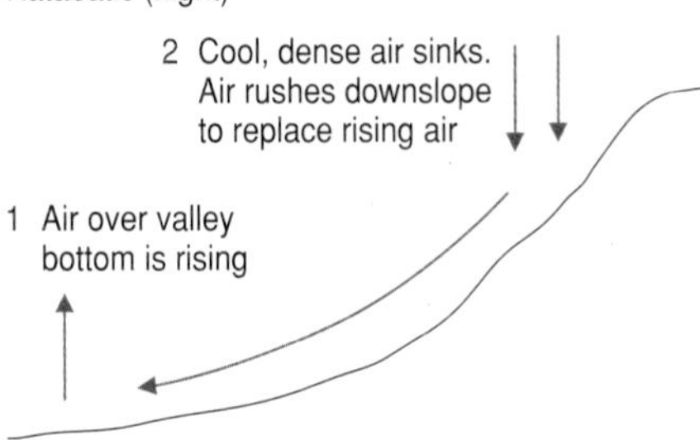

Fig 2.13 Anabatic and katabatic winds. These valley winds caused by differential heating of the valley slopes and sides are similar to on-shore and off-shore breezes but can be much stronger.

The effect of topography

The aspect (orientation) of a slope will determine the amount of solar radiation and therefore the amount of light and heat that slope receives. In the northern hemisphere, south-facing slopes receive more solar radiation than north-facing slopes. On northern slopes, this will result in weaker plant growth, lower evaporation rates and lower rates of snow melting.

During the day, the higher parts of mountains and the valley sides, especially those facing the sun, will heat up more than the valley floor. The air above the valley sides will therefore rise, and air from the valley floor will move upslope to replace the rising air. This is called an **anabatic wind**. A **katabatic wind** blows downslope during the evening when the valley sides cool quickly, and the cool dense air sinks back into the valley floor (Figure 2.13).

Air flow will be affected by topography. As wind blows through a narrower part of a valley or through a gap in a hedge the velocity will increase in order to allow the passage of the increased mass of air. Velocity will also increase as air is forced over a mountain range. Friction with the rough ground will cause the wind to slow down and spread out. Therefore low pressure and eddies form in the lee of a mountain (Figure 2.14). The presence of high terrain may also cause clouds to form which could then give rise to orographic rain.

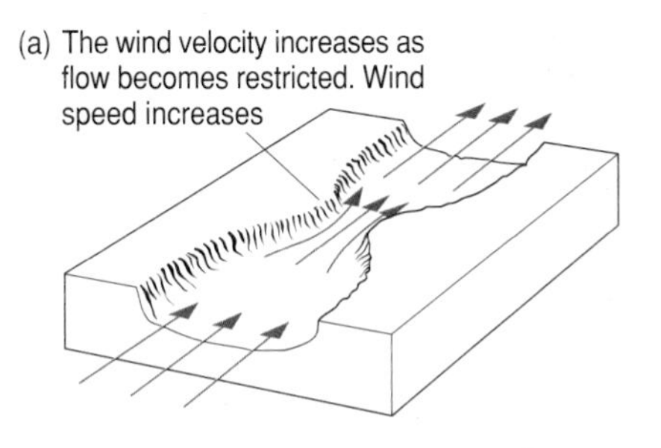

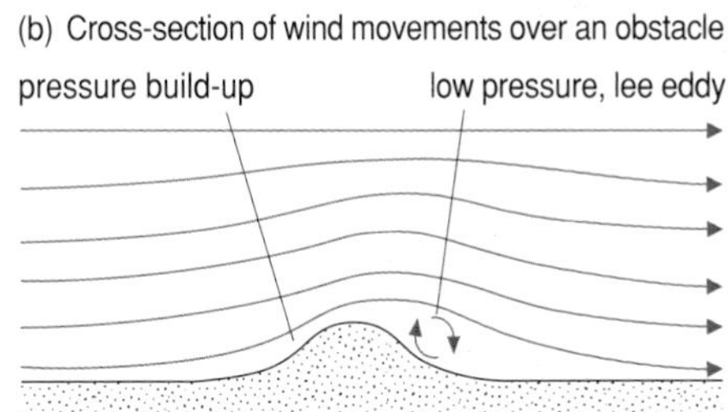

Fig 2.14 The impact of topography on wind.

The effect of land use

Land use affects the amount of radiation absorbed by the land, resulting in differential heating, which in turn will affect surface winds and cloud formations. Differential heating of the land and sea causes **on-shore** and **off-shore breezes**. During the day, air above the land will rise because the land heats up faster than the sea. An **on-shore breeze** will therefore occur, as air moves from the sea to replace the rising air. During the night, air will rise at a faster rate over the sea, which has retained more heat than the land. The wind will therefore blow as an **off-shore breeze** (Figure 2.15). Remember that wind blows from areas of high pressure (falling air) to areas of low pressure (rising air). Wind will also be faster over lakes and the sea than over land because there is much less frictional resistance.

The differential heating of land and water results in a temperature-buffering effect near large water bodies. During the day a lake will have a cooler temperature, and at night the lake will remain warm compared to the surrounding land. A similar effect happens throughout the changing seasons and, as a result, vegetation near lakes has a longer growing period than vegetation growing further away.

As a result of evaporation, the air above water contains more moisture than surrounding air, and so clouds are more likely to develop if this air rises.

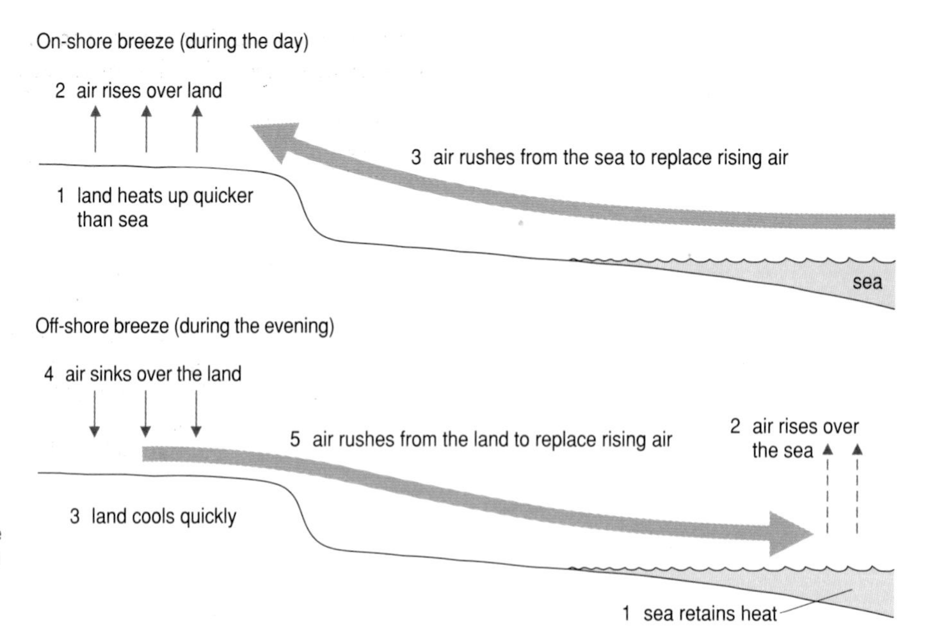

Fig 2.15 On-shore and off-shore breezes.

In coastal areas, sea breezes develop because of the differential heating of the sea and the land. The breezes flow from areas of high pressure to areas of low pressure.

The effect of vegetation

Wind speed lessens as it gets closer to the earth's surface because of friction with the ground. The zone of low wind speed is called the **friction layer**. The height to which the friction layer extends varies depending upon the roughness of the ground.

The presence of tree cover will greatly reduce wind speed, especially at low levels. Air trapped below a forest canopy will have a higher humidity, owing to **evapotranspiration**, and will heat up and cool down at a slower rate than the surrounding air, resulting in reduced convection. Tropical rainforests are so named because they have such a high humidity.

By planting trees and hedgerows as a **wind-break**, humans can manipulate the local climate so that the wind does not damage crops or erode the soil (Chapter 14). A low-density, permeable barrier, such as a hedgerow, does not stop the wind but simply slows it down below the speed which can cause damage (Figures 2.16 and 2.17). Behind a permeable barrier, the wind only returns to its previous velocity after a horizontal distance which is 20 to 30 times the vertical height. Behind an impermeable barrier, such as a wall, the initial velocity is regained at a horizontal distance 10 to 15 times the

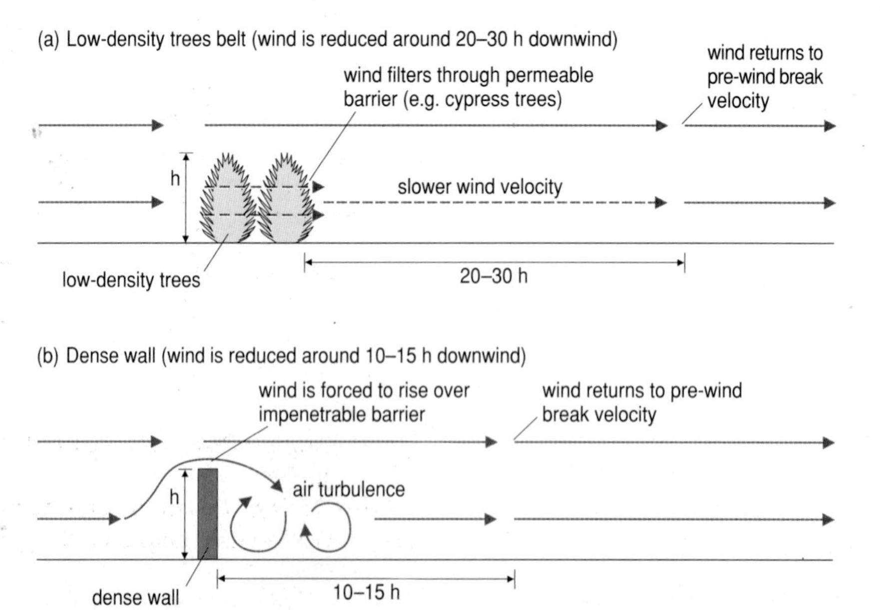

Fig 2.16 The effect of wind-breaks on wind velocity.

The wind speed behind the impermeable barrier in **(b)** quickly returns to dangerous levels, whereas the permeable barrier in **(a)** keeps the wind speed low for a considerable distance.

vertical height. Using vegetation as a wind-break also has the advantage that it is cheaper, less visually intrusive and may act as a wildlife habitat.

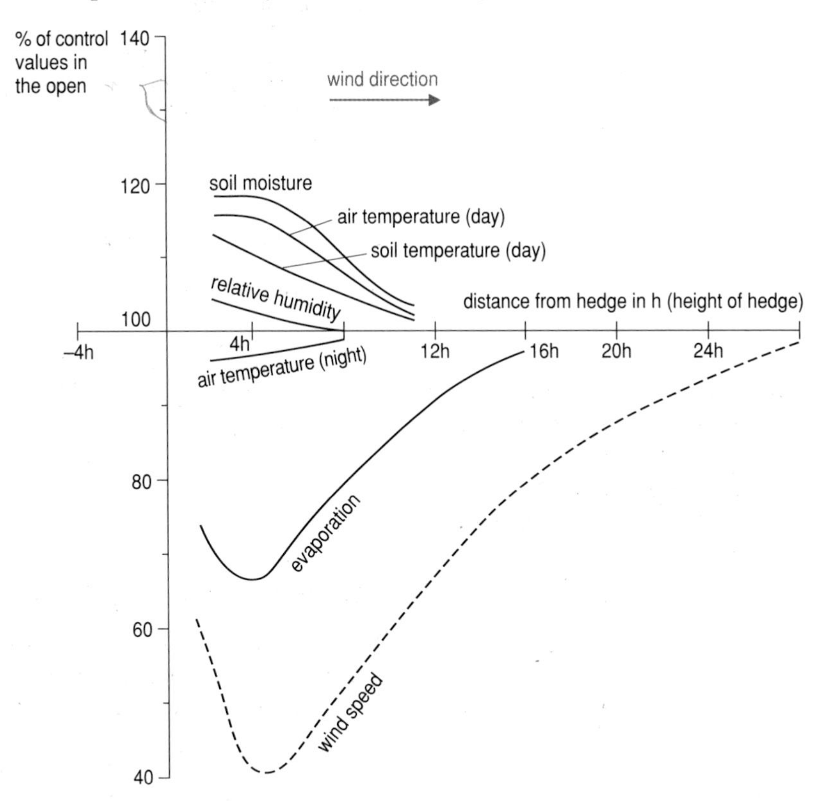

Fig 2.17 The microclimatic effect behind a hedge.

Figure 2.17 illustrates some of the other advantageous effects which a hedge may create. There will be an increase in soil moisture content, humidity, soil temperature, a decrease in evaporation, and the air temperature will be higher during the day and lower at night. The belt will also retain snow for longer and thus protect the soil from frost damage.

QUESTION

2.6 The diagram shows a tree shelter which is often used to help establish young trees. The graph shows the results of an investigation to study the effects of tree shelters on early growth of oak trees.

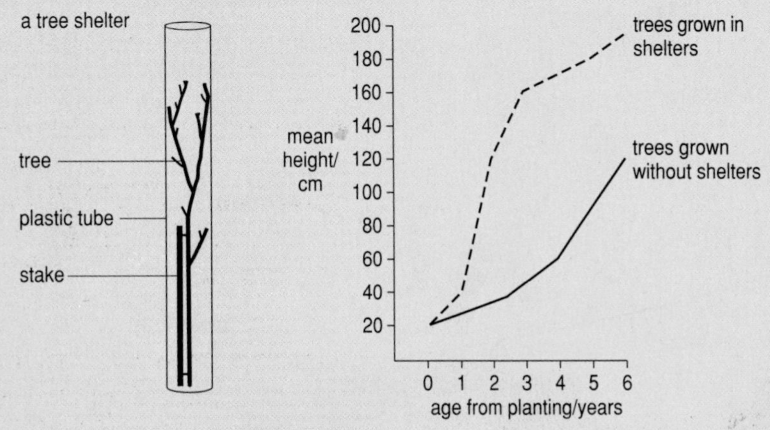

(a) Suggest **three** reasons for the increased growth rate of trees in shelters.

(b) During hot, sunny days plants may be seen to wilt. Explain how wilting may help plants to survive.

AEB June 1995

Vegetation blocks the ground surface from direct solar radiation. The soil temperature will therefore be generally lower than in areas which are not vegetated. Deforestation may result in dramatic changes to the local climate, the most serious of which is increased soil erosion as a result of increased wind, surface run-off and rainsplash (Chapter 14).

Urban climates

The characteristics of urban areas, such as little vegetation cover, high buildings, the generation of dust and the radiation of heat, combine to form a very different climate to that of the surrounding rural areas. Some of the effects are beneficial and others are undesirable (Table 2.3).

Table 2.3 Advantages and disadvantages of urban climates

Advantages	Disadvantages
lower heating bills	increased air conditioning bills
reduced wind speeds	increased wind turbulence
less frost	increased number of thunderstorms
less snow	pollution
quicker melting of snow	greater cloud cover, therefore less sunshine
overall reduced atmospheric humidity	high temperatures cause aggression
	particulate smogs
	photochemical smogs

Building materials such as brick, stone and tarmac have a large specific heat capacity with a low albedo, and therefore absorb heat during the day which is gradually released during the night. At the same time, heat is being released by industry and domestic uses. These two effects result in average temperatures that are around 1°C higher than the surrounding rural areas. This **urban heat island effect** is greatest during calm anticyclone conditions because the heat is otherwise lost through strong winds.

At dawn, urban areas are warmer than the neighbouring countryside. This increased temperature results in lower heating bills, fewer frosts, less snow and quicker melting of snow. The absence of bodies of water and vegetation usually causes lower humidity.

Urban areas emit considerable amounts of dust which may combine with fog at ground level to form an '**urban smog**' or act as condensation nuclei, accelerating cloud formation. Thus in urban areas there is less sunshine, the rainfall is more intense and there are an increased number of thunderstorms. A cocktail of invisible harmful gases, such as ozone (O_3), carbon monoxide (CO) and nitrous oxides (NO_x), may also develop in urban areas (this is called a **photochemical smog**). The production of these chemicals is mainly associated with pollution from vehicle exhausts, and therefore the effect is worse in hot, dry, sunny cities such as Athens, Rio de Janeiro, and Mexico City. The problem, which is discussed in more detail in Chapter 10, has become so bad in Athens that the authorities have tried to ban all non-essential traffic from the city centre.

SUMMARY ASSIGNMENT

The graph shows the average annual deposition of solar radiation on earth.

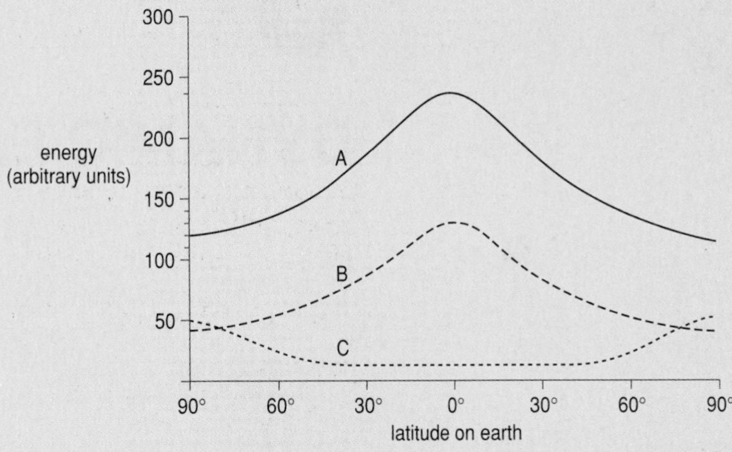

Curve A: radiation at the top of the atmosphere
Curve B: absorbed by the earth's surface
Curve C: reflected by the earth's surface

1. From the graph:
   (a) estimate what percentage of solar radiation reaching the atmosphere is absorbed by the earth's surface at the equator
   (b) explain why the earth's albedo increases between latitudes 60° and 90°
   (c) suggest an explanation for the fact that the curves **A** and **B** are very similar.
2. Give **two** explanations for the way in which the amount of solar radiation reaching the earth's surface changes with latitude.
3. Explain why the atmosphere is heated mainly by the earth rather than by direct solar radiation.
4. Draw a diagram to illustrate the temperature and pressure relationships which produce on-shore breezes.
5. Distinguish between the terms *weather* and *climate.*
6. Describe, in each case, two important characteristics of
   (a) a depression or cyclone
   (b) an anticyclone.

AEB June 1992

Chapter **3**

DYNAMIC EARTH

Geology touches every aspect of our lives. Every year we use millions of tonnes of rocks to build our towns and cities, roads and airports. Our beaches and coastlines are continuously created and destroyed by the processes of deposition and erosion. It is the disintegration of rocks which has given us the soil upon which most terrestrial life depends. More dramatically, earthquakes and erupting volcanoes are unwelcome reminders that the earth is a dynamic planet.

> **LEARNING OBJECTIVES**
>
> After completing the work in this chapter you will be able to:
>
> **1.** describe the structure of the earth
>
> **2.** describe the processes of weathering and erosion
>
> **3.** explain the different ways by which igneous, sedimentary and metamorphic rocks are formed.

3.1 INSIDE THE EARTH – INTERNAL PROCESSES

The earth was formed billions of years ago and its interior has been cooling down ever since. The interior of the earth can now be separated into three major concentric zones, the **core**, the **mantle** and the **crust** (Figure 3.1).

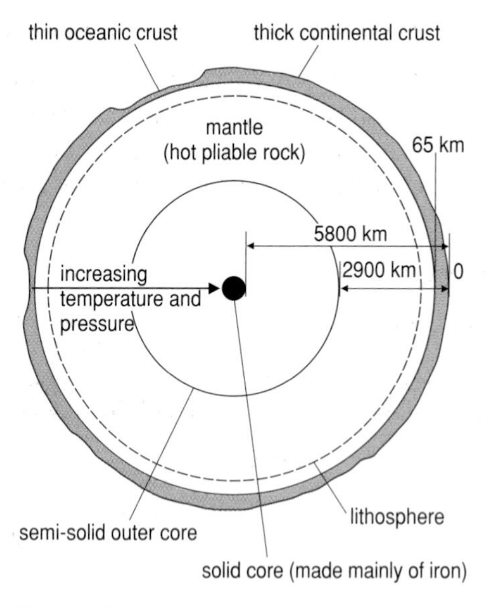

Core: Solid due to extremely high pressure

Outer core: Semi-solid as a result of the radioactive decay of unstable elements such as uranium

Mantle: Rocks are pliable due to high temperature. The mantle contains a greater proportion of lighter elements such as silicon and aluminium. The outer region is known as the lithosphere. The outermost part of the lithosphere is called the crust

Crust: There are two types of crust – continental and oceanic. Continental crust is thicker but much less dense than oceanic crust. According to the theory of plate tectonics, the crust is broken up into a mosaic of slowly moving plates

Fig 3.1 Structure of the earth.

The crust, which makes up just 1% of the weight of the planet, can be thought of as the skin on a bowl of rice pudding. There are two types of crust: oceanic crust and continental crust. **Oceanic crust** is thin but dense and makes up the ocean floor. **Continental crust** is the crust from which the great land masses of the continents is formed. It is much less dense and it therefore 'floats' upon the ocean crust.

According to the theory of **plate tectonics**, the crust is broken up into a mosaic of huge, drifting plates (Figure 3.2). It is the movement of these vast plates, sometimes grinding together, sometimes pulling apart, which results in volcanic eruptions and earthquakes at the plate boundaries, and these give rise to some of our most dramatic landforms, such as the mountain ranges of the Andes and the Himalayas. Beneath the sea even more dramatic features are produced, including deep ocean trenches, such as the Marianas Trench in the western Pacific Ocean, and the longest chains of volcanoes on earth, such as those running down the middle of the Atlantic Ocean.

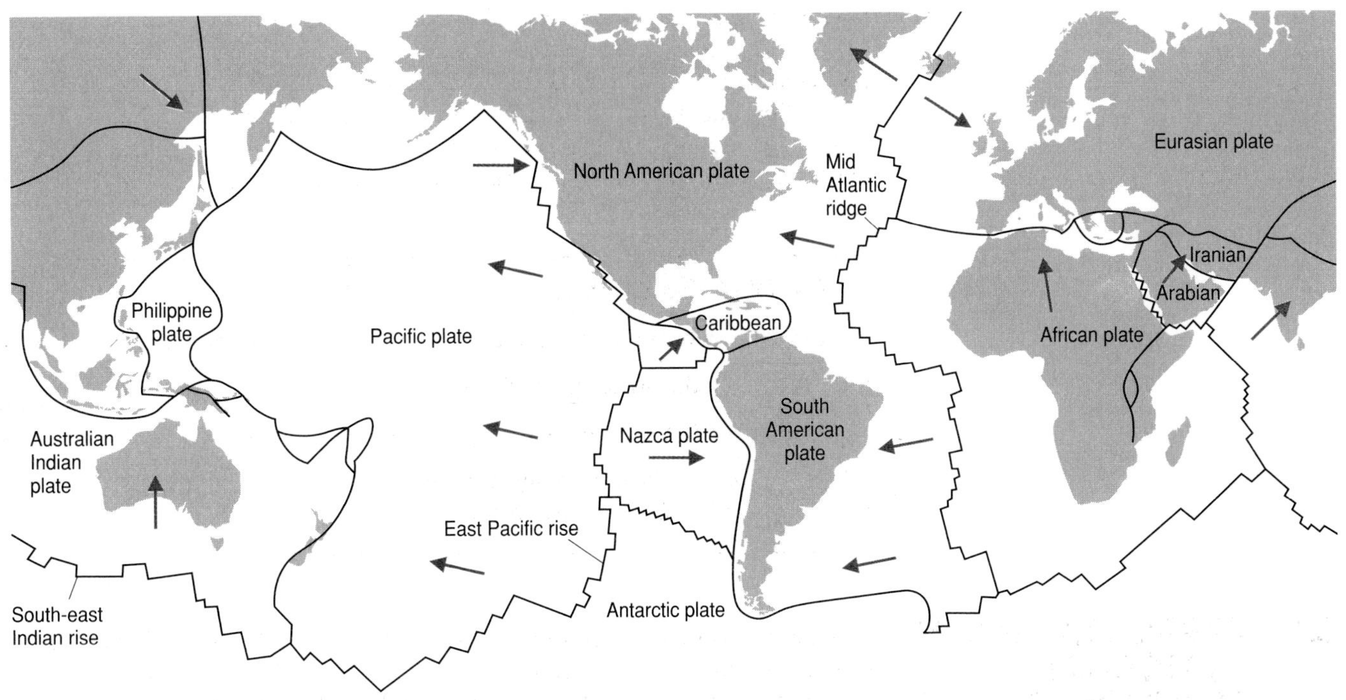

Fig 3.2 Plate boundaries.

There are three types of plate boundary.

Divergent (constructive) plate boundaries

When two plates move away from each other, molten material rises up into the rift to form new rocks. In this way, the Mid-Atlantic ridge formed between the Eurasian and Atlantic plates (Figure 3.3 a).

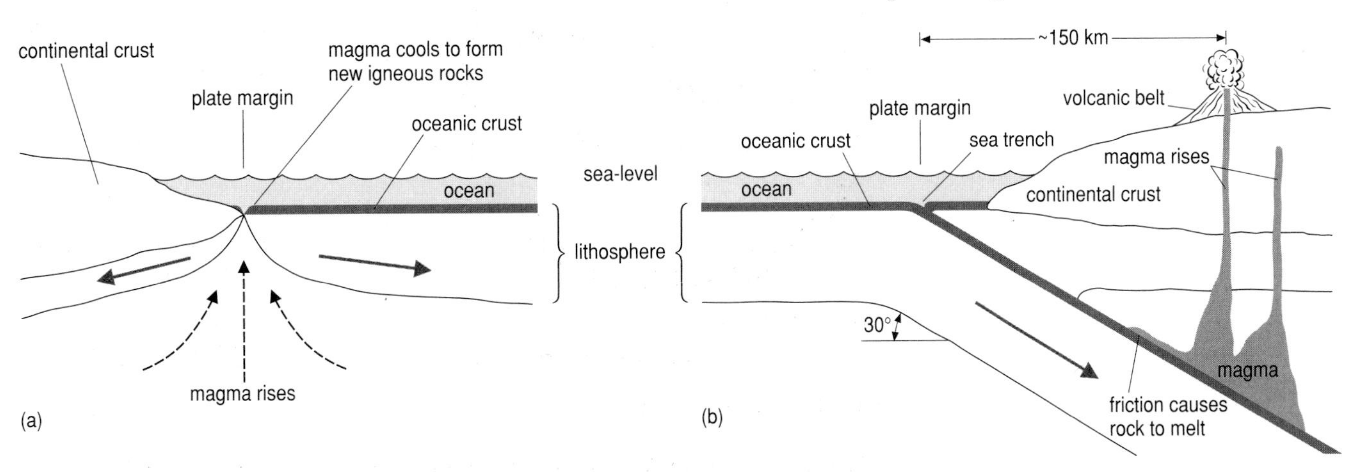

Fig 3.3 Convergent and divergent plate boundaries.

(a) Cross-section of a divergent (constructive) plate margin. New igneous rocks form when magma rises between two plates that are moving apart. This new rock forms a mid-ocean ridge on either side of the plate margin. The North American and Eurasian plates, for example, are moving apart at a rate of 1.8 cm per year.

(b) Cross-section of a convergent plate boundary. The more dense oceanic crust descends below the continental crust. As the crust descends, friction causes rock to melt, and it rises up into the continental crust and forms a volcanic chain.

Convergent (destructive) plate boundaries

When an oceanic and a continental plate collide, the denser oceanic plate is pushed down (**subducted**) beneath the continental plate. A trench up to 8 km deep usually forms at the boundary of the two plates (Figure 3.3 b). The subducted plate melts because of the higher temperatures at depth, and the molten material may then rise and erupt through the surface to form a volcano or even a chain of volcanoes. The release of tension as the subducted plate disintegrates may also cause earthquakes, which give rise to **tsunamis** (tidal waves).

Conservative plate boundaries

Conservative plate boundaries occur when plates slide past each other along a fracture (or fault) in the outer part of the crust (the lithosphere – Figure 3.1). Earthquakes occur as the plates grind and jerk past one other. San Francisco is situated almost directly above the San Andreas fault, which marks the boundary between the North American and Pacific plates.

(a) Aerial photograph of San Andreas fault as it crosses Carrizo Plain. The fault is the valley-like scar running from bottom right to top centre.

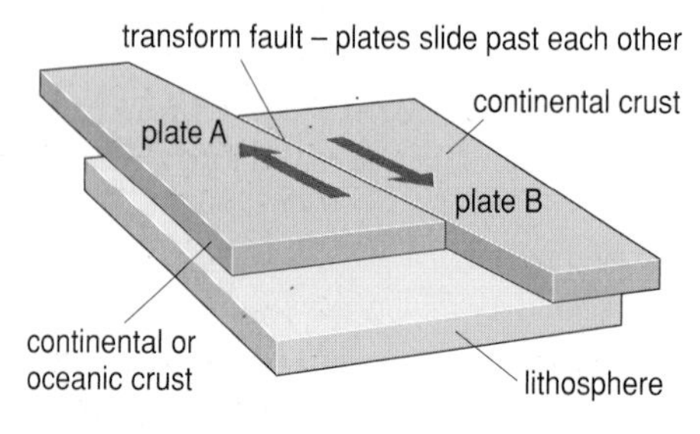

Fig 3.4 The San Andreas fault.

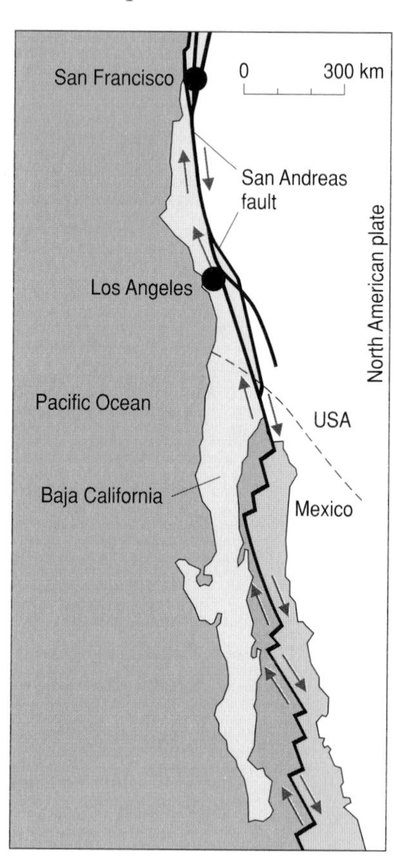

(b) Plan view of the San Andreas fault.

The North American and Pacific plates are sliding past each other at a rate of 5.6 cm per year. This causes tension which, when released, gives rise to earthquakes.

QUESTIONS

Figure 3.2 shows a plan view of the plates that cover the earth.

3.1 Contrast what is happening to the size of the Pacific and Atlantic Oceans because of plate tectonic movements.

3.2 Suggest a possible reason why Iceland has formed as a new land mass from the middle of a deep ocean.

3.2 INSIDE THE EARTH – EXTERNAL FEATURES

The rocks that protrude from the surface of the earth are continually subjected to change due to the processes of weathering and erosion, along with mass movement (movement of material downslope under the influence of gravity) and transportation. **Weathering** is the breakdown of rocks *in situ*, and the harder the rock, the slower it will weather. **Erosion** is the breakdown of rocks involving transportation, that is, rocks moved away from their point of origin by wind, waves, streams, rivers or glaciers. There are three types of weathering. **Mechanical weathering** breaks rocks up but does not alter their chemical composition. **Chemical weathering** involves the chemical change and decomposition of rocks as they react with air and water. **Biological weathering** results from the activities of plants and animals. The major forms of mechanical, chemical and biological weathering are summarised in Table 3.1 a. Mechanisms of erosion are illustrated in Table 3.1 b. Weathering produces a material called **regolith**. Such material may move downslope simply under the forces of gravity (mass movement).

Table 3.1 (a) Mechanisms of weathering

Mechanical	
release of pressure	when an overlying rock is removed, the release of pressure may cause the underlying rock to expand and fracture
freeze–thaw	when water in pores and crevices freezes, it expands (by around 10%), possibly fracturing the rock
temperature change	minerals within rocks may expand and contract at different rates, possibly causing lines of weakness and fracturing
Chemical	
solution	some minerals, e.g. rock salt, are water soluble and will dissolve in and be removed by rainwater
hydration	the incorporation of water into the crystal structure of a mineral may make it easier to degrade by erosive processes
hydrolysis	the breakdown of minerals by the hydrogen and hydroxyl ions of water e.g. hydrolysis of orthoclase feldspar $4KAlSi_3O_8 + 4H^+ + 18H_2O \rightarrow Si_4Al_4O_{10}(OH)_8 + 4K^+ + 8Si(OH)_4$ the potassium ions and silicic acid may then be removed in solution
oxidation	the chemical combination of oxygen with a mineral, making it easier to separate by erosive processes
reduction	the chemical removal of oxygen from a mineral, making it easier to separate by erosive processes
carbonation	carbonic acid, formed when carbon dioxide dissolves in rainwater, will chemically attack limestone to form calcium hydrogen carbonate (bicarbonate) which will dissolve and be removed $H_2O + CO_2 \rightarrow H_2CO_3$ $H_2CO_3 + CaCO_3 \rightarrow Ca(HCO_3)_2$
Biological	
root penetration	roots may penetrate crevices and physically shatter rocks
trampling	the feet of humans and any other animal may physically damage rocks
humic acids and carbon dioxide	produced by soil organisms and chemically attack rock material

Table 3.1 (b) Mechanisms of erosion

Mechanism	
abrasion	scraping of any hard material against a rock, e.g. when material carried by a river scrapes along its bed or sides
attrition	collision of rocks, e.g. when a wave tosses a pebble against a cliff
hydraulic action	the sudden injection of water into crevices causes the air in the crevices to become compressed, increasing pressure which may lead to disintegration and removal of parts of the rock
ice plucking	movement of a glacier after freezing onto loose debris transports it away

Mineral resources

The earth's crust is composed of minerals and organic material. Minerals are elements or inorganic compounds which have a crystalline structure and are made up of an orderly arrangement of atoms or ions (atoms carrying a charge). Some minerals, such as gold and sulphur, are made up of just one element, while others, such as apatite ($Ca_5(PO_4)_3F$), from which phosphate fertilisers are produced, contain several elements. Some of the most important rocks in industry and the main mineral which they contain are shown in Table 3.2.

Table 3.2 Industrially important minerals

Rock/source of mineral	Main mineral
limestone	calcium carbonate $CaCO_3$
salt	halite NaCl
gypsum	gypsum $CaSO_4 . 2H_2O$
china clay	kaolinite $Al_4Si_4O_{10}(OH)_8$

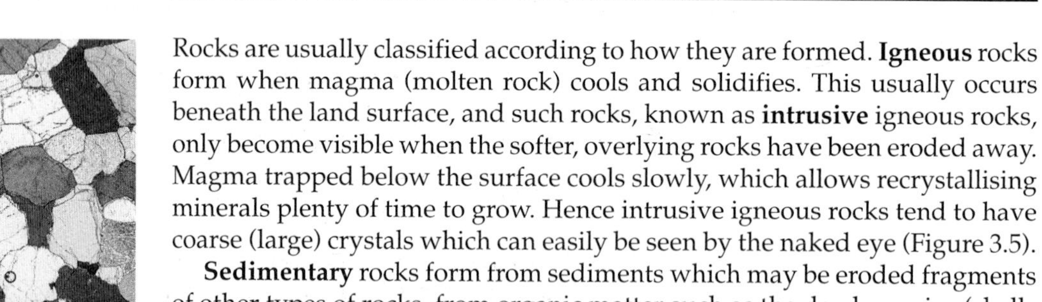

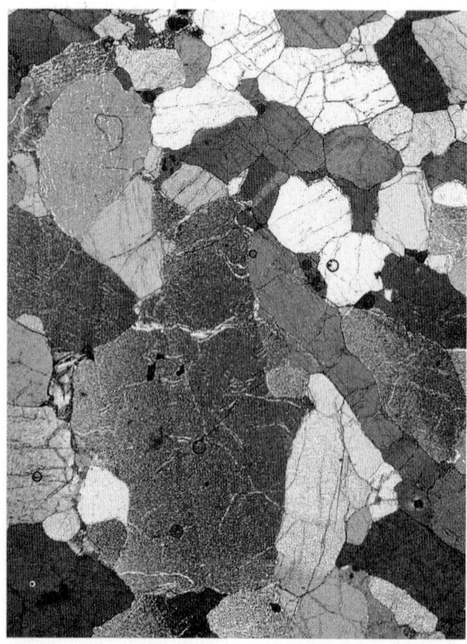

Fig 3.5 Polarised light micrograph of a thin section of monzonite, an igneous rock, showing the crystal structure.

Rocks are usually classified according to how they are formed. **Igneous** rocks form when magma (molten rock) cools and solidifies. This usually occurs beneath the land surface, and such rocks, known as **intrusive** igneous rocks, only become visible when the softer, overlying rocks have been eroded away. Magma trapped below the surface cools slowly, which allows recrystallising minerals plenty of time to grow. Hence intrusive igneous rocks tend to have coarse (large) crystals which can easily be seen by the naked eye (Figure 3.5).

Sedimentary rocks form from sediments which may be eroded fragments of other types of rocks, from organic matter such as the dead remains (shells or skeletons) of marine organisms, or from chemical precipitation of minerals out of solution. Eroded material from pre-existing rocks can accumulate in many different environments. Deposition may occur in terrestrial environments, such as deserts and glacial areas, and in aquatic environments, such as lakes (lacustrine), rivers (fluvial) and the sea (marine). The most common sedimentary rock is limestone (calcium carbonate – $CaCO_3$), which is formed in marine environments and which is a source of many fossils.

Sediments may accumulate over thousands of years and may eventually form deposits hundreds of metres thick. The tremendous pressure of the material may cause the sediments to bind together, forming rocks such as sandstone or limestone, in a process known as **cementation**.

Fig 3.6 Cross-section through a limestone block. This limestone has developed through the accumulation of microscopic animals whose shells were made of calcium carbonate.

Metamorphic rocks form when an igneous or sedimentary rock is subjected to great heat and pressure, as may happen, for example, at a plate boundary. Table 3.3 lists common examples of the three major rock types. Over 90% of rocks in the crust are igneous, but due to the effects of weathering over 75% of the surface is covered by sedimentary rocks.

Table 3.3 Rock types

Rock type	Origin	Example
igneous	magma	granite, basalt
metamorphic	igneous rock	granite→gneiss
	sedimentary rock	limestone→marble, mudstone→slate
sedimentary	mechanical breakdown	conglomerate, sandstone
	organic accumulation	coal
	chemical precipitation	gypsum

QUESTION

3.3 The diagram below shows stages in the formation of evaporite deposits.

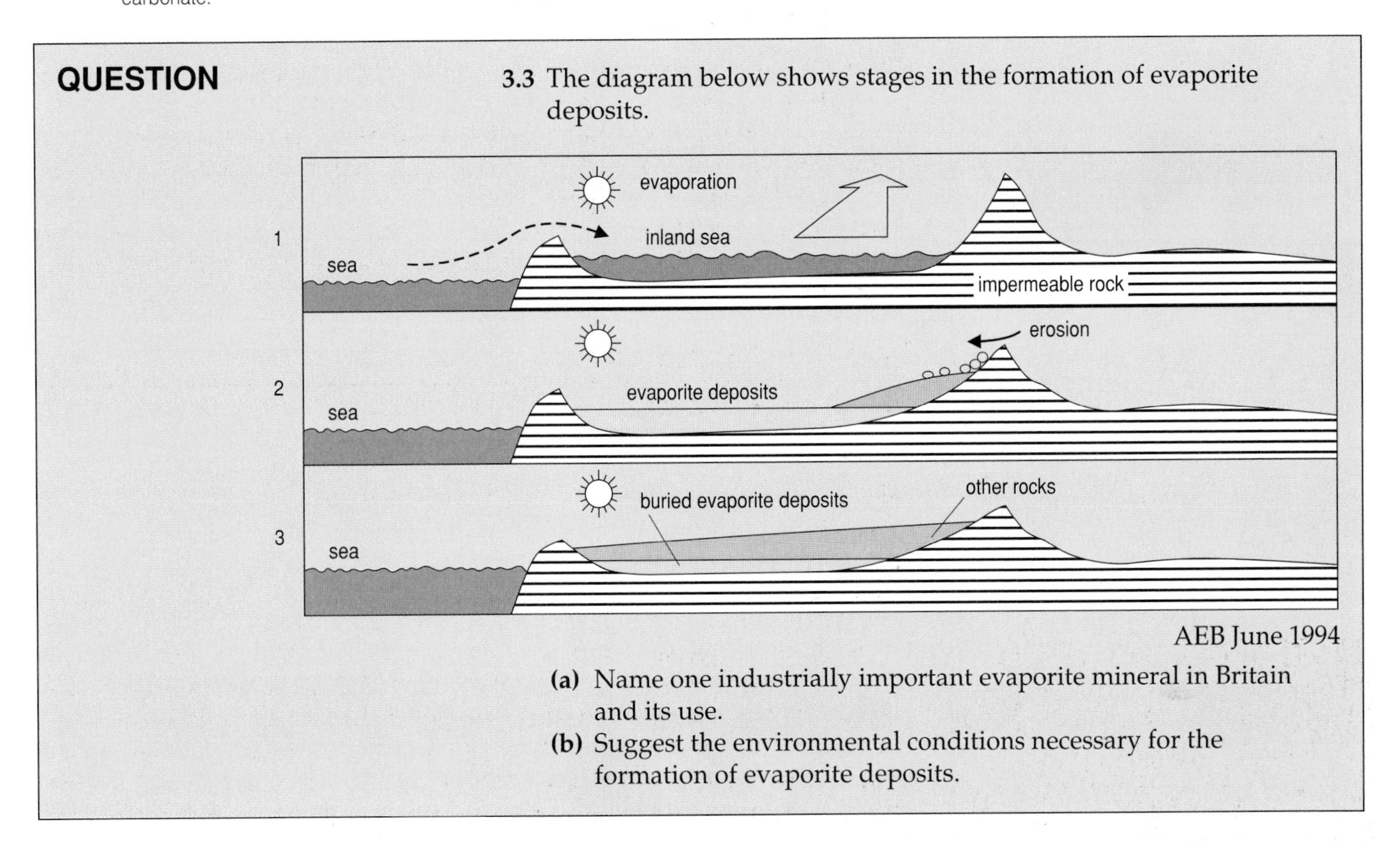

AEB June 1994

(a) Name one industrially important evaporite mineral in Britain and its use.

(b) Suggest the environmental conditions necessary for the formation of evaporite deposits.

As we have seen, rocks can be created, destroyed or dramatically changed. These processes form the basis of the rock cycle (Figure 3.7).

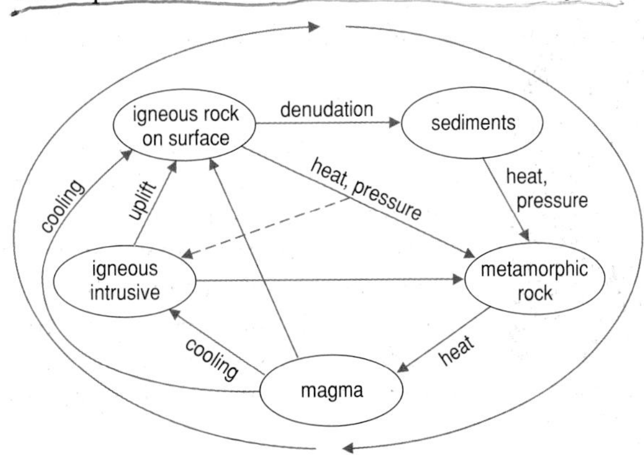

Fig 3.7 The rock cycle. Within the cycle there are three stages in the formation of sedimentary rock, namely **denudation** (weathering and erosion of rock material), **transportation** and **deposition**.

QUESTIONS

3.4 Describe the links to and from metamorphic rock in the rock cycle.

3.5 Name the two main energy sources that drive the rock cycle.

The geological time-scale

The geological time-scale is split into eons, eras, periods and epochs, each of which is subdivided into unequal time units on the basis of the fossils within or their relative age. Each subdivision is given a name derived from a Greek or Latin term, a geographical region or a rock type.

Eon	Era	Period	Epoch	Age (million years)
Phanerozoic	Cenozoic	Quaternary	Holocene	0.01
			Pleistocene	1.6
		Tertiary	Pliocene	5.3
			Miocene	23.7
			Oligocene	36.6
			Eocene	57.8
			Palaeocene	66.4
	Mesozoic	Cretacous		144
		Jurassic		208
		Triassic		245
	Palaeozoic	Permian	Divisions not shown	286
		Pennsylvanian		320
		Missisippian		360
		Devonian		408
		Silurian		438
		Ordovician		505
		Cambrian		570
Proterozoic		No divisions used		2500
Archaean				3800
Hadean				approx. 4650

Fig 3.8 The geological time-scale.

The oldest rocks on earth are from the Archaen era (4 billion years ago). Life on land started 300 to 400 million years ago, but the earliest multicellular life started in the Precambrian Age, 700 million years ago.

At present, we are in the Quaternary period and the Holocene epoch (Greek for 'wholly recent'). The boundary of the Pleistocene and the Holocene epochs is given as the end of the last Ice Age, which was 10 000 years ago.

3.3 LANDFORM DEVELOPMENT

Landforms are created by the effects of weathering, erosion, mass movement and tectonic movement on the rocks around. The important geomorphological regions in the UK are glacial, river and marine systems, although there are many others throughout the world (for example deserts, high mountain ranges, salt lakes and savanna plains).

Glacial landforms

Glacial landforms (Figure 3.9) develop from erosion or deposition by ice and glacial meltwater. In Britain, the last Ice Age was at its peak around 18 000 years ago, extending as far south as London, and ended around 10 000 years ago (during the Pleistocene geological period).

Erosional landforms occur in upland areas, such as the Brecon Beacons, Snowdonia and the West Highlands of Scotland, where the climate is cold enough for snow to accumulate.

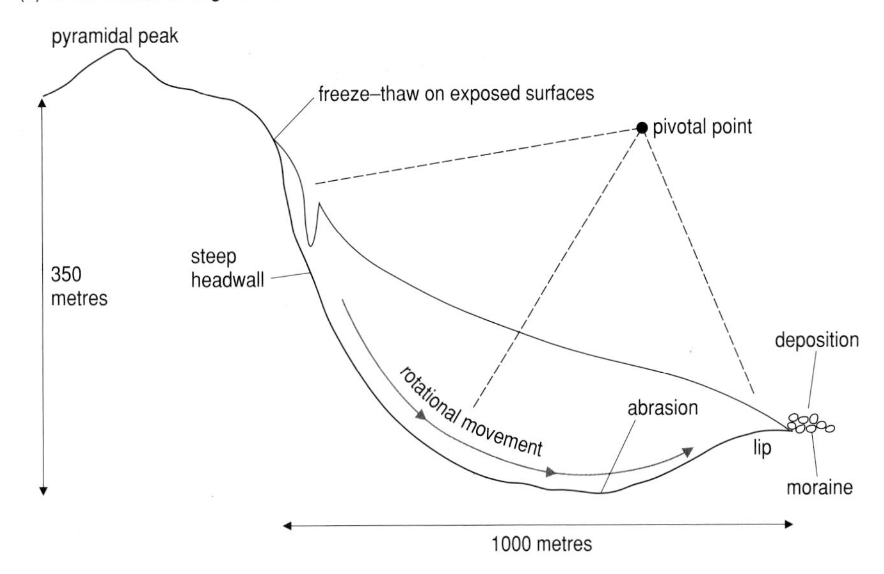

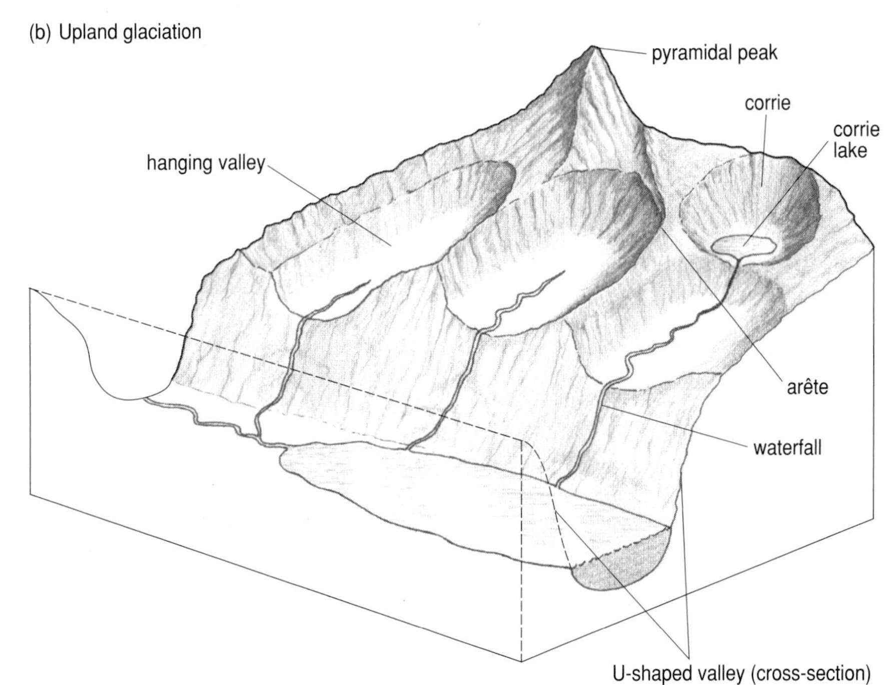

Fig 3.9 Glacial landforms.

Corries are 'amphitheatre-shaped' depressions formed by a small glacier. Formation starts when snow accumulates in a hollow. This ice rotates under its own weight, scours the underlying rock, and creates a hollow depression at the base and a steep headwall. The corrie may fill with water to form a lake.

An **arête** is a sharp 'knife edge' ridge that develops when two corries erode into each other. **A pyramidal peak** is left between the corries.

The cross-section of a glaciated valley is characteristically U-shaped, whereas a river valley has a V-shaped profile.

The valley bottom, which may contain a lake once the glacier has melted, will become filled with sediments and thus develop a flat bottom. A hanging valley is a tributary valley which is left stranded high up on the sides of a main valley because the main valley erodes at a much faster rate.

Depositional landforms occur at the edge of these valley glaciers, and also in much of eastern Britain, where vast ice sheets extended from Scandinavia.

The material deposited directly by ice is called **till** and will contain boulders, sands and clays that may have been transported great distances. An individual boulder deposited on a different rock type is called an **erratic**. However, erratics are very small depositional features compared to moraines, drumlins and outwash plains (Figure 3.10). A **moraine** is a mound of till that was deposited at the end of the glacier. **Drumlins** are large, elongated till deposits that are aligned in the direction of ice flow.

Deposition can also result from meltwater coming off glaciers. **Outwash plains** are thin, flat fluvio-glacial deposits (materials deposited from rivers flowing out of a glacier), which can cover considerable areas (Chapter 14). Meltwater streams underneath the glacier can sometimes become filled with sediment so that when the glacier melts away, a long sinuous ridge of material is left standing. This is called an **esker**.

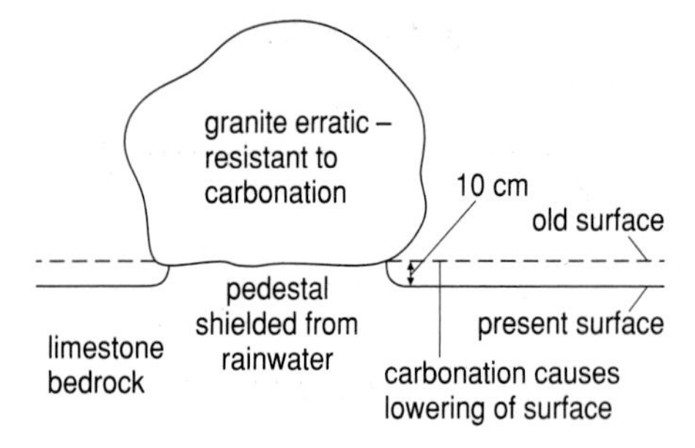

(a) An erratic.
The erratic acts as an 'umbrella', shielding the limestone from the effects of carbonation. By measuring the distance from the present surface to the bottom of the erratic, we can infer that carbonation has lowered the surface by approximately 10 cm in 10 000 years.

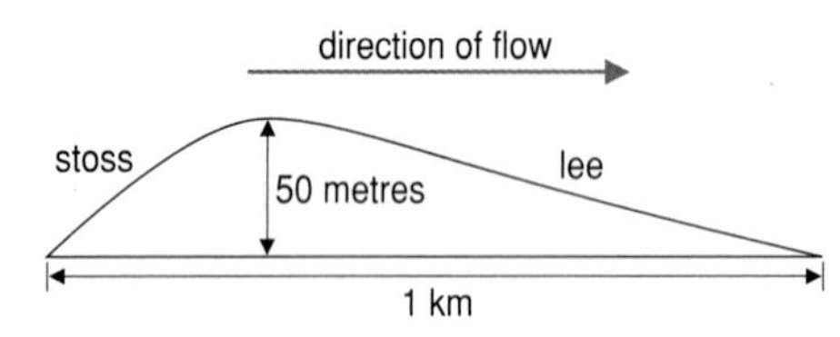

(b) A drumlin.
Drumlins are massive depositional features aligned in the direction of flow that often occur in large numbers or 'swarms' (for example in Upper Ribblesdale, North Yorkshire).

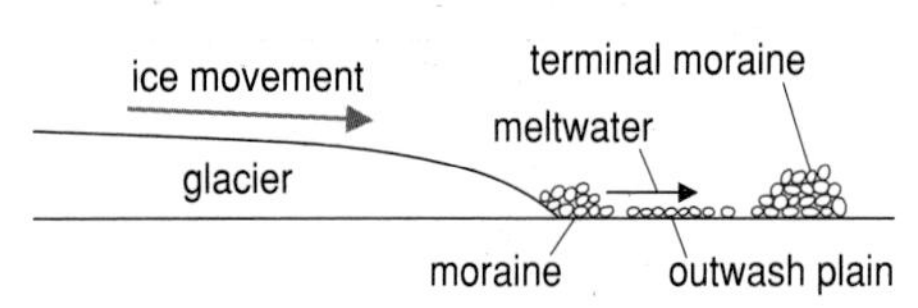

(c) A moraine.
The longer the snout of the glacier remains stationary, the more material can be deposited and the larger the moraine will be.

Fig 3.10 Erratics, drumlins and moraines.

River systems

Rivers act as an essential resource but may also represent an environmental hazard (Chapter 11). The major uses of river systems include: damming valleys for water storage, direct extraction for water supplies, transportation, waste disposal, leisure (for example canoeing or fishing) and as a wildlife habitat (Chapter 11). Hazards to people include flooding and soil erosion. Rivers are important agents of erosion. Material can be moved within a channel by the mechanisms shown in Figure 3.11.

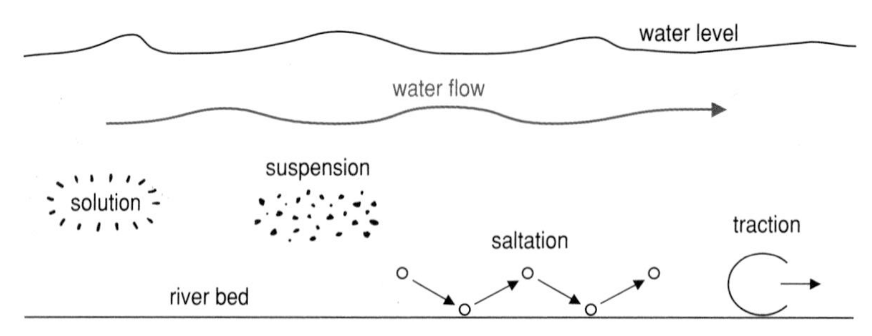

Fig 3.11 Sediment transport in a river.
- Solution – material is dissolved within the water column.
- Suspension – fine particles may be light enough to be carried completely within the water column.
- Saltation – larger particles bounce along the bottom, sometimes in suspension, sometimes hitting the river bed.
- Traction/rolling – large pebbles that are too heavy to be picked up by the current can be rolled along the bottom.

River systems typically have a **V-shaped valley cross-section** and a smooth **graded long profile (**Figure 3.12). At the top of the long profile there are numerous waterfalls, the channel is narrow and relatively straight, and has a steep gradient, a low volume and a high erosion rate. Towards the bottom of the long profile the river **meanders** (bends), has a low gradient and a high volume. At a meander, the water tries to go straight ahead, and therefore erosion occurs on the outside of a bend. Deposition occurs on the inside of the bend where the velocity of the stream is not strong enough to carry material (Figure 3.13).

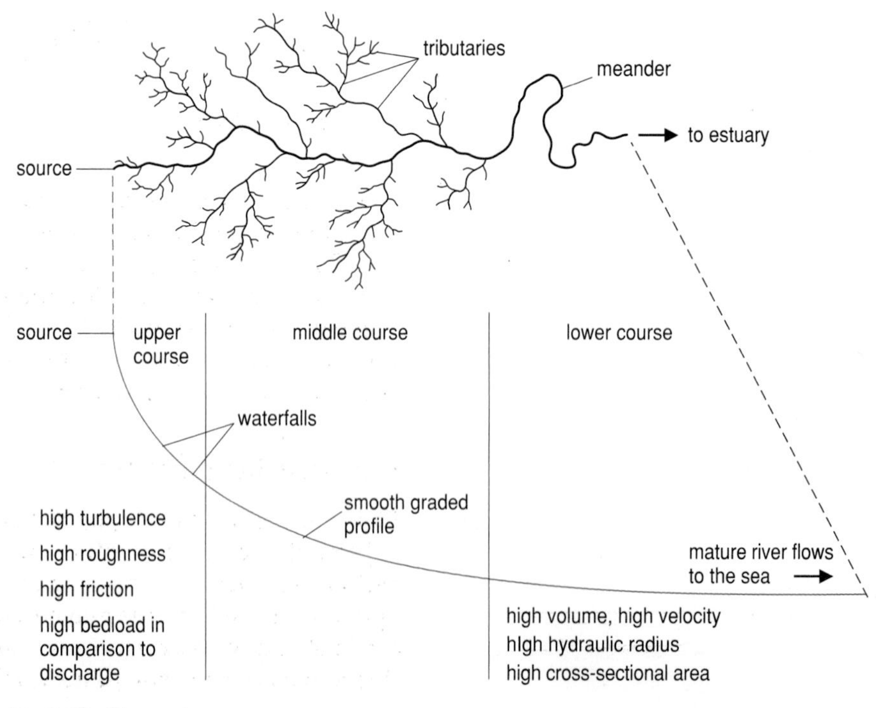

Fig 3.12 River systems.

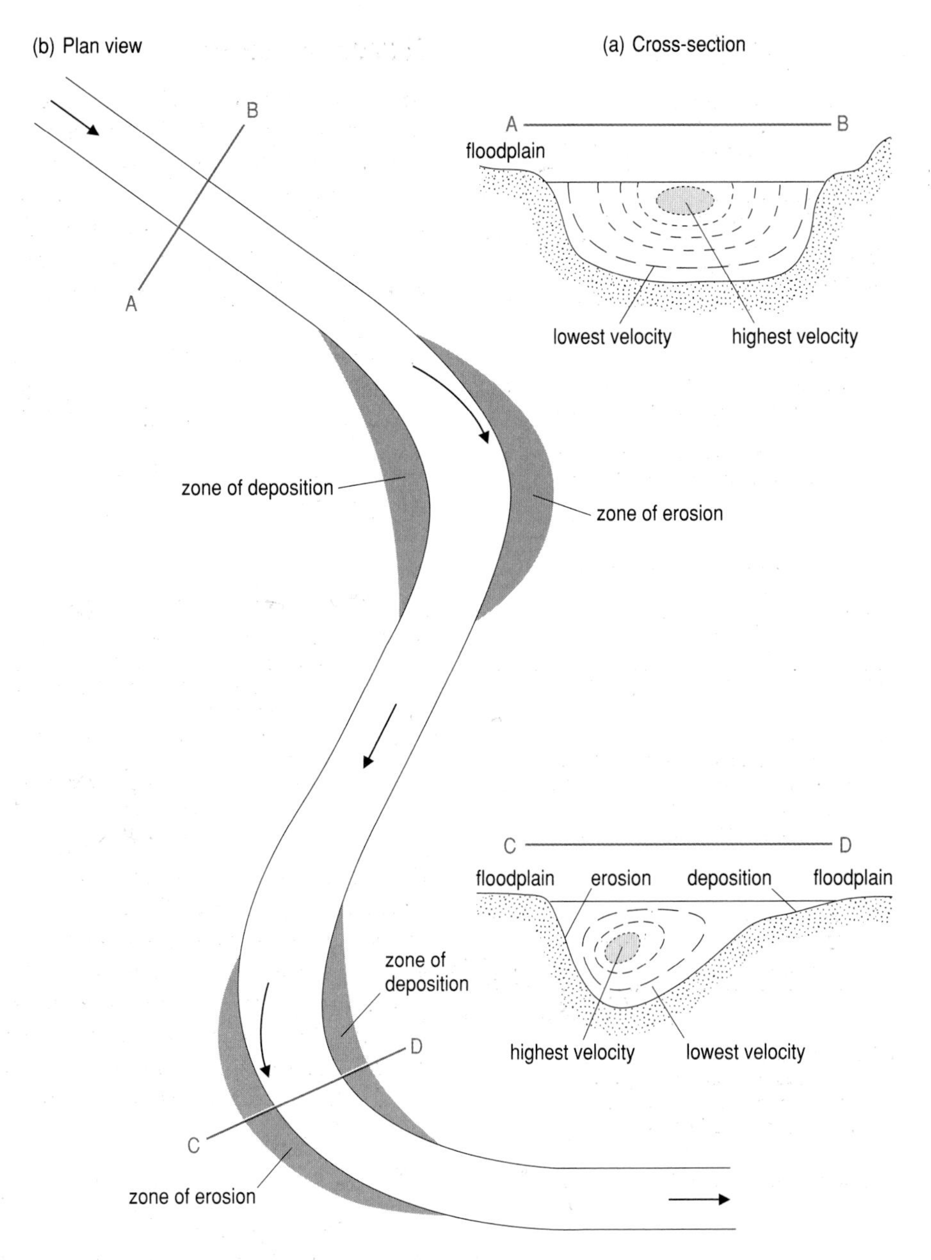

Fig 3.13 River channels.

Due to these areas of erosion and deposition, the river channel is constantly migrating across the valley bottom. The flat area on either side of the channel (the **floodplain**) is fertile because of silt deposition during floods, although some farmers have installed drainage channels and pumping systems to prevent damage caused by these floods. Altering the hydrological regime in this way can damage nearby plant communities that have developed specialist characteristics as a result of the seasonal flooding.

Coastal landforms

Over half of the world's population live in coastal zones, and such areas are subject to intense economic and industrial activity. Coastal erosion is a serious problem for many parts of the world, threatening properties, agricultural land and the beaches, which annually attract millions of tourists. Understanding coastal processes may be even more important in the future due to the potential rise in sea-levels (Chapter 1).

QUESTION

3.6 Suggest why large human settlements have often developed on coastlines.

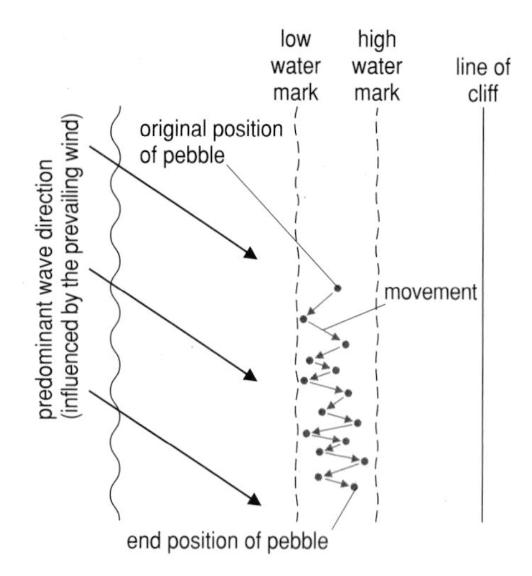

Fig 3.14 Long-shore drift – a plan view.
Material is transported along the shoreline through the action of long-shore drift.

Material is transported along the coastline by the process of **long-shore drift** (Figure 3.14). In this process, material is moved towards and along the beach by the forward impact of the waves (swash), and back towards the sea as the wave retreats (backwash). The force of waves is rarely perpendicular to the shoreline, and therefore material is moved in the predominant wave direction.

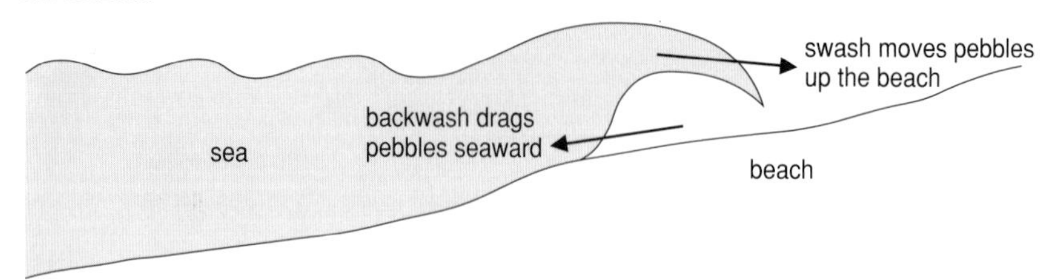

Fig 3.15 Cross-section of a breaking wave.
A wave is constructive when material is moved towards the beach, and destructive when material is moved back down towards the sea.

A **saltmarsh** is a flat coastal depositional landform inhabited by halophytic (salt-tolerant) plants. Saltmarshes are important resources for waste disposal, conservation, recreation and food, and, by spreading out the force of the sea in a complex drainage network, they help to prevent cliff erosion.

The most obvious coastal landform is the **sea cliff**. Cliffs erode when stormwater forms a wave-cut notch by removing material at the base. This notch gradually enlarges until the land above becomes too heavy and slides down onto the beach below (Figure 3.16). The landslide is removed by long-shore drift and another wave-cut notch then forms. The coastline is also eroded by wind, mass movements, freeze–thaw and chemical weathering.

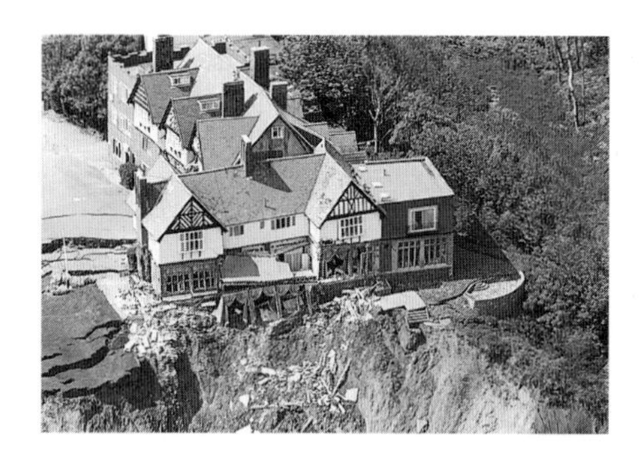

Fig 3.16 Sea cliff erosion resulting in the collapse of Holbeck Hall Hotel.

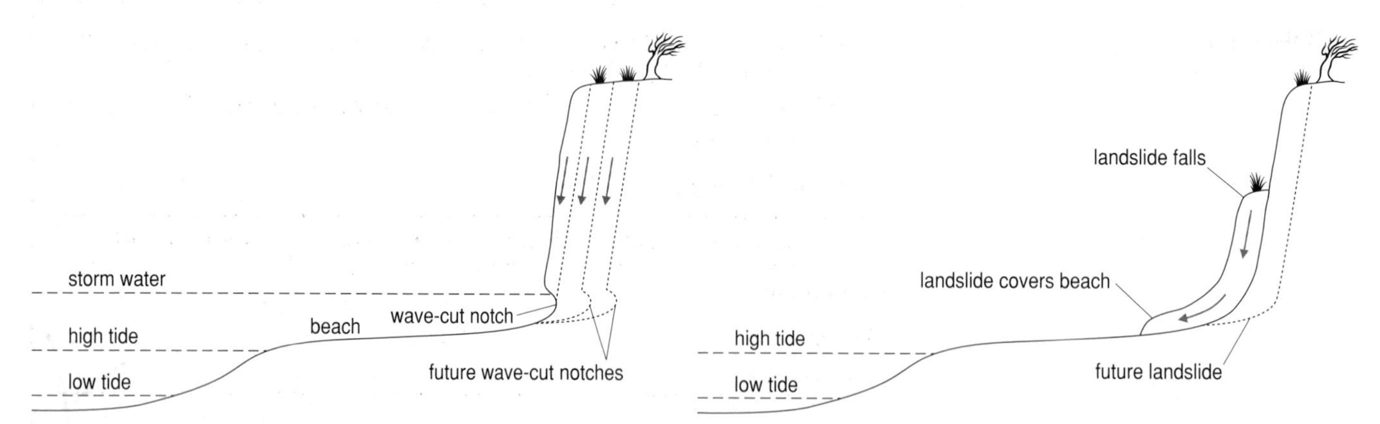

Fig 3.17 The effects of coastal landslides.
Numerous management techniques have been implemented to prevent both the actions of long-shore drift and cliff erosion. However, management can be very expensive and is therefore only applied to areas which merit the expense. Many sea fronts have been protected by huge concrete walls and large boulders, which will take longer to be destroyed by the sea. Groynes, which are a familiar site on many beaches, are used to trap sediment moving by long-shore drift (Figure 3.18).

However, management may create other problems or may simply move the problems further along the coast. For example, damage to an important

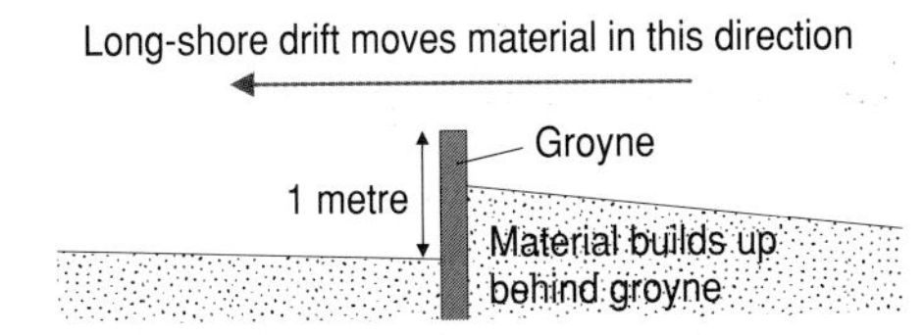

Fig 3.18 Cross-section of a groyne.
Groynes, designed as a barrier to prevent long-shore drift, are a common sight on tourist beaches. This preserves the sand on the beaches, which helps the local tourist industry.

coastal bird sanctuary could be caused by trapping sediment on a tourist beach, thus depriving the sanctuary of sediment which will increase erosion and prevent deposition. There are many measures that can be used to counteract coastal processes.

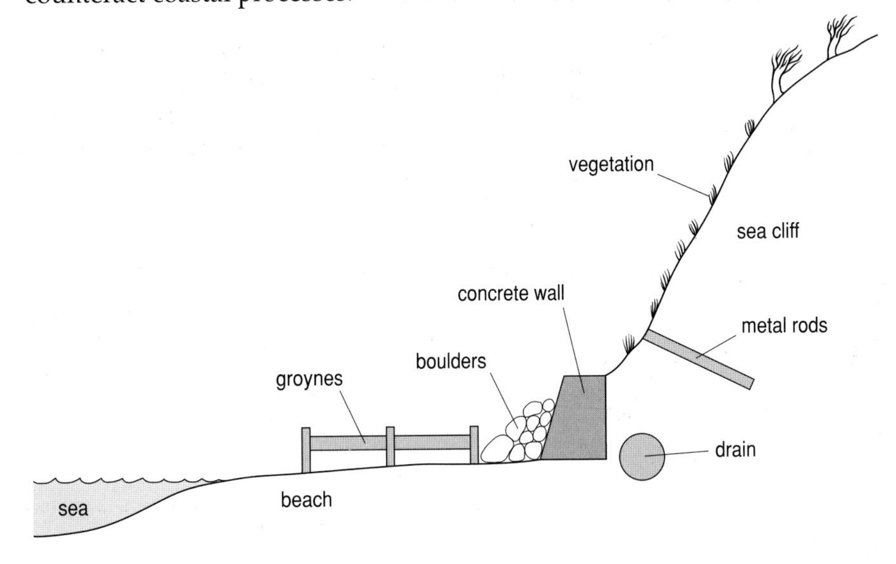

Fig 3.19 Combating coastal erosion.

Sea cliff erosion

Wave action produces a wave-cut notch at the base of a cliff. This notch will enlarge until the material above falls as a landslip onto the beach. The slip will gradually be removed by long-shore drift and a new wave-cut notch will form at the base of the cliff.

QUESTION

3.7 Explain how each of the measures in Figure 3.19 helps to prevent coastal erosion.

SUMMARY ASSIGNMENT

1. **(a)** Explain how biological weathering on a highly jointed rock will aid physical disintegration.
(b) Suggest three factors which could influence the rate of movement of rock debris on a slope.

2. Complete the table below.

Type of plate boundary	Movement of plates	Named example
convergent		
divergent		
conservative		

Chapter 4

SOILS

The soil is one of our most valuable resources. Without it, most of life on earth could not occur, yet it is a resource which many people take for granted.

LEARNING OBJECTIVES

After completing the work in this chapter you will be able to:

1. describe the components of a fertile soil
2. describe the process of soil development
3. explain why different types of soil develop in different areas.

4.1 SOIL COMPONENTS AND PROPERTIES

Soil is a complex mixture of air, water, mineral salts, weathered minerals, organic matter and millions of living organisms, most of which are microscopic (Figure 4.1).

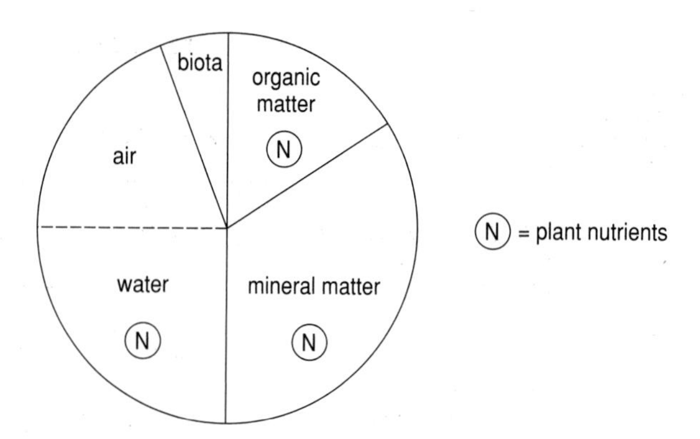

Fig 4.1 Soil components in agricultural loam.

Table 4.1 Soil texture

Type of particle	Size of particles / mm
stones	> 2.0
sand	0.02–2.0
silt	0.002–0.02
clay	< 0.002

Soil texture

Soil texture refers to the coarseness or size of the mineral particles in a soil and is determined by the relative proportions of sand, silt and clay particles in the soil (Table 4.1).

Soil texture is important because it affects the structure of the soil, that is, how the particles fit together, as well as the nutrient and water availability in the soil. It is classified by plotting the relative proportions of sand, silt and clay on a triangular graph (Figure 4.2).

Sandy soils have many pore spaces and are consequently free-draining and easy to cultivate. Clay soils, however, are usually much less free-draining, heavier and more difficult to cultivate. The ideal soil texture for agriculture is loam (Chapter 14) because its clay content (20%) holds nutrients and water, its sand content (40%) ensures good drainage and large pore spaces for aeration, and its silt content (40%) helps to hold the clay and sand together.

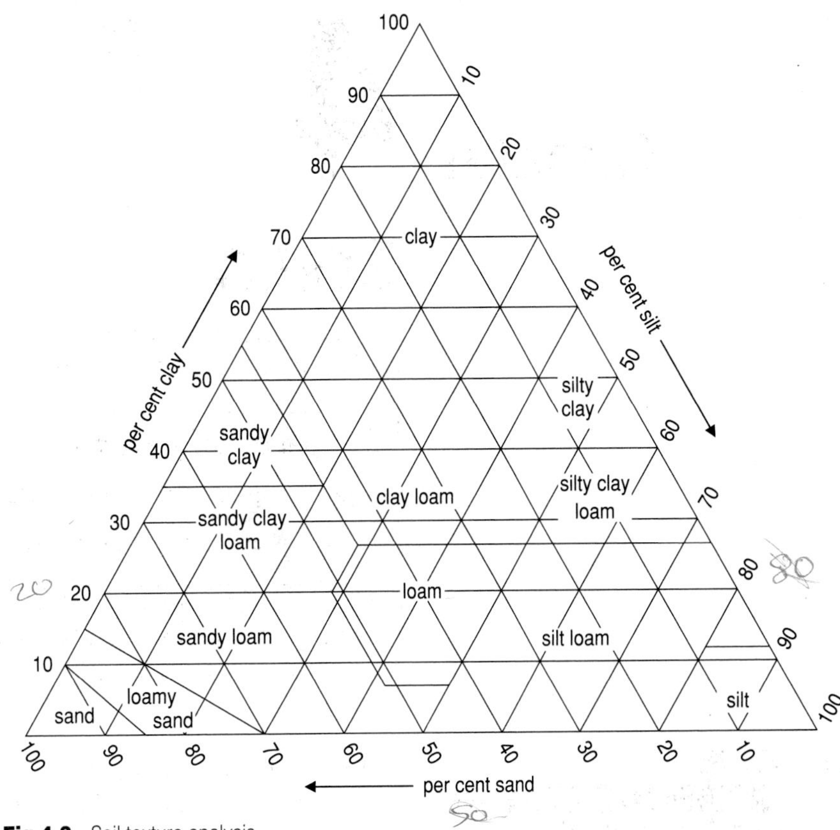

Fig 4.2 Soil texture analysis.

Soil structure

Type of structure /ped	Size of structure /mm	Description of peds	Shape of peds	Location /horizon–texture	Agricultural value
crumb	1 to 5	Small individual particles similar to breadcrumbs Porous		A horizon – loam soil	The most productive. Well aerated and drained – good for roots
granular	1 to 5	Small individual particles Usually non-porous	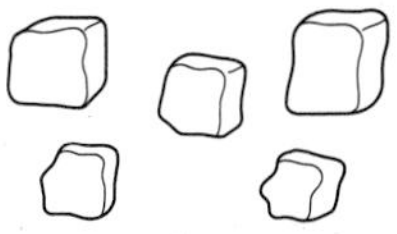	A horizon – clay soil	Fairly productive. Problems with drainage and aeration
platy	1 to 10	Vertical axis much shorter than horizontal, like overlapping plates preventing flow of water	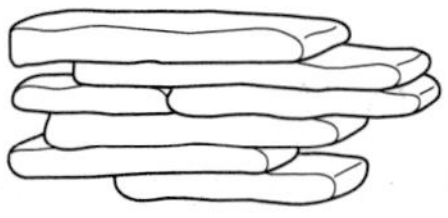	B horizon – silts and clays or when compacted by farm tractors	The least productive. Hinders water and air movement. Restricts roots
blocky	10 to 75	Irregular shape. Horizontal and vertical axes about equal May be rounded or angular but closely fitting	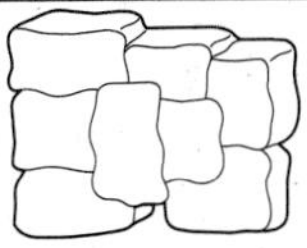	B horizon – clay-loam soils	Productive. Usually well drained and aerated
prismatic	20 to 100	Vertical axis much larger than horizontal. Angular caps and sides to columns		B and C horizons – often limestones or clays	Usually quite productive. Formed by wetting and drying. Adequate water movement and root development
columnar	20 to 100	Vertical axis much larger than horizontal Rounded caps and sides to column		B and C horizons – alkaline and desert soils	Quite productive (if water available)

Fig 4.3 Soil structure.

Soil structure describes how the soil components fit together into aggregates or peds (Figure 4.3). It is the arrangement of the peds which determines the size and nature of the soil pores, and this largely influences aeration, drainage and how easy it is for plant roots to penetrate the soil. Peds are held together by the stickiness of their clay component, as well as by humus, roots, fungal hyphae and inorganic 'cements', such as iron and aluminium hydroxides.

The most fertile soils have a **crumb ped** structure. This means that they are well aerated and drained, having a structure of small individual particles similar to bread crumbs. Platey ped formations are not fertile because they hinder the infiltration of water and air. Clayey soils have a high water retention and poor drainage because their small, flat clay platelets fit together neatly. Sandy soils have large pore spaces and therefore have good infiltration but low water retention (Chapter 11).

QUESTIONS

4.1 What is the difference between soil texture and soil structure?

4.2 How does soil texture influence soil structure?

4.3 What effect will a heavy tractor moving over a soil have on the soil structure?

Soil minerals

Primary soil minerals are those minerals that were present in the parent material (underlying rock or sediment) and have entered the soil in an unaltered state. They tend to be weather-resistant, insoluble minerals, such as quartz.

Secondary soil minerals are those minerals that were not present in the original parent material and have been formed in the soil through the process of chemical weathering. Secondary minerals, such as aluminosilicates and iron, and aluminium and manganese hydroxides, are found in the clay components of soils.

Organic material

The breakdown of plant litter and animal faeces into humus is termed humification and is accelerated in hot and wet conditions. Humus is an extremely complex mixture containing, amongst other things, organic acids, polysaccharides, amino acids and phosphates. Besides providing a source of essential elements, such as nitrogen, sulphur and phosphorus (Chapter 8), humus may become bound to clay minerals, iron and aluminium hydroxides, to form stable aggregates which improve the physical properties of the soil for plant growth (Chapter 14).

Soil organisms (biota)

Each gram of soil usually contains millions of bacteria, and it is these, along with what may seem fantastic numbers of fungi, insects and earthworms, which are responsible for the return of nutrients to the soil from decomposing plant and animal material.

Earthworms drag leaves into their burrows and eat them. The organic material which passes out of the other end of the worm is chemically quite different from the leaf that was eaten. Moisture content and pH are both increased, as is the surface area of the material. All of these changes make it easier for bacteria and fungi to break down the organic material.

Leguminous plants, such as peas (*Lathyrus* sp.), clover (*Trifolium* sp.) and alder (*Alnus* sp.), fix atmospheric nitrogen from the soil air as part of the nitrogen cycle (Chapter 8). A plentiful supply of oxygen is essential for the activities of aerobic bacteria, which are responsible for the breakdown of organic matter and the release of nutrients back into the soil.

Soil water

The amount of water in the soil depends on the water-holding capacity and the infiltration capacity, both of which are influenced by the soil's texture and structure. Soil water exists in four different forms:

- Free water (or gravitational water) collects in large pores, is under the influence of gravity, and thus can freely drain away.
- Capillary water is held in medium-sized pores against the force of gravity by capillary action.
- Hygroscopic water exists as a thin layer attracted by surface tension to every soil particle.
- Combined water is chemically bound to a soil particle.

Neither combined water nor hygroscopic water is available to plants. Gravitational water is only available while it is infiltrating the soil, and thus capillary water is the major water source for plants.

Soil air

Air is present in the pore spaces which are not completely filled with water. The amount of air in a soil is therefore dependent upon the texture, structure and the amount of soil water. Most of the animals which live in the soil need oxygen to survive, as do the roots of plants. If no air is present (anaerobic), perhaps because the soil is waterlogged and all the air spaces in the soil are filled with water, plant roots will be unable to grow as they will have insufficient oxygen. Under these conditions of low oxygen, the rate of decomposition of organic matter by micro-organisms will be very slow.

Soil pH

The pH of the soil has a major effect on the availability of soil nutrients. Many of the important elements which plants need are most easily available in slightly acidic soils. The pH of very acidic soils may be increased by the addition of lime.

4.2 SOIL DEVELOPMENT

To illustrate the process of soil formation we will consider the formation of soils on newly formed sand-dunes (Figure 4.4).

There are six main factors which affect soil development.

- Parent material, that is, the underlying rock or sediments which are weathered to provide mineral matter. If the parent material is a very hard rock, such as igneous granite, then weathering and soil formation will proceed slowly (Chapter 3).
- Climate has a major effect on the rate of weathering; generally the hotter and wetter the climate, the faster weathering proceeds. The rate of plant growth will be affected by climate, and so will the amount and rate at which water and nutrients are removed from the soil. The rate of decomposition, which returns nutrients and organic matter back to the soil, is largely determined by water availability and temperature. In fact, water movement within the soil controls most of the other processes which occur in the soil.

- Relief plays an important role because aspect and altitude will affect the climate (Chapter 2), and gradient will affect the drainage characteristics.
- Soil organisms greatly affect the chemical properties of the soil. Earthworms burrowing in the soil mix and aerate it and increase its organic content. The invertebrates, bacteria and fungi which break down organic matter return minerals to the soil, often helping to determine the pH of the soil in the process (Chapter 8).
- Time is an important factor. The rate of development largely depends upon the climate but, in general, it may take 400 years to develop 1 cm of a temperate soil.
- Humans may be viewed as a sixth factor. The way they cultivate soils will affect the soil's development and composition (Chapter 14).

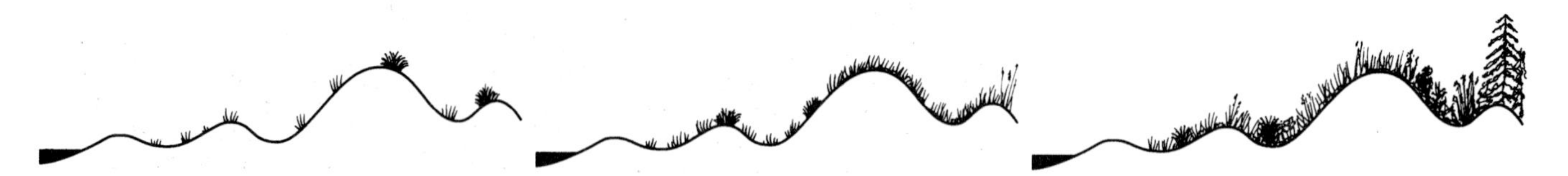

(a) Initially all there is is sand. This is basically silica. The dune is exposed, cold and windy, and sand is blown back and forth. Conditions for plant growth are poor; the many air spaces between the coarse grains of sand allow water to drain away almost immediately, and very few mineral nutrients are available. Only the seeds of specialised salt-tolerant (halophytic) plants such as marram grass are able to germinate above the high watermark. The roots of the grass slowly penetrate the sand and take hold.

(b) The growth of the marram grass causes **physical** and **chemical** changes in the dune. The plant shields the sand from the wind, reducing sand blow and wind speeds around the surface of the soil, which in turn means that temperatures are slightly higher than those experienced in non-vegetated parts of the dunes. The marram grass therefore creates a microclimate.

As some plants die, **organic matter** or **humus** is added to the sand. Decomposers break down this organic matter, releasing nutrients which can then be absorbed by plant roots. The organic matter improves the structure of the soil and also helps to hold water molecules in the sand, giving plant roots a chance to absorb them.

(c) As the physical and chemical conditions improve, the seeds of other plant species are able to germinate and grow. Such species absorb nutrients from deep below the dune, which are released into the surface layers when the plants die. The microclimatic effect becomes more marked, the roots bind the sand and the water, humus and nutrient content continue to increase. The shifting sands of the dune are becoming soil.

Fig 4.4 Soil development on sand-dunes.

QUESTION

4.4 Explain how the nature of the parent material may affect the development of soils on sand-dunes.

These are the general factors which will affect the rate and type of soil development in a given area. However, the soil should be thought of as an extremely complex system, and it is only by examining in more detail the inputs, outputs and internal processes of the soil that we can obtain a realistic picture of what is happening.

4.3 SOIL TYPES

The major inputs to and outputs from a soil are shown in Figure 4.5. Minerals are supplied by the parent material and from the decomposition of dead plants and animals. These minerals can be moved up or down through the soil by water and may be removed by plant roots. Gases enter the soil from the atmosphere and from the soil organisms, and these gases may then chemically alter the minerals, making them more or less available to plants. The relative importance of these processes will be greatly affected by the general factors discussed in Section 4.2, Soil development. They differ greatly from one area to another, and for this reason different soil **types** develop in different areas. Some of the major processes which help to determine soil type are summarised in Table 4.2.

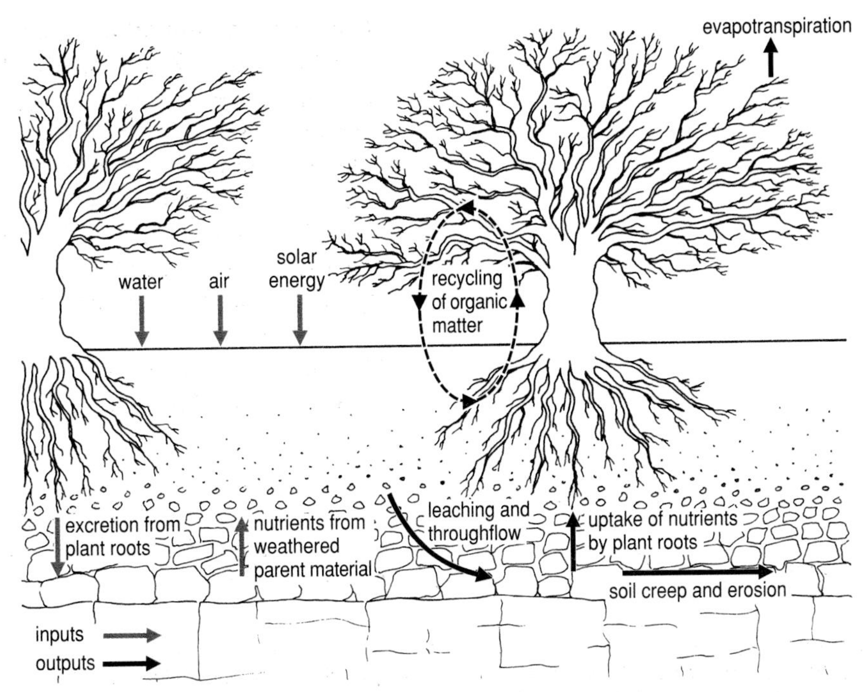

Fig 4.5 The soil system.

Table 4.2 Soil-forming processes

Process name	Details of process	Environmental implications
humification	decomposition of organic matter into humus	essential for soil fertility
chelation	the release of acids during humification, which attack the clay soil components	loss of iron and aluminium via cheluviation (a form of leaching)
organic sorting	redistribution of organic matter by animal activity (especially worms)	improves soil fertility throughout the profile
translocation (Figure 4.6)	movement of any component of the soil from one place to another by suspension or solution	can be detrimental or beneficial
eluviation	removal of particles in suspension	may reduce fertility in upper horizons
leaching	removal of nutrients in solution	removes nutrients from the topsoil which decreases fertility, but can prevent domination of the upper layers by salt and other unwanted components
podsolisation (Figure 4.8a)	extreme form of leaching where the A horizon has become bleached	translocated material forms in very concentrated layers (e.g. iron pans), which impedes drainage, causing high water content in the upper layers
waterlogging	soil profile is saturated with water	anaerobic conditions reduce numbers of soil biota
gleying	seasonally waterlogged profile	
calcification	leaching does not remove all the calcium which, due to evaporation, forms nodules as it rises up the profile	very fertile, as calcium is essential for plant growth, but calcium can form on the surface in drier climates
ferralitisation (Figure 4.8c)	the formation of clay, iron and aluminium complexes (sesquioxides), due to high rates of chemical weathering	soil is poor because nutrients are easily taken up by the plants or are quickly leached away
salinisation	evaporation causes sodium ions to dominate the upper layers	the surface becomes too base-rich for plants (salt crust)

By digging soil pits, we can see that most soils are divided into horizontal layers called **horizons**. These horizons, which develop as a result of the processes listed in Table 4.2, may differ from each other in colour, texture, structure, and organic and nutrient content. Individual soil types have characteristic horizons (Figure 4.7).

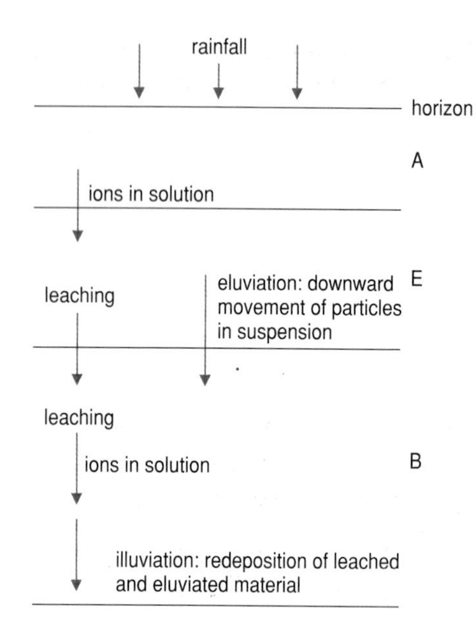

Fig 4.6 Translocation processes in soil.

The uppermost layer is an accumulation of organic material (the **leaf litter** – Horizon L), which decomposes into the **fermentation layer** (Horizon F) and finally the **humus layer** (Horizon H). These three layers consist almost entirely of organic matter and are grouped together into the **surface organic layers**.

Beneath the surface organic layers is the topsoil (Horizon A). An **eluvial layer** (Horizon E) is produced by leaching (the removal of nutrients in solution).

The subsoil (**illuvial layer** – Horizon B) is produced from the accumulation of material moved from the topsoil. The translocated materials may accumulate together or in separate layers.

Beneath the subsoil is a zone of weathered parent material called the **regolith** (Horizon C), beneath which is the original parent material (**bedrock** – Horizon D).

A particular soil profile may not contain all of the possible horizons, and the depth of any horizon will vary from place to place. The boundary between horizons may also be indistinct, and transitional horizons may occur. For example, the Horizon AC is a transitional zone which lacks the characteristics of both Horizons E and B. Examples of important soil profiles are shown in Figure 4.8.

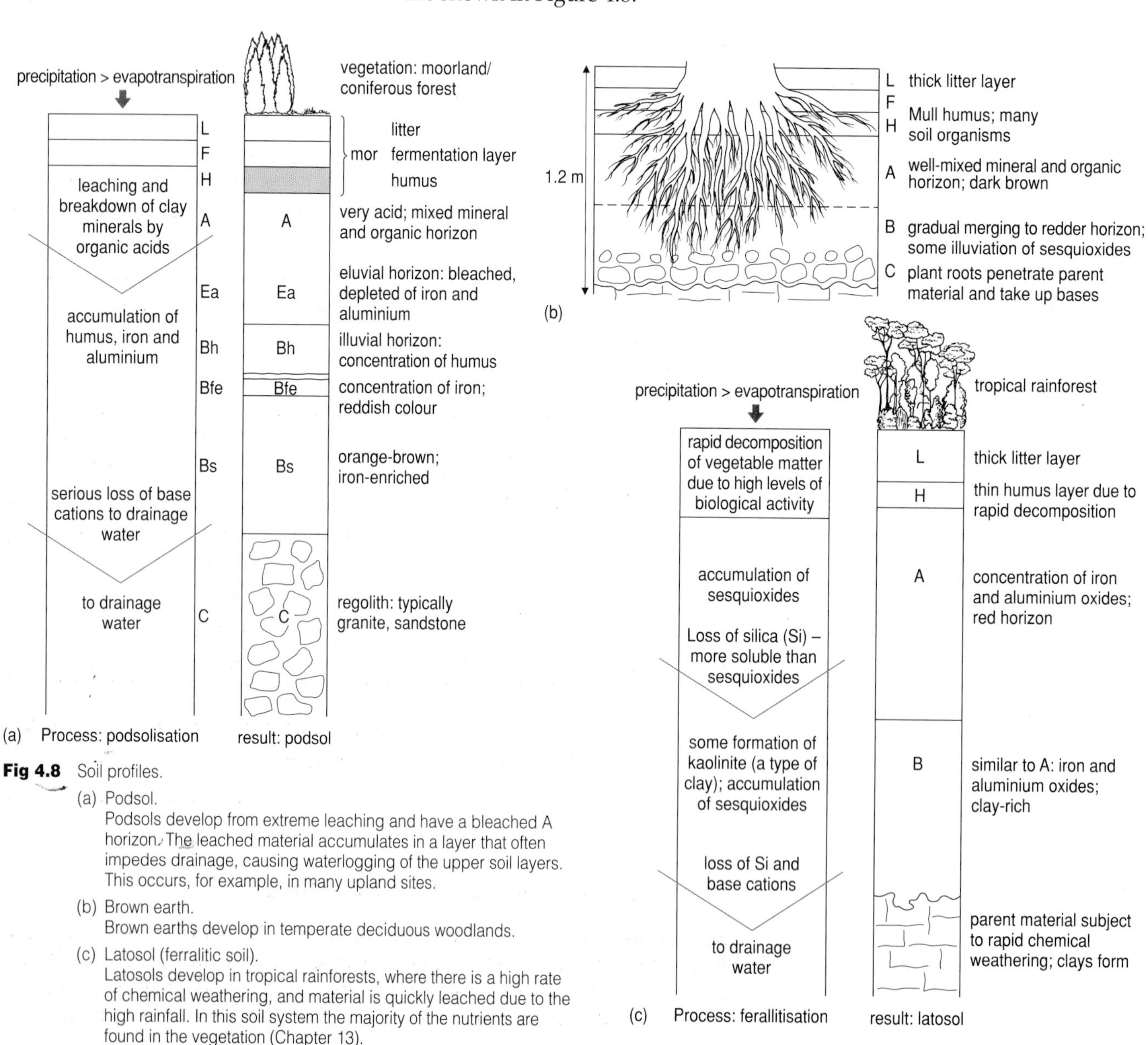

Fig 4.8 Soil profiles.

(a) Podsol.
Podsols develop from extreme leaching and have a bleached A horizon. The leached material accumulates in a layer that often impedes drainage, causing waterlogging of the upper soil layers. This occurs, for example, in many upland sites.

(b) Brown earth.
Brown earths develop in temperate deciduous woodlands.

(c) Latosol (ferralitic soil).
Latosols develop in tropical rainforests, where there is a high rate of chemical weathering, and material is quickly leached due to the high rainfall. In this soil system the majority of the nutrients are found in the vegetation (Chapter 13).

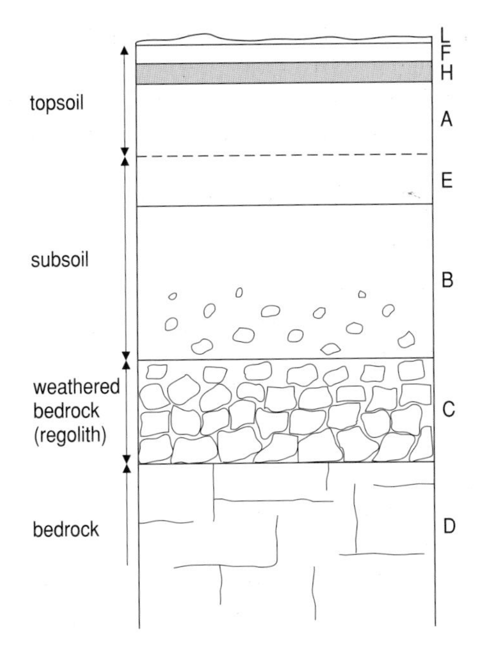

Fig 4.7 Soil horizons.

Figure 4.9 shows a cross-section of an English chalk downland. This is a good illustration of how different soil profiles can develop with a change in relief even on the same bedrock.

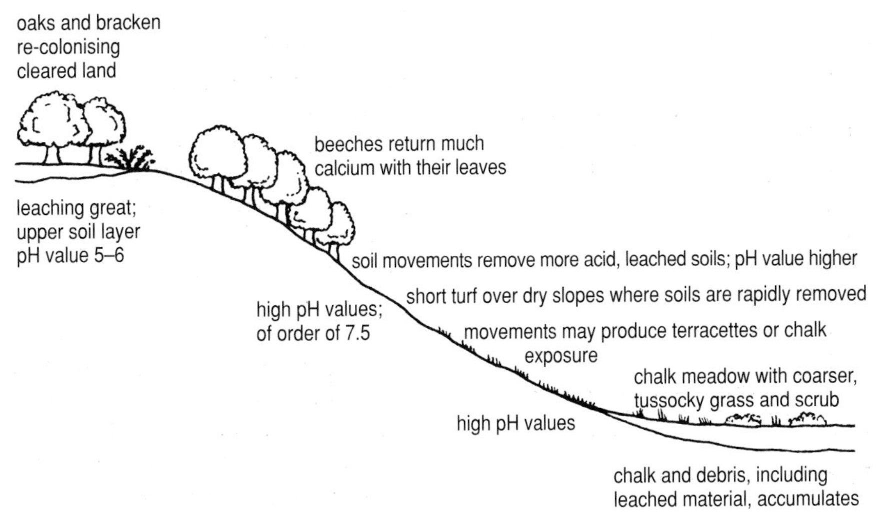

Fig 4.9 Cross-section of an English chalk downland.

In all chalkland soil profiles, leaching removes calcium from the upper soil, and therefore alkalinity increases with depth. Soils at the hilltop have a low pH because leaching is more efficient at altitude. Some of the slopes may be too steep (scarp slopes) for a soil to develop because the chalk is eroded away. The eroded material accumulates at the foot of the slope, and so this area has a very high pH value.

QUESTION

4.5 The table below lists some of the properties of soils with different textures.

Table 4.3 Soil properties and textures

	Water infiltration rate	Water-holding capacity	Nutrient status	Aeration	Ease of working
clay	poor	good	good	poor	poor
silt	medium	poor-medium	medium	poor-medium	medium
sand	good	poor	poor	good	good
loam	medium	medium	medium	medium	medium

(a) Explain why root growth of plants may be very poor in clay soils.
(b) How could a farmer improve a sandy soil?

Zonal, intrazonal, and azonal soils

Zonal soils are those which develop mainly as a result of climate, that is, the effects of the other factors have been masked by the effect of climate. Zonal soils are the best way of classifying soils on a world scale, because the more localised changes in parent material, biota and relief are ignored.

Intrazonal soils develop as a result of a dominant local factor. For example, a **rendzina** is a thin calcareous soil which develops from a limestone bedrock. Hydromorphic soils are those in which water is continually present, such as gleys and peats. Halomorphic soils are those that have developed through the process of salinisation, that is, the build-up of salts in soils.

QUESTION

4.6 The cross-profile and the table show the results of a field survey on a hillslope in Wales.

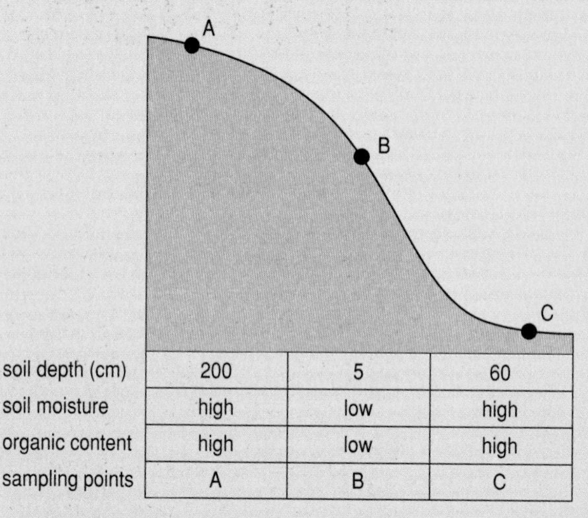

soil depth (cm)	200	5	60
soil moisture	high	low	high
organic content	high	low	high
sampling points	A	B	C

(a) Suggest why hydromorphic soils are present both at A and C in the cross-profile.

(b) Suggest why the soil depth is low at B but high at C.

Azonal soils are young soils which have not yet developed a characteristic soil profile. Azonal soils can be divided into lithosols, regosols and alluvial soils. Lithosols develop through high-altitude weathering (for example from scree). Regosols develop on unconsolidated material (for example sand-dunes or volcanic ash). Alluvial soils form on alluvium (for example floodplains).

SUMMARY ASSIGNMENT

1. Describe the components of a fertile soil.
2. Complete the table below by suggesting what effect the ploughing in of crop residues would have on each of the soil characteristics.

Characteristic	**Effect of ploughing in the remains of a harvested crop**
soil mineral content	
soil organic matter	
soil organisms	
soil water-holding capacity	

3. List the advantages and disadvantages of the zonal soil classification.

Chapter 5

AUTOTROPHIC NUTRITION AND PRODUCTIVITY

Every living organism requires a source of energy, and nutrition is the process by which all organisms obtain their energy source – food. This chapter will consider those organisms (autotrophs) which can make their own complex organic molecules from relatively simple, inorganic molecules.

LEARNING OBJECTIVES

After completing the work in this chapter you will be able to:

1. describe the principles of autotrophic nutrition
2. understand the importance of photosynthesis
3. understand the purpose of the C_4 pathway
4. explain why productivity varies greatly between ecosystems
5. describe how farmers and foresters can use this knowledge to increase productivity.

Autotrophic nutrition involves the conversion of inorganic raw materials such as carbon dioxide and water into complex organic compounds such as carbohydrates. This process requires energy. Autotrophs can be split into two types, depending on the energy source they use.

Photoautotrophs use sunlight energy for this process. All green plants are photoautotrophs. Most photoautotrophs trap sunlight in the green pigment, chlorophyll, and use the energy to convert carbon dioxide and water into carbohydrates such as glucose and starch. This is called photosynthesis.

Chemoautotrophs do not use sunlight as the source of energy for organic synthesis; instead they use the energy released from inorganic chemicals such as ammonia (NH_3), hydrogen sulphide (H_2S) or iron (Fe) for this conversion. This is called chemosynthesis. All chemoautotrophs are bacteria.

In contrast, **heterotrophs** such as sheep, camels and humans are unable to make their own organic compounds to use as food and they must therefore take them in, ready-made by the autotrophs. All heterotrophs therefore must rely completely on autotrophs either directly or indirectly (that is, eating animals which have eaten plants) for their energy.

5.1 PHOTOSYNTHESIS

The sun provides the ultimate source of energy for life on earth and, by trapping it and using it to produce food, green plants form the start of most food chains (Chapter 7). Photosynthesis also helps to regulate atmospheric carbon dioxide levels and therefore tropospheric temperatures (Chapter 1). It also provides the oxygen which is essential for aerobic organisms (Chapter 6) and for the formation of the protective ozone layer in the stratosphere (Chapter 1).

QUESTION

5.1 'Every last gram of humanity's food comes from photosynthesising plants.'
How can this statement be true for:
(a) sausages
(b) fish and chips?

5.2 THE REQUIREMENTS FOR PHOTOSYNTHESIS

The basic process of photosynthesis can be summarised by the following equation:

$$\text{carbon dioxide} + \text{water} \xrightarrow[\text{by chlorophyll}]{\text{sunlight energy absorbed}} \text{carbohydrates} + \text{oxygen}$$

Chlorophyll

The green pigment, chlorophyll, absorbs light from the visible part of the absorption spectrum (Figure 5.1a). Chlorophyll is really a mixture of several different pigments – chlorophyll a, b, carotene and xanthophyll. Each of these pigments absorbs slightly different wavelengths. The **absorption spectrum** shows that chlorophyll absorbs a lot of red and blue light but only a little green light. Most of the green light is reflected, which is why chlorophyll and the leaves containing chlorophyll appear green.

(a) Absorption spectrum: measured using a spectrometer

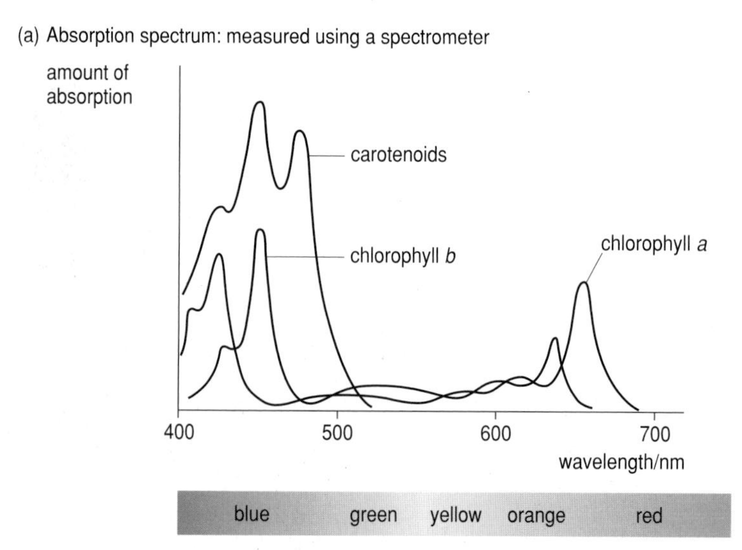

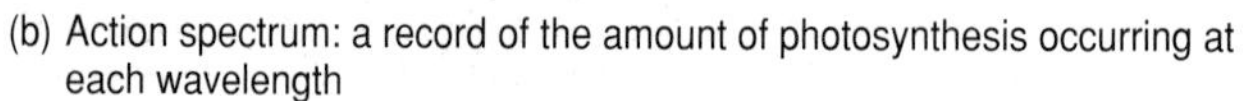

(b) Action spectrum: a record of the amount of photosynthesis occurring at each wavelength

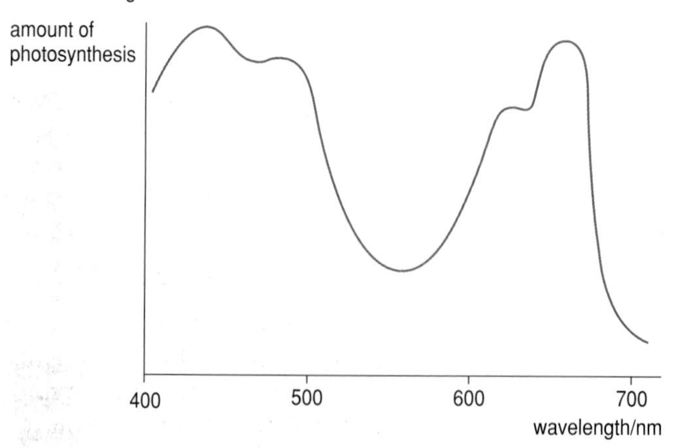

Fig 5.1 The absorption spectrum and the action spectrum of chlorophyll.

To find out whether or not the wavelengths that are absorbed are those that actually power photosynthesis, we can shine individual wavelengths at an aquatic plant and use the volume of oxygen given off as a measure of the rate of photosynthesis. Provided all other variables are kept constant, the more effective the wavelength, the more oxygen will be produced. We can then plot a graph showing the volume of oxygen given off against each wavelength. This graph is known as an **action spectrum** (Figure 5.1b).

Putting the action spectrum and absorption spectrum together, we can see that there is a close fit; that is, the wavelengths that are absorbed most are the ones which stimulate photosynthesis the most. Since each of the individual pigments absorbs a slightly different part of the spectrum, the total amount of light absorbed is greater than if chlorophyll were a single substance. This means that photosynthesis can proceed faster.

Chlorophyll is contained in chloroplasts that are found in the photosynthetic cells. Figure 5.2 shows the structure of a chloroplast as seen under an electron microscope. The membranes within the chloroplast have two important functions:

- They hold the chlorophyll molecules in a position where they can absorb maximum light energy. Stacking the chlorophyll molecules on the grana (singular = granum) increases the surface area over which light can be absorbed.
- They internally divide the chloroplast, allowing very different chemical reactions to occur in different parts. The chemical reactions which occur on the membranes are quite different from, and therefore need quite different conditions to, the ones which take place in the stroma.

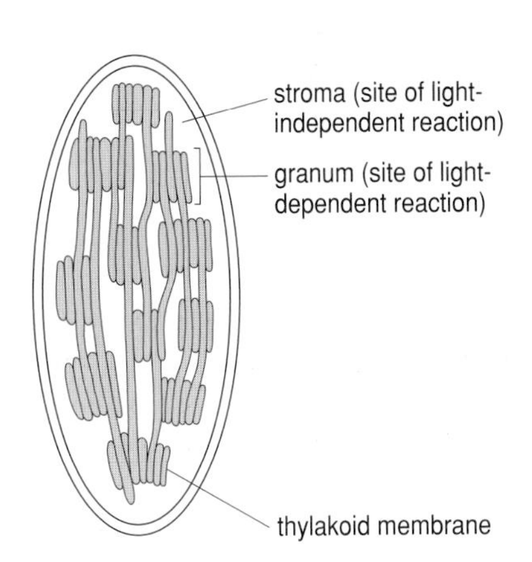

Fig 5.2 The chloroplast.

QUESTION

Fig 5.3 A deciduous woodland in winter and summer.

5.2 Suggest why evergreen species such as Sitka spruce are faster growing than deciduous species such as oak.

Carbon dioxide (CO_2)

As discussed in Chapter 1, the atmosphere contains 0.035% of gaseous carbon dioxide. This is absorbed by plants and provides the carbon needed to make carbohydrates, polysaccharides, fats and proteins, which make up the body of the plant. The carbon dioxide diffuses into the leaves of the plant through microscopic pores called **stomata**.

Water

Water is essential for every chemical reaction in the plant, which all occur in solution. Water is absorbed by the very fine root hairs and is then transported through the plant in numerous microscopic tubes which collectively make up the **xylem**. Water is directly involved in the chemical reactions of photosynthesis but also allows plant cells to maintain their shape (turgidity). The xylem extends all the way through the plant – from the roots, up through the stem, and into the leaves via petioles (leaf stalks) and veins in the leaves. This means that no cell is very far from a source of water. Water diffuses out of the leaf through the stomata in a process called **transpiration**. Water lost in this way is replaced by water drawn up through the xylem in the stem. The water moving up the plant is called the **transpiration stream**. Figure 5.4 summarises the routes by which light, carbon dioxide and water reach the sites of photosynthesis – the chloroplasts.

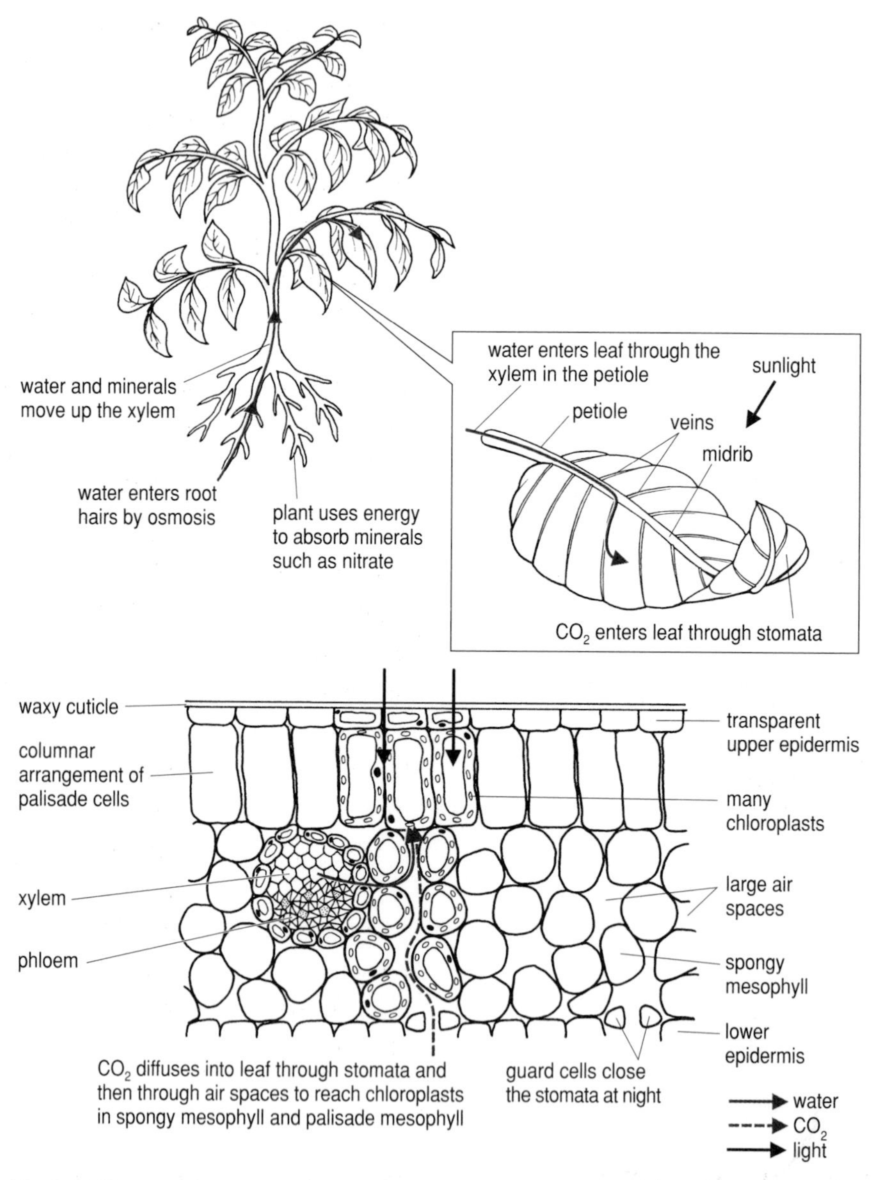

Fig 5.4 The principle of photosynthesis.

QUESTION

5.3 Outline the ways in which the structure and arrangement of the leaf help to promote photosynthesis.

5.3 WHAT HAPPENS DURING PHOTOSYNTHESIS?

Photosynthesis may be split into two stages:

Light-dependent stage

This stage needs light and occurs on the grana of the chloroplasts. The chlorophyll molecules arranged on the grana absorb light energy and use it to synthesise molecules of **ATP** (adenosine triphosphate).

Light also hits and splits water molecules (photolysis). This produces oxygen, which diffuses out of the leaf via the stomata, and hydrogen ions which play an important role in the light-independent stage.

Light-independent stage

The light-independent stage does not need light energy and occurs in the stroma of the chloroplast. The carbon dioxide, which has diffused into the leaf via the stomata and then into the stroma of the chloroplasts, is now reduced; it combines with hydrogen to form carbohydrates. The hydrogen used in this reaction (or, more accurately, the hydrogen ions) came from the water which was split in the light-dependent stage. This process of reduction, which occurs in several steps, requires energy. The energy is supplied by the ATP molecules which were also formed during the light-dependent stage. In other words, two of the products of the first stage are used up in the second. The carbohydrates produced may provide a source of energy for the plant, or, in combination with other elements such as nitrogen, may be used to make proteins and other molecules which the plant needs (Table 5.1).

Table 5.1 Which nutrients does a plant need and why?

Element	Form absorbed by plant	Used for
calcium	Ca^{2+}	cell walls
magnesium	Mg^{2+}	making chlorophyll and proteins
phosphorus	$H_2PO_4^-$	proteins, DNA, ATP
potassium	K^+	stomatal movement
nitrogen	NO_3^- or NH_4^+	making proteins
sulphur	SO_4^{2-}	enzyme activity
boron	H_3BO_3	regulates metabolism
iron	Fe^{2+} or Fe^{3+}	enzyme activity
copper	Cu^+ or Cu^{2+}	leaf growth
zinc	Zn^{2+}	root/stem growth
molybdenum	MoO_4^{2-}	nitrogen fixation
nickel	Ni^{2+}	enzyme activity
manganese	Mn^{2+}	enzyme activity
chlorine	Cl^-	regulates water potential

5.4 C_4 PLANTS

The mechanism outlined in Figure 5.4 is called the C_3 system because the first stable product of photosynthesis (in the light-independent stage) is a sugar containing only three carbon atoms. However, some plants, such as maize, use a slightly different system where the first product is a four-carbon compound. Such plants are termed C_4 plants (Table 5.2).

Table 5.2 Common C_3 and C_4 plants	
Common C_3 and C_4 plants used in food production	
C_3	C_4
wheat	maize
rice	sugar cane
potato	millet

The purpose of the C_4 mechanism

C_4 plants are much more efficient at capturing carbon dioxide than C_3 plants and are able to continue to photosynthesise even when the concentration of CO_2 is very low. The C_4 system works very effectively under conditions of low CO_2 levels, high light intensities and high temperatures, which is why many C_4 plants are found in the tropics and arid regions and are widely grown as staple food crops (Figure 5.5).

Fig 5.5 C_4 plants, such as maize, are much more efficient at capturing CO_2.

QUESTION

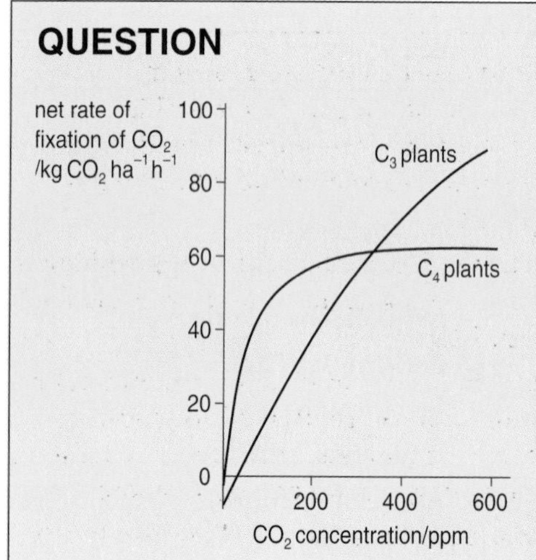

5.4 The graph shows the effect of carbon dioxide concentration on net carbon dioxide fixation rates in C_3 and C_4 plants.

(a) Under normal atmospheric conditions (CO_2 concentration = 0.035%), which type of plant would be photosynthesising most efficiently?

(b) A commercial grower was considering increasing the CO_2 concentration artificially inside one of his greenhouses to 500 ppm. The greenhouse contained both C_3 and C_4 plants. Suggest how the greenhouse owner might try to increase CO_2 levels in the greenhouse.

(c) If the owner succeeded in raising the level of carbon dioxide to 500 ppm, what would be the effect on the C_3 and C_4 plants?

(d) How would the greenhouse owner calculate whether or not this was an economically worthwhile proposition?

5.5 LIMITING FACTORS

The rate of photosynthesis is affected by temperature, light intensity and wavelength, carbon dioxide concentration, water and mineral availability and pollution. These factors may affect the rate individually or interact together. If we are to maximise photosynthesis and hence food production, it is essential that we have some understanding of how and why these factors affect the rate of photosynthesis.

Figure 5.6 a shows the effect of increasing light intensity on the rate of photosynthesis of an aquatic plant kept at 20°C. The rate is again measured by the volume of oxygen released. Other variables such as temperature and carbon dioxide have been kept constant.

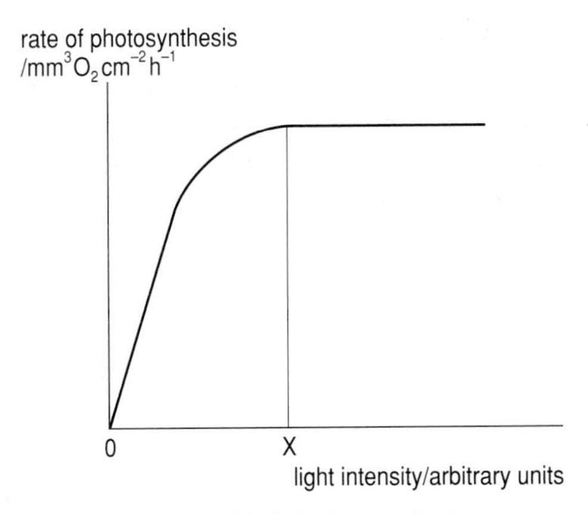

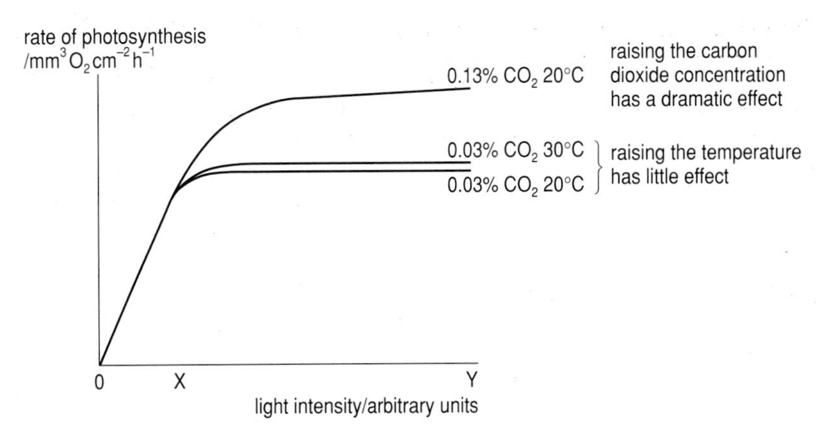

Fig 5.6 Factors which limit photosynthesis.

- As light intensity increases between 0 and X units, the rate of photosynthesis increases. Between 0 and X, it is the light intensity which limits the rate of photosynthesis; that is, light is the limiting factor.
- However, as light intensity increases above X, there is no further increase in the rate of photosynthesis. This indicates that some other factor is limiting the rate. In theory, this could be any of the other factors such as temperature or CO_2, etc. In fact, both of these may be restricting the rate of photosynthesis, but one will have a greater restrictive effect than the other.

To help us identify the overall limiting factor, the investigation can be repeated over the same range of light intensities, but firstly at a higher temperature (with carbon dioxide concentration kept constant), and secondly at a higher carbon dioxide concentration (with temperature kept constant).

Table 5.3 Minimising the effect of limiting factors on photosynthesis to increase food production

Limiting factor	Function / effect	Response
water	solvent for all chemical reactions; transport of minerals and sugars around plant; supporting role in cells and tissues; source of hydrogen atoms to reduce CO_2	irrigate in arid areas; timed sprinklers in greenhouses; add organic matter to soil to increase its water-holding capacity; shelterbelts reduce wind speed and therefore rate of water loss from plants
carbon dioxide	source of carbon for synthesis of carbohydrates, fats, proteins, etc.	impossible to regulate in the field, but greenhouses may contain a heating system based on fossil fuels which release CO_2
light	provides energy for the chemical reactions of photosynthesis; splits water to provide hydrogen atoms for the reduction of CO_2	artificial light may be provided if competition for light is severe or to control light/dark regimes; some trees may be removed to leave more light for those remaining – thinning, e.g. in a forest
temperature	affects the rate of activity of enzymes, which control the rate of photosynthesis (especially light-independent reaction) and the speed of seed germination	greenhouse temperature may be thermostatically controlled; soil temperature may be increased by adding mulches to the surface
mineral availability	may be divided into macro- and micro-nutrients	artificial fertilisers may be added; mineral availability often determined by soil pH, therefore lime may be added to increase the pH; legumes, e.g. clover or peas, may be planted which will fix gaseous nitrogen
pollution	many effects, e.g. acid rain may damage epidermis of leaf	pollution control strategies
pests	herbivores and pathogens may attack plant, causing damage and disease	pest-control measures

Over the light intensity X to Y, an increase in carbon dioxide has a greater effect on the rate of photosynthesis than an increase in temperature. We can therefore say that carbon dioxide concentration was the limiting factor. The **law of limiting factors** can now be stated as 'the limiting factor of a chemical process is that factor which is nearest its minimum value'.

All people involved in plant production, whether they be market gardeners, tulip-growers or foresters, need to have some understanding of limiting factors, since by manipulating light regimes, carbon dioxide concentrations and water supply, they can increase productivity. Table 5.3 summarises some of the approaches used to minimise the effect of limiting factors.

5.6 CHEMOSYNTHESIS

Chemosynthetic bacteria convert inorganic raw materials into organic compounds using energy released from the oxidation of compounds and elements. As such, they are extremely important in mineral recycling (Chapter 8). Some of the most important groups of chemosynthetic bacteria are described below.

Bacteria in the nitrogen cycle

Chemosynthetic bacteria are largely responsible for the cycling of nitrogen between the atmosphere, soil, plants and animals (Chapter 8). Ammonia is released from the rotting remains of dead animals and plants. This is used as a source of energy by *Nitrosomonas*, which oxidises the ammonia into nitrites. Another type of bacteria, *Nitrobacter*, oxidises the nitrite into nitrate. This can then be absorbed from the soil by plant roots and used to make proteins or nucleic acids such as DNA. Other chemosynthetic bacteria release nitrogen back into the atmosphere or convert gaseous nitrogen into nitrates. The nitrogen cycle is discussed fully in Chapter 8.

Bacteria which oxidise sulphur compounds

Some bacteria such as *Thiobacillus* can use sulphur or sulphate (SO_4^{2-}) as a source of energy to produce sulphuric acid. This knowledge can be used in pest control. For example, the fungus which causes potato scab cannot tolerate acid conditions. If an affected area is treated with sulphur and *Thiobacillus*, the bacterium will oxidise the sulphur to sulphuric acid which will destroy the pest. Some species of bacteria actually live in hot, acid springs where they oxidise volcanic hydrogen sulphide.

Bacteria which can oxidise sulphur and iron compounds can be used to leach ores to obtain valuable metals (Chapter 15). For example, a solution of the appropriate bacteria can be sprayed over columns of broken-up ores which might contain sulphides of iron, lead, nickel and zinc. The bacteria will oxidise the sulphides to produce sulphuric acid. The acid will dissolve the metals which can then be recovered by precipitation. A similar process is now being used to extract copper from very low grade ores.

5.7 PRODUCTIVITY

Productivity may be defined as the rate at which energy or organic matter is built up into tissues (**assimilated**) by a crop, per unit area, per unit of time. Units might therefore be expressed as kJ ha^{-1} y^{-1}.

Gross Primary Productivity (**GPP**) is the total amount of energy or organic matter captured or **fixed** by green plants. However, not all of this is available for the animals that eat the plants, the herbivores, because some of the energy is used by the plant itself in staying alive. The energy lost in this way is called the respiratory loss. The amount of energy left after respiration is called the **Net Primary Production** (**NPP**). A simple equation can be used to show this relationship:

Gross Primary Production (GPP) – Respiration (R) = Net Primary Production (NPP)

It is NPP which is actually available for animals to eat, so farmers and foresters are also interested in minimising R. How farmers try to do this is discussed in Chapter 6.

It is worth noting that although oceans cover two-thirds of the earth's surface, they contribute only one-third of global NPP. Marine NPP is mainly determined by light intensity and nutrient availability but, as Table 5.4 shows, there are wide variations in the productivity of different parts of the ocean. There is also a fundamental difference in the way humans use the productivity of the land and oceans. On land, humans have concentrated on developing and utilising NPP directly through crops, whereas in the oceans we have exploited **secondary production**, that is, the productivity of the fish which eat the marine plants. Phytoplankton pie is not yet on the menu! We will return to this theme in Chapter 10.

Global patterns of NPP

NPP varies greatly between different types of plant community (Table 5.4). By rearranging our equation we can see that:

$$NPP = GPP - R$$

GPP is influenced by temperature, carbon dioxide concentration, water and nutrient availability, light intensity and wavelength, and pollution. The amount of energy used by a plant community in respiration is largely determined by the plant biomass which the photosynthesising tissues have to support. Hence the figure for R may be 70% in tropical forests, 60% in temperate forests, 40% in crops and 30% in microscopic marine plants (phytoplankton).

Table 5.4 Net annual primary production and plant biomass for the earth dry weight

Ecosystem type	Area / 10^6 km^2	Net Primary Productivity per unit area / g m^{-2} or t km^{-2}		World Net Primary Production / 10^9t	Biomass per unit area / kg m^{-2}	
		Normal range	Mean		Normal range	Mean
tropical rainforest	17.0	1000–3500	2200	37.4	6–80	45
temperate deciduous forest	7.0	600–2500	1200	8.4	6–60	30
boreal forest	12.0	400–2000	800	9.6	6–40	20
tundra and alpine	8.0	10–400	140	1.1	0.1–3	0.6
desert and semi-desert shrub	18.0	10–250	90	1.6	0.1–4	0.7
cultivated land	14.0	100–3500	650	9.1	0.4–12	1
swamp and marsh	2.0	800–3500	2000	4.0	3–50	15
total continental		**149**		**773**	**115**	**112.3**
open ocean	332.0	2–400	125	41.5	0–0.005	0.003
upwelling zones	0.4	400–1000	500	0.2	0.005–0.1	0.02
continental shelf	26.6	200–600	360	9.6	0.001–0.04	0.01
algal beds and reefs	0.6	500–4000	2500	1.6	0.04–4	2
estuaries	1.4	200–3500	1500	2.1	0.01–6	1
total marine		**361**		**152**	**55.0**	**3.033**

SUMMARY ASSIGNMENT

1. Using the information given in this section and from your own knowledge:
   (a) Explain why NPP may vary with latitude.
   (b) Suggest reasons for the NPP values of:
      (i) tropical rainforests
      (ii) boreal forest
      (iii) tundra and alpine
      (iv) continental shelves
      (v) estuaries.
2. Explain the role of each of the following in photosynthesis:
   (a) light
   (b) water
   (c) chlorophyll.
3. Plant geneticists hope someday to be able to incorporate the C_4 mechanism of photosynthesis into C_3 plants. What would be the advantages of this?

Chapter 6

HETEROTROPHIC NUTRITION AND RESPIRATION

In Chapter 5 we saw that photoautotrophic organisms (producers) use the energy from sunlight to convert carbon dioxide and water into organic molecules. **Heterotrophic organisms** cannot do this and have to consume some organic molecules ready-made into their bodies. To do this they feed upon plants or other animals and are therefore called consumers. Rabbits, foxes and eagles are all consumers. The release of energy is called **respiration** and occurs in every living cell of every living organism. Both producers and consumers release the energy they need from the oxidation of organic compounds such as glucose or fat. Farmers use their knowledge of nutrition and respiration to try to ensure that their livestock obtain and retain as much energy as possible, as this energy can then be used for growth.

LEARNING OBJECTIVES

After completing the work in this chapter you will be able to:

1. explain the need for and the process of digestion
2. describe the adaptations of some herbivores to digest cellulose
3. describe the process of respiration
4. understand the concept of homeostasis
5. explain how farmers may use their knowledge of these processes to increase production.

6.1 NUTRITIONAL REQUIREMENTS

Humans need six major constituents in their diet, in the correct proportions, in order to obtain a **balanced diet**. The six components, along with fibre, are described in Table 6.1.

Table 6.1 Components of a balanced diet

Component	Use
carbohydrates	provide a fuel for all of the millions of chemical reactions which occur in the body; sugars and starches are examples of carbohydrates
fats	like carbohydrates, fats may be used to fuel chemical reactions, but they also provide a longer-term energy store and store certain vitamins (A, D, E and K); fats which are stored under the skin have an insulating function
proteins	provide a source of amino acids which are the building blocks from which proteins are made; tissue replacement and the formation of new tissues during growth need a continuous source of new amino acids

vitamins	organic compounds needed in minute quantity for normal health; many are needed to allow enzymes – the biological catalysts which speed up chemical reactions – to work properly
minerals	elements such as iron, calcium and sodium are needed for many chemical reactions in the body and for the production of, for example, pigments
water	essential for all life, as every chemical reaction in every cell must occur in solution; 70% of the human body mass is made of water; water lost through urine or sweat must be replaced in food or drink
fibre	this is mainly cellulose from plant material; fibre has no nutritional value but helps to prevent diseases of the digestive system and constipation

Fig 6.1 The effect of famine in Sudan, Africa.

An unbalanced diet

A large proportion of people on the planet have insufficient food to supply either their energy or protein requirements or both. **Malnutrition** occurs when an individual's diet lacks specific, essential components or when a particular component cannot be absorbed or used. Although malnutrition affects adults in many parts of the world, it is perhaps an even more serious problem in young, growing children who need large quantities of energy and protein. **Kwashiorkor** develops in children who have sufficient energy but insufficient protein. Symptoms include pot-belly and poor muscle development. **Marasmus** is a condition caused when children have both insufficient energy and protein. When the body has insufficient energy sources, it begins to break down its own proteins and the body wastes away.

Conversely, the most common form of malnutrition in the western world is **obesity**, caused by excessive intake of energy in relation to energy expenditure. Obesity increases the likelihood of coronary heart disease and high blood pressure.

6.2 THE PRINCIPLES OF HUMAN DIGESTION

The whole process of human digestion can be summarised under the following headings:

(a) **Ingestion** The taking into the body of large, complex, insoluble organic food substances through the mouth.

(b) **Digestion** The breaking down of these large insoluble organic molecules into smaller, simpler, soluble molecules so that they can be absorbed across the wall of the digestive tract and into the body. Digestion occurs in the mouth, stomach and small intestine.

(c) **Absorption** The movement of the small molecules across the wall of the digestive tract into the blood or lymph system.

(d) **Assimilation** The small soluble building blocks now need to be reassembled into the complex molecules which the body requires. Sugars like glucose may be linked together to form the carbohydrate glycogen, which is stored in the liver and muscles. Amino acids may be reassembled (or assimilated) into proteins, for example to form muscles.

(e) **Egestion** The removal of faeces from the body, that is, undigestible fibre along with dead cells and dead bacteria.

Every chemical reaction within cells produces waste products which, if they were allowed to accumulate, would be toxic. Excretion is the removal from the body of these waste products. The processes of egestion and excretion are

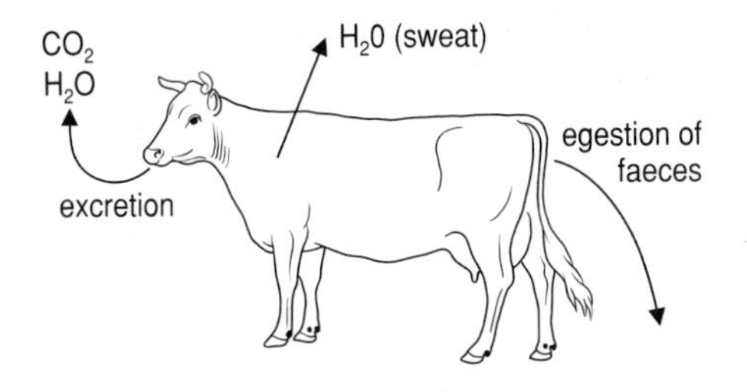

Fig 6.2 The processes of egestion and excretion.

illustrated in Figure 6.2. Some excretory materials pass out from the liver in the bile into the small intestine and are removed from the body in the faeces with which they combine in the intestine.

Digestion of a sandwich

(a) Ingestion Imagine eating a cheese and tomato sandwich made with wholemeal bread which contains starch, protein, fat, fibre and vitamins.

(b) Digestion

(i) Starch is partly broken down into maltose in the mouth. Maltose is broken down to glucose in the small intestine. Any remaining starch is also broken down to glucose in the small intestine.

(ii) Long chains of proteins are split into smaller chains in the stomach. The smaller chains are broken down into amino acids in the small intestine.

(iii) Fats are broken down into fatty acids and glycerol in the small intestine.

(c) Absorption Small, soluble products of digestion (glucose, amino acids, fatty acids and glycerol) are absorbed in the small intestine.

(d) Assimilation Most of the absorbed substances then pass in the blood to the liver. Regulation of glucose, amino acids and fats occurs in the liver. This ensures the levels of these substances remain constant in the blood. Surplus digestive products are removed from the general blood system in the liver.

(e) Egestion The fibre from the bread, along with the cellulose cell walls of the tomato and its seeds, are removed from the body in the faeces. Despite having no nutritional role, it is clear that fibre (cellulose, lignin and other non-digestible polysaccharides) is an essential part of the diet. Fibre appears to protect against cancers of the intestine. It also prevents constipation.

Different parts of the digestive system are specially adapted to carry out their functions effectively (Figure 6.3).

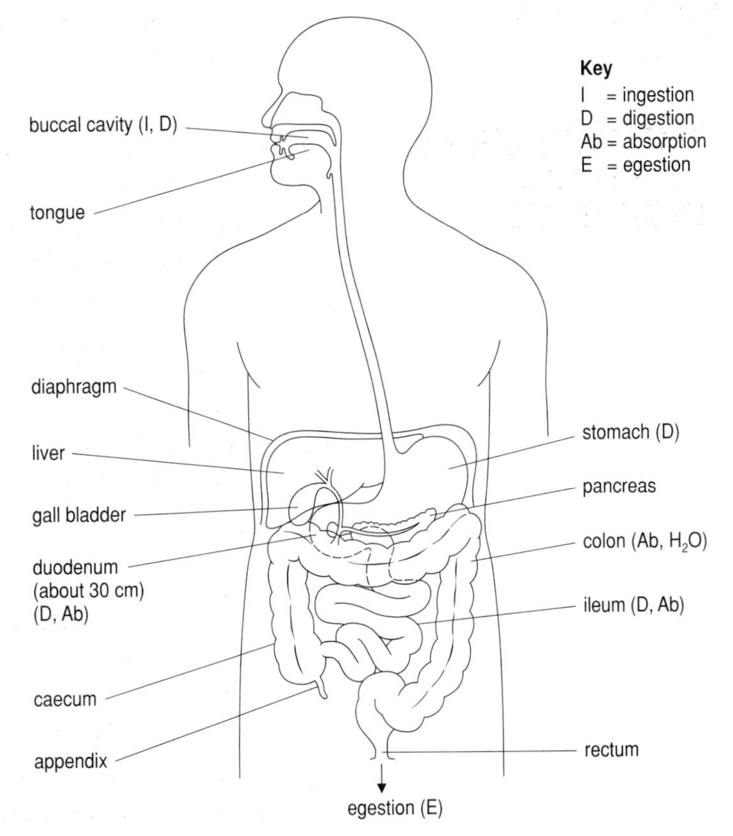

Fig 6.3 The human digestive system.

CASE STUDY

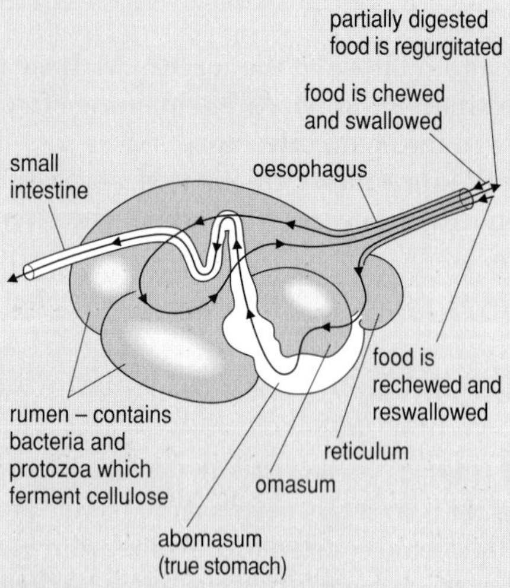

The problem with eating plants

Plant cell walls contain cellulose which humans cannot digest. **Ruminants** such as cows and sheep, however, have specially adapted extra stomachs containing bacteria which can break down cellulose. They regurgitate their food (bring the food back up to the mouth) from these stomachs and chew it again before swallowing it a second time.

- Chewed grass is swallowed and passes into the **rumen** and **reticulum**.
- The rumen and reticulum contain anaerobic bacteria which digest the cellulose, releasing glucose. This is fermented into organic acids which are absorbed into the blood to form the main source of energy for the animal. Carbon dioxide and methane are belched out, adding to the enhanced greenhouse effect!
- The remains of the grass is then regurgitated into the mouth and chewed again (chewing the cud), effectively offering a second chance to chew the grass and break down the plant cell walls. When swallowed, the food then passes through the omasum to the true stomach and then to the rest of the digestive system. The rest of the food is therefore digested in the usual way.

Rabbits also have a special mechanism for dealing with cellulose. Cellulose digestion is carried out by anaerobic bacteria in the **caecum** (part of the large intestine) and appendix. The partially digested material is egested at night as soft pellets, and these are immediately eaten. The products of cellulose digestion are then absorbed into the blood as the contents of the soft pellets pass through the digestive system. Waste material is then egested as hard pellets during the daytime.

partially digested food is regurgitated
food is chewed and swallowed
small intestine
oesophagus
food is rechewed and reswallowed
rumen – contains bacteria and protozoa which ferment cellulose
reticulum
omasum
abomasum (true stomach)

Fig 6.4 The digestive system of a cow.

Every chemical reaction that occurs in the body requires a source of energy. Carbohydrates, fats and proteins are the usual source of that energy and the process by which energy is released from substances like carbohydrates and fats is called **respiration**.

QUESTION

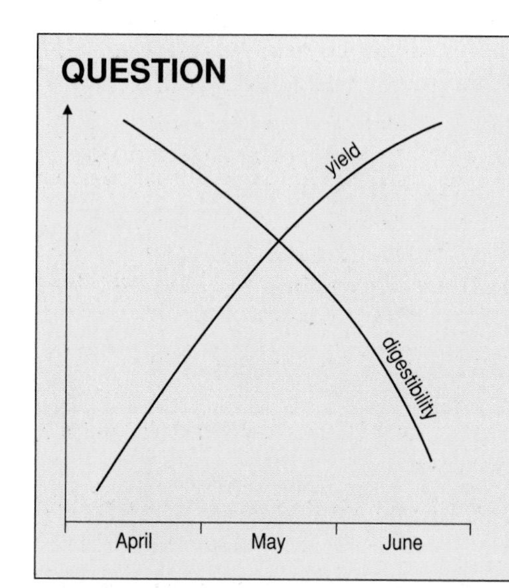

6.1 The graph shows the relationship between the digestibility and yield of a grass crop in Britain.

(a) From the graph:

(i) describe the relationship between digestibility and yield

(ii) explain why most British farmers cut their grass crop in mid-May.

(b) State **two** environmental factors which affect the yield of a grass crop.

(c) Explain how the complexity of organic molecules within the grass determines the digestibility.

AEB June 1991

6.3 RESPIRATION – THE RELEASE OF ENERGY FROM FOOD

In animals and plants, the usual respiratory substrates (materials from which energy is released) are glucose, fats and proteins. Whatever the substrate, the process of breakdown occurs in a large number of steps, each of which is controlled by a specific catalyst called an **enzyme**. Many of these steps release a small amount of the total energy in the substrate molecule and this energy is then used to make molecules of **ATP** (adenosine triphosphate). This energy-carrying molecule can then be used whenever energy is needed in a cell. However, ATP does not capture all or even most of the energy which is released from the substrate during respiration. About 60% is lost as heat. Respiration proceeds most efficiently when oxygen is available. This is known as aerobic respiration; anaerobic respiration releases a much smaller amount of energy from the substrate. The processes of aerobic and anaerobic respiration can be summarised in the equations below:

Aerobic: glucose + $O_2 \longrightarrow CO_2 + H_2O$ + energy

Anaerobic: glucose $\longrightarrow$ lactate + energy (in animals)

glucose $\longrightarrow$ ethanol + carbon dioxide + energy (in plants, for example in plant roots in waterlogged conditions)

In anaerobic respiration, less energy is released for ATP synthesis. The remaining energy that is not released is retained in the molecules of lactate and ethanol.

6.4 HOMEOSTASIS

Homeostasis is the maintenance of a constant internal environment, for example the composition of tissue fluid and body temperature. Organisms that have homeostatic mechanisms have an advantage because they provide their cells with relatively constant conditions, independent of changes in the external environment. This allows them to function efficiently.

Temperature regulation

Organisms use two different strategies for surviving temperature variations. **Ectotherms** (or **poikilotherms**) derive their body heat from the environment and so have specialised enzymes, cell membranes and biochemical functions to tolerate a wide range of body temperatures. Ectotherms include single-celled organisms, plants, all invertebrates, and some vertebrates such as fish, amphibians and reptiles. Some ectotherms have some control over their body temperature mainly by using behavioural responses, for example basking in the sun. To raise body temperature, bees and moths generate heat through rapidly contracting flight muscles. Conversely, if body temperature becomes too high, the insect's heart beats faster and extra haemolymph (blood) flows to hairless parts of the body (abdomen) and heat radiates away. Such adaptations allow insects such as bumblebees to fly on cool, damp days.

Endotherms (or **homeotherms**), in contrast, produce most of their body heat internally, and monitor and maintain a constant body temperature through a variety of mechanisms. Only birds and mammals belong to this group. For example, in humans, if the brain detects that the body is too hot, the following mechanisms operate to lower temperature:

- Metabolic rate is reduced. This reduces the amount of heat generated by respiration within the body.
- Blood vessels near the skin carry more blood (vasodilation), which increases heat loss to the air by conduction.
- Sweating increases. Evaporation of the sweat released onto the skin results in a loss of heat.

These mechanisms continue until the body returns to normal temperature (36°C to 37°C). Similarly, if body temperature becomes too low:

- Shivering and increased metabolic activity occurs.
- Blood vessels transport less blood near the surface of the skin (vasoconstriction), so less heat is lost to the air by conduction.
- Sweating is reduced or ceases.

Unlike animals, flowering plants cannot move away from unfavourable environments, and so most are regional; different species are adapted to different temperature regimes, ranging from Antarctica (where species can survive freezing) to tropical rainforests. In addition, many plants are seasonal and avoid the harshest times of the year by only growing at times when temperatures are suitable. For example, flowering plants such as daffodils are bulbs during winter, which is a dormant period, and emerge in the spring to reproduce as the temperature increases.

Water regulation

Osmoregulation is the process that controls the relative amounts of water and dissolved solutes in organisms. This is essential to maintain cell structure and functions, as all biochemical reactions occur in solution.

Invertebrates have several mechanisms to achieve some level of osmoregulation. For example, all insects, such as locusts from arid regions, have a hard impervious cuticle (outer coating) which prevents water loss through evaporation and water gain. However, insects can still dry out and die under very dry conditions.

Osmoregulation in mammals ensures that the total volume of blood plasma (the liquid portion of blood) and the concentration of dissolved substances in the plasma remain constant. This occurs by controlling the amount of water (and salts) gained and lost by the body. Drinking replaces lost water and is controlled by thirst. Drinking can compensate for the loss of water and salts from the body which occurs in urine, sweat and expired air. However, water loss in urine is actively regulated by the kidneys and is controlled by anti-diuretic hormone (ADH) produced by the brain. Loss of water from the body results in high levels of ADH, causing the kidney to retain more water in the blood and produce a smaller volume of urine. Excessive gain of water produces low levels of ADH, increasing water loss at the kidneys, and so larger volumes of water are lost in urine.

Plants show an enormous range of adaptations to prevent water loss. The leaves of cacti are reduced to form spines (reducing the surface area for water loss), and the stem is swollen as a water storage organ. All these external structures have a thick waxy cuticle (outer coating) and a reduced number of stomatal pores. In addition, the roots are spread out laterally to absorb all available moisture. Plants which show modifications for water conservation are known as **xerophytes**.

6.5 MAXIMISING FOOD PRODUCTION

In any form of food production, it makes sense to try to minimise energy losses; energy loss is always an economic cost to a farmer. However, some losses are inevitable and this has implications for the sort of agriculture they might choose to practise (Figure 6.5).

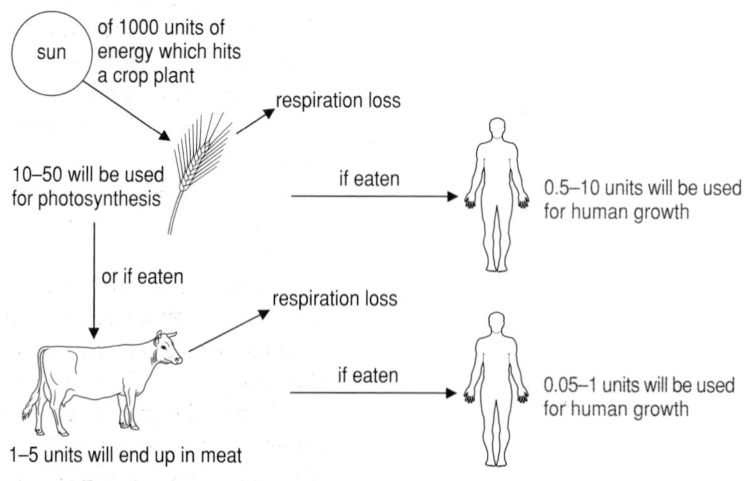

Fig 6.5 Energy losses in food production.

Table 6.2 Energy yield from different forms of food production

Food/Crop	Energy yield per unit area / kJ acre^{-1}
potatoes	29.0
rice	16.5
grains	13.25
milk	4.5
butter	2.0
eggs	1.25
beef	1.0

Some energy will be lost as heat through respiration at each stage of the food chain, and it therefore makes sense to keep the food chain as short as possible; in other words, a given area of land can produce more food if we use it to grow plants rather than support animals (Table 6.2). This is one argument for vegetarianism.

Any form of agriculture can be seen as a system of inputs and outputs. A farmer provides inputs of energy in the form of fertilisers, cattle feed, heating and lighting, and receives outputs in the form of harvested crops, milk and meat. This is illustrated in Figure 6.6.

Chapter 5 considered how primary production, for example the growth of our crops, can be maximised. **Secondary production** refers to the output of meat and milk which comes from the animals which have often been fed on these crops. We can use our understanding of digestion and respiration to try to increase secondary production.

Firstly, agricultural scientists have done much to improve the digestibility of animal feed, thereby increasing the amount of energy and nutrients which the animal can absorb from it. Our knowledge of respiration has allowed some farmers to develop systems which minimise energy losses through this process. The amount of energy needed by a heterotroph will be affected by its age, sex and level of activity. The more active an animal, the more energy it will require. This is made available through respiration. Farmers can use this knowledge to try to minimise energy losses when they rear animals. By tethering animals or housing them in small spaces, movement can be restricted, reducing energy loss. More of the food will therefore be converted to body tissue and the animal will put on more weight.

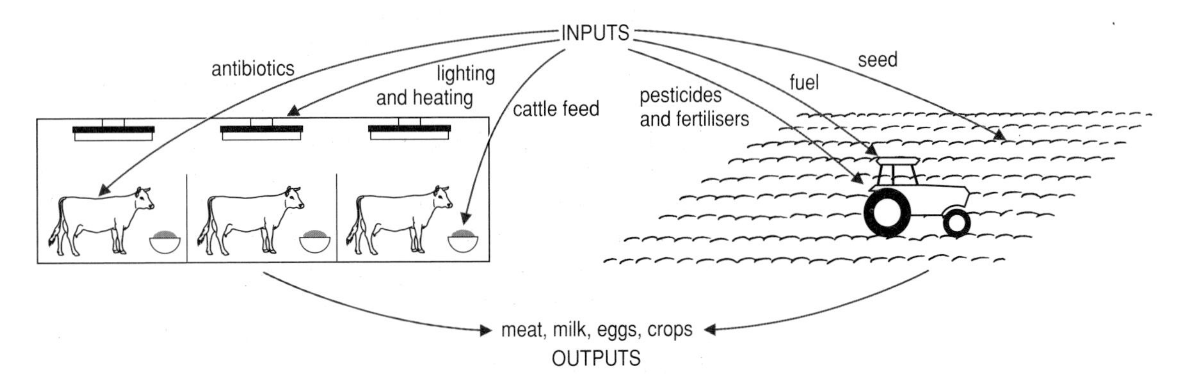

Fig 6.6 Inputs and outputs of modern farming.

Mammals, such as livestock, and birds are examples of **homeotherms (endotherms)**, animals which, through a variety of mechanisms, try to maintain their internal body temperature at a constant level, usually well above that of the surrounding environment. The important point here is that homeotherms use up most of their energy intake to simply keep warm. This knowledge can enable farmers to keep the surrounding temperature close to that of the animals' internal temperature so that little energy is lost trying to keep warm. This results in the animal using more energy to produce useful products, such as meat or milk. However, the artificial heating of stock sheds uses up expensive fossil fuels. Extra output is only gained by increased inputs.

Besides concerns about animal welfare, it is principles such as these which encourage some people to become **vegetarian**. Certainly, the environmental implications of eating meat are varied and can affect almost every aspect of the biosphere. How eating meat in Birmingham may unwittingly help to accelerate the enhanced greenhouse effect, add to acid rain problems over Sweden and increase flooding in Brazil, amongst other things, will be discussed in Chapter 14.

SUMMARY ASSIGNMENT

1. Explain the adaptations which each of the following have used in order to increase their digestion of cellulose:
   (a) rabbits
   (b) cows.

2. What are meant by the terms 'respiration' and 'homeostasis'?

3. The organisms in any ecosystem can be grouped into autotrophs and heterotrophs, according to how they obtain their food. Heterotrophs, in turn, can be classified according to the type of food they eat.
   (a) Outline how autotrophs make their food.
   (b) The figure below shows a general model of heterotrophic nutrition for an animal. The data relate to food ingested from a square metre of land over one year. With reference to the diagram, answer the questions.

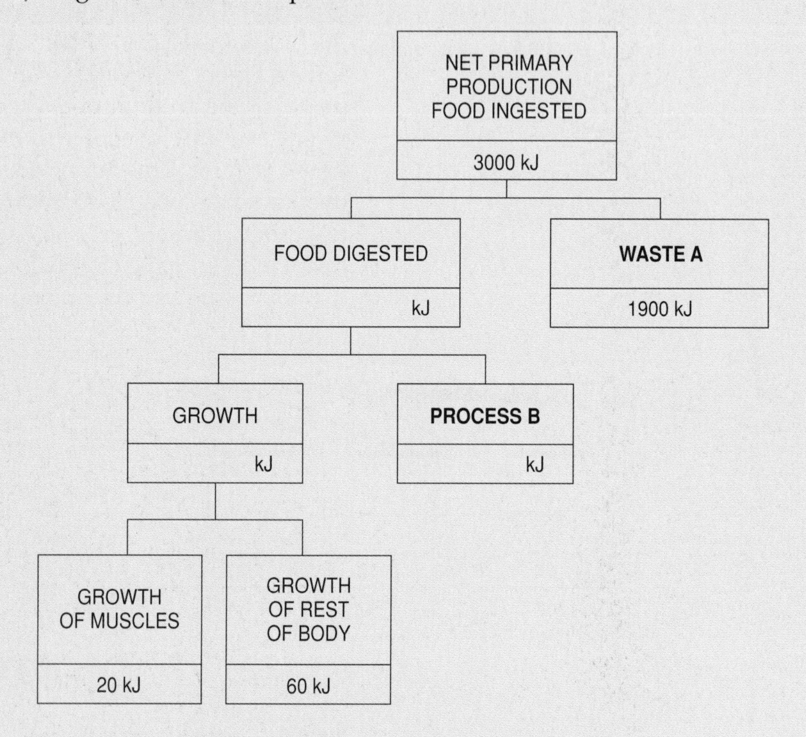

   (i) Explain what is meant by 'net primary production'.
   (ii) Identify waste A. Identify process B.
   (iii) State the general purpose of process B.
   (iv) Calculate the missing values in the diagram for:
   Food digested
   Growth
   Process B.
   (v) Calculate the percentage gross growth efficiency of the animal.

   (c) In relation to the diagram shown, outline the main reason for **each** of the following practices which may be used in some intensive animal production systems:
   (i) feeding the animals on a carefully controlled diet
   (ii) keeping the animals in a confined space.

JMB June 1992

Fig 6.7 Battery hens.

4. **(a)** Name three factors which are being controlled in the battery system.
   (b) Explain how control of these factors may increase meat production.
   (c) Name two other animals which can be kept in similar systems.

5. Details of four agricultural food chains, K, L, M and N, are given in the table below.

Four agricultural food chains

Food chain	J of human food per J × 10^5 solar radiation	Edible protein per hectare
K food → crop → humans	200–250	42
L food→ crop → livestock → humans	15–30	10–15
M intensive grassland → livestock → humans (meat)	5–25	4
N grassland and crops → livestock → humans (milk)	30–50	17

(a) In terms of conversion of solar energy to human food, which chain is most efficient and which is least efficient?
(b) Give two examples of a food crop appropriate to chain K.
(c) What kind of livestock might figure in chain L?

AEB June1992

Chapter **7**

ECOSYSTEMS

Humans share the earth with millions of other kinds of organisms. Uniquely, the earth has a protective, oxygen-rich atmosphere and an abundance of water, and it is this which has allowed life to develop. The part of the earth including air (atmosphere), water (hydrosphere) and minerals (lithosphere) where life can exist is called the **biosphere**.

Within the biosphere, organisms interact. By providing oxygen, green plants enable animals to aerobically respire (Chapter 6), a process that in turn releases the carbon dioxide which the plants use in photosynthesis (Chapter 5). Plants provide food and habitats but rely upon animals and micro-organisms to release the nutrients which they need to grow. Animals such as insects are also frequently involved in the reproductive processes of plants (for example bees pollinate flowers).

The biosphere existed long before humans came along. In fact, in terms of geological time, we are the latest of latecomers, and just as dependent as any other organism on the complex inter-relationships which support life on the planet. If we are to play our part and maintain the biosphere, we need to understand it.

LEARNING OBJECTIVES

After completing the work in this chapter you will be able to:

1. explain the structure of the biosphere
2. describe the categories of organisms which are found in all ecosystems
3. describe feeding relationships within an ecosystem
4. explain why energy and biomass decrease at each level in an ecosystem
5. describe some of the non-feeding relationships within an ecosystem
6. outline the principle of ecological succession.

7.1 STRUCTURE OF THE BIOSPHERE

A **species** is one particular type of organism; the individuals of a particular species can interbreed to produce living, fertile offspring. *Homo sapiens* (humans) is just one species, *Rananculus repens* (creeping buttercup) is another.

A **population** consists of all the members of a species living in a particular area at the same time. All of the different populations of organisms which live together, for example the grasses, trees, insects, deer and birds which live in a part of a wood, are called a **community**. The interaction of the community with the physical environment is termed an ecological system or **ecosystem**.

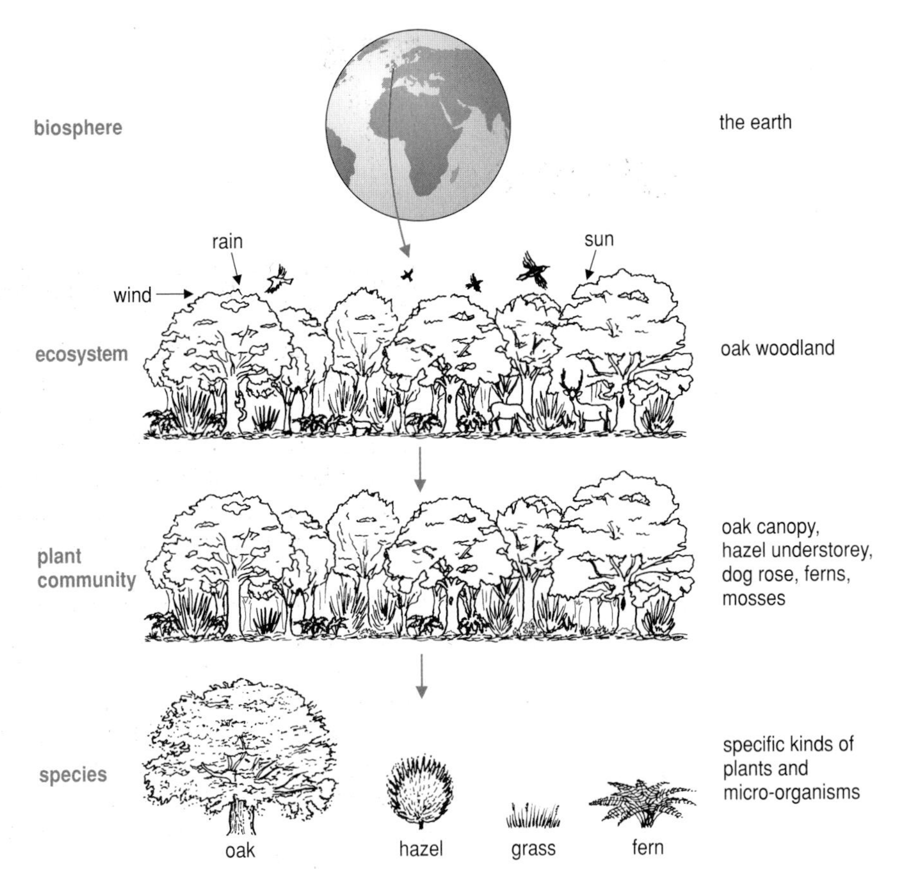

Fig 7.1 Structure of the biosphere.

7.2 STRUCTURE OF AN ECOSYSTEM

The living organisms in an ecosystem – the plants, animals and micro-organisms – are referred to as the biota or the **biotic factors**. The non-living chemical and physical parts of the environment are referred to as the **abiotic factors**.

Categories of organisms

Although there are many types of ecosystem, they all have the same basic categories of organism: primary producers, consumers, detritivores and decomposers. Typical examples of each category are shown in Figure 7.2, which depicts an oak woodland ecosystem.

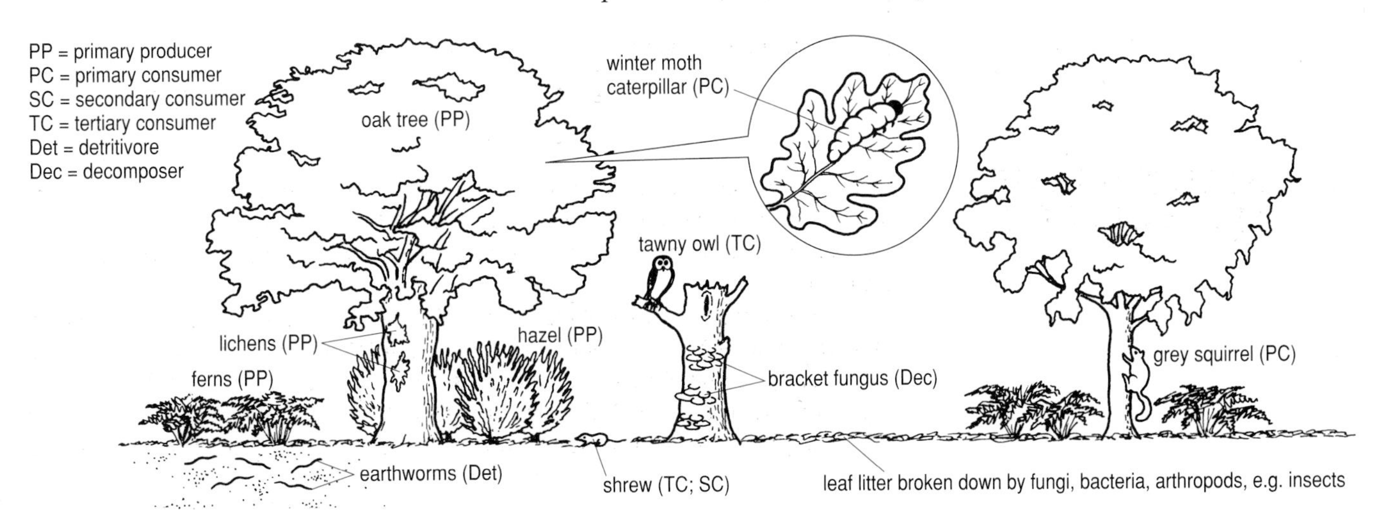

Fig 7.2 Biota of an oak woodland ecosystem.

Producers

Producers are mainly green plants which use the sun's energy during photosynthesis to convert water and gaseous carbon dioxide into carbohydrates, releasing oxygen as a waste product (Chapter 5). The producers then use these carbohydrates, plus minerals such as nitrogen absorbed from the soil, to make all of the other complex organic compounds they require. Plants are a source of food for other organisms in the ecosystem.

Consumers

Primary consumers obtain their energy in the form of complex organic molecules by feeding on the producers (Chapter 6). This category includes organisms as diverse as rabbits and blue whales. Organisms that feed on the primary consumers are called secondary consumers, and those that feed on the secondary consumers are called tertiary consumers (Figure 7.2).

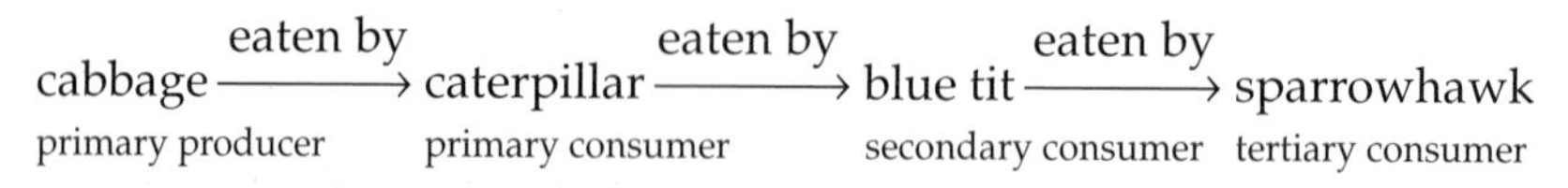

Fig 7.3 Predators in action. Lionesses chasing gazelle.

Note that an organism may fit into more than one category; a blue tit may feed on seeds when it is acting as a primary consumer, but if it also eats insects, it is acting as a secondary consumer.

Those primary consumers that only feed on plants are called herbivores, and secondary, tertiary or higher order consumers that kill other organisms are called carnivores. Consumers that eat both plants and animals are called omnivores. When an animal such as a lion attacks and kills an antelope, the lion is said to be acting as a predator and the antelope that is killed is called the prey. Parasites are plants or animals that live on or in another organism, called the host, causing it harm but usually without killing it. The fungus that causes athlete's foot is a parasite on humans. Fungal parasites also cause blight and smuts in crop plants.

Detritivores and decomposers

Detritus includes partly broken-down dead plant and animal material, such as fallen leaves and berries, animal faeces and dead animals. The organisms that feed on this dead material are called detritus feeders or **detritivores**. An earthworm is a detritivore. Decomposers such as bacteria or fungi normally digest every type of dead organism and their waste products. Bacteria and fungi that are involved in this breakdown are known as **saprophytes**.

Decomposers play a vital role in an ecosystem. By breaking down (rotting) the dead remains of other organisms, they release nutrients back into the soil (Chapter 4). These nutrients can then be absorbed by plants, and so the nutrient cycle continues.

Feeding relationships, food chains and food webs

A **food chain** is a sequence of organisms, each of which feeds on a type of organism from the preceding food level. For example:

oak tree ⟶ leaf-eating insect ⟶ shrew ⟶ fox

In a food chain, the direction of the arrows signifies the transfer of energy (energy flow). However, food chains do not give an accurate picture of the very complex feeding relationships that usually exist. Shrews, for example, eat many other things besides leaf-eating insects, and foxes may eat a wide variety of animals. Consequently, almost all food chains within an ecosystem are interconnected and form a web of feeding relationships, known as a food web (Figure 7.4).

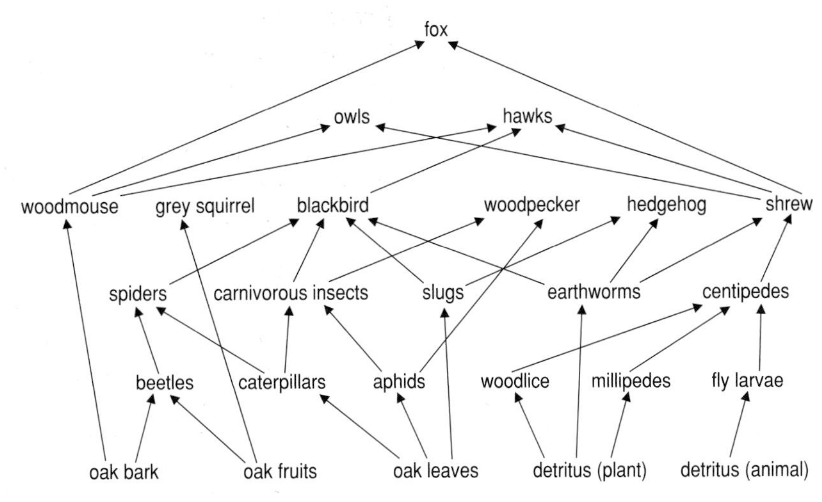

Fig 7.4 Part of a food web of an oak woodland.

The organisms within a food chain or web are organised into different feeding or trophic levels (TL) as follows:

secondary consumer	e.g. blue tit	TL3
primary consumer	e.g. caterpillar	TL2
primary producer	e.g. cabbage	TL1

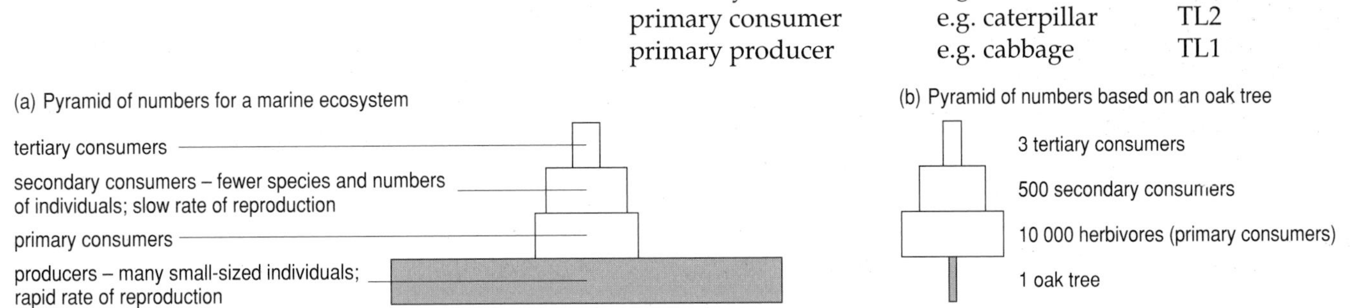

Fig 7.5 Pyramid of numbers.

If the number of organisms at each trophic level in the food web are shown diagrammatically, the result is often a pyramid known as a **pyramid of numbers** (Figure 7.5). Thus it takes a large number of primary producers to support a smaller number of primary consumers, which in turn provide food for an even smaller number of carnivores. However, this is problematic when we consider the number of organisms at each trophic level based upon a tree, as we have not taken into account the size of the organisms. One tree may support thousands of herbivores, so here the 'pyramid' of numbers appears as in Figure 7.5 b. This is called an inverted pyramid. A more reliable ecological pyramid is a **pyramid of biomass** which, as its name suggests, shows the mass of biological material at each trophic level in the food web (Figure 7.6 a and b).

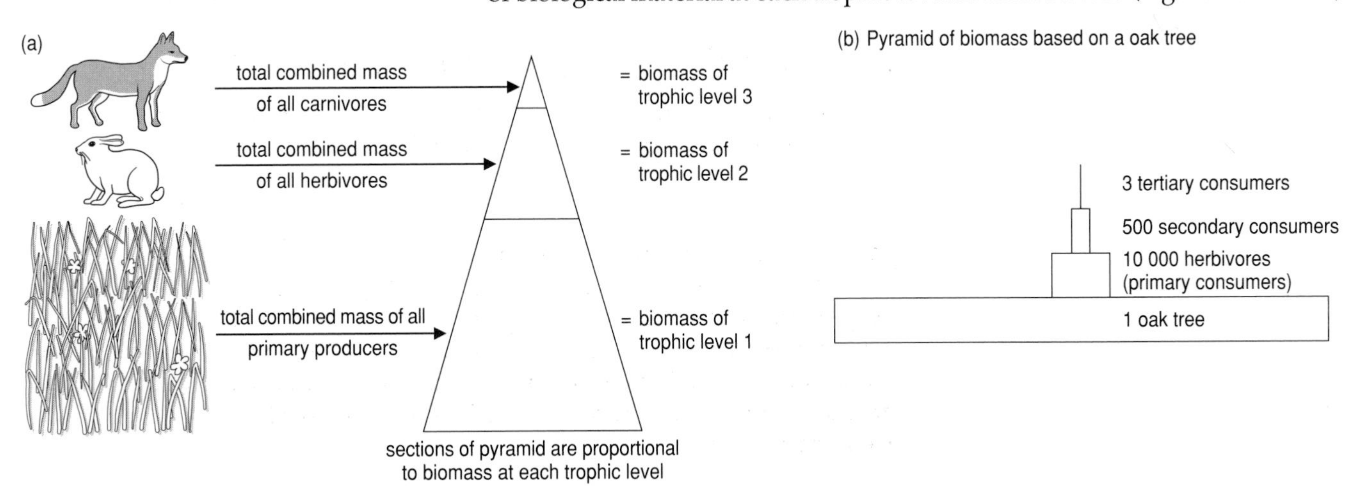

Fig 7.6 A pyramid of biomass.

(a) Generalised pyramid of energy, showing how the energy flowing through a trophic level decreases as you go up a food chain

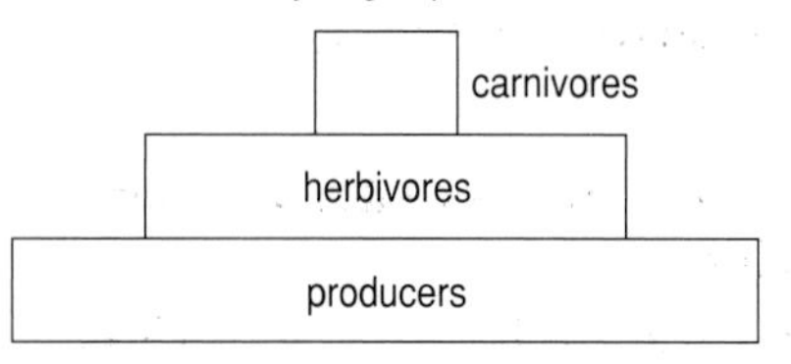

(b) Pyramid of energy from the Arctic tundra, Devon Island, Canada. Units are kJ m^{-2} yr^{-1}

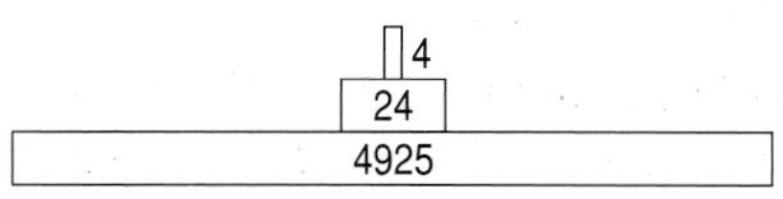

Fig 7.7 Pyramids of energy

The most reliable ecological pyramid is a pyramid of energy. This represents the amount of energy flowing from one trophic level to another. It is usually expressed in kilojoules per metre squared per year. Since energy is lost at every trophic level, these always form upright pyramids.

Why does biomass decrease at each trophic level?

Biomass decreases at each trophic level because the transfer of energy from the primary producer to the primary consumer and between the consumers is inefficient. Energy is lost at each stage, that is, at each of the transfers. Look at the food chain below.

green plant ⟶ rabbit ⟶ fox

The green plants contain a certain amount of energy in the form of carbohydrates, fats and proteins, etc. The rabbits, however, only eat a proportion of the plant. For example, the roots are left behind to die and their remains are eaten by detritivores or broken down by decomposers. This potential source of energy has therefore been lost to the rabbits, which cannot use it to make new rabbit tissue (biomass). The rabbit also loses energy, because it cannot digest or break down all of the material it consumes. Undigested material is egested as faeces, which can be thought of as pellets of energy, which again support the detritus and decomposer food chains. Finally, most of the energy that is obtained from the grass is simply used as fuel for respiration to keep the rabbit alive or is lost as heat when energy is transformed, during respiration, from one form to another. This energy cannot be used to increase biomass.

These losses, through not all of the organism being eaten, through faeces, through loss of energy as heat or because it has been used as a fuel in respiration, occur between each trophic level. In reality then, it may take 100 kg of grass to make 10 kg of rabbit which supports 1 kg of fox. For example, energy losses are represented by the 90 kg (90%) in energy transfer from grass (producer) to rabbit (primary consumer).

QUESTION

7.1 The figure illustrates a simplified food chain associated with tropical grasslands.

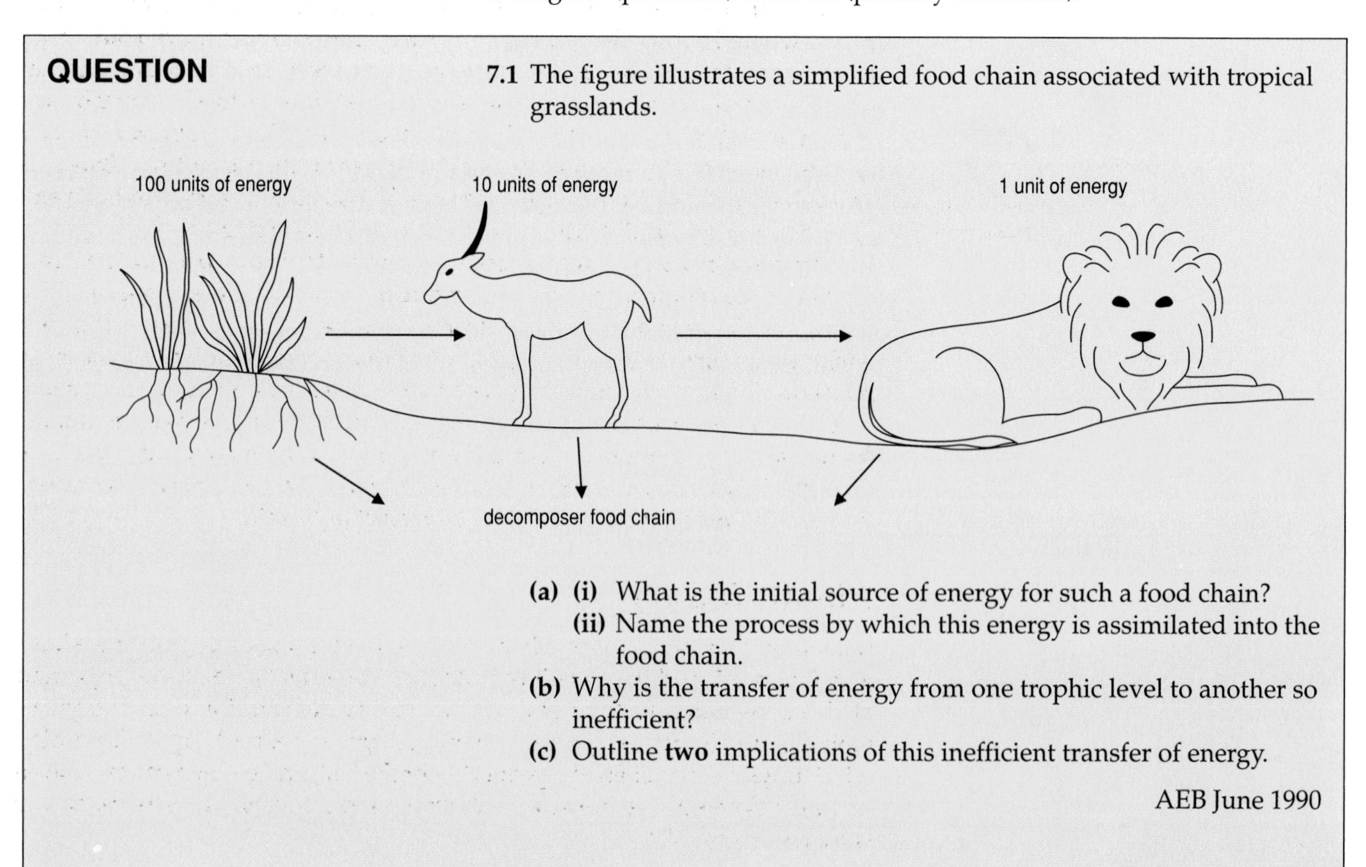

(a) **(i)** What is the initial source of energy for such a food chain?
(ii) Name the process by which this energy is assimilated into the food chain.

(b) Why is the transfer of energy from one trophic level to another so inefficient?

(c) Outline **two** implications of this inefficient transfer of energy.

AEB June 1990

Non-feeding relationships

Although food webs link all the individuals of an ecosystem, there are some important relationships that do not depend on feeding alone.

Fig 7.8 Root nodules on the roots of broad bean caused by nitrogen-fixing bacteria, *Rhizobium* sp.

Symbiosis

This is where two organisms live together in an intimate situation that is of benefit to them both. Here are two examples:

- Lichens are made up of a fungus and an alga. The fungus attaches itself to a surface such as a tree, rock or gravestone and provides a means of attachment and some protection for the alga. The alga, a producer, photosynthesises and produces sugars for the fungus.
- Some bacteria that belong to the genus *Rhizobium* live only in the roots of leguminous plants such as clover. The bacteria invade the roots of plants of the legume family, for example peas, beans and clover, causing swellings or **root nodules** to form. Inside these nodules the bacteria convert nitrogen gas, which has diffused into the roots from the soil air, into ammonia. This process is called **nitrogen fixation**. The plant is able to use this ammonia to make amino acids, which are the building blocks of proteins. In this way, the bacteria provide the plant with a usable source of nitrogen. In return, the bacteria are provided with the sugars which the plant has produced by photosynthesising.

Competitive relationships

The place where a particular organism or population lives is called its **habitat**. The habitat of red squirrels, for example, is pine woodland; for water beetles, their habitat might be a pond. Different types of woodland, for example, offer very different habitats and therefore support very different populations of plants and animals. However, even when species occupy the same habitat, there may be very little competition between them because different species occupy different **niches**.

An animal's niche refers to its role in a habitat – what it feeds on and when, where it finds shelter, etc. For example, two bird species could both feed on insects, but if one feeds exclusively on bark beetle larvae and another on leaf hoppers, then they will not be in competition for food. Similarly, one species that limits itself to feeding on tree tops will not compete with a species that feeds by pecking through the litter on the forest floor. Bats and swallows may both feed on flying insects, but one feeds during the day and the other at night, and therefore they occupy different niches.

Competition may be **intraspecific**, between two members of the same species, or **interspecific**, between members of two different species. For example, oak saplings may compete amongst themselves (intraspecific) and with young birch trees (interspecific) for available light or for water or nutrients. Such competition may lead to one species eliminating the other.

These two forms of competition also exist in the animal kingdom. Robins from adjacent gardens may aggressively compete for territory (intraspecific). Two species of barnacles, *Balanas balanoides* and *Cthamalus stellatus*, compete for space on many rocky shores (interspecific) in Britain.

Abiotic factors

Remember that an ecosystem has two components: the biota, which we have just discussed, and the non-living, abiotic components. The most important of these are rainfall (total amount as well as distribution over the year), temperature (daily and annual averages and extremes), light, humidity, wind, physical and chemical properties of the soil, and oxygen concentration in water. Some of the influences of abiotic factors in an oak woodland are shown in Figure 7.9.

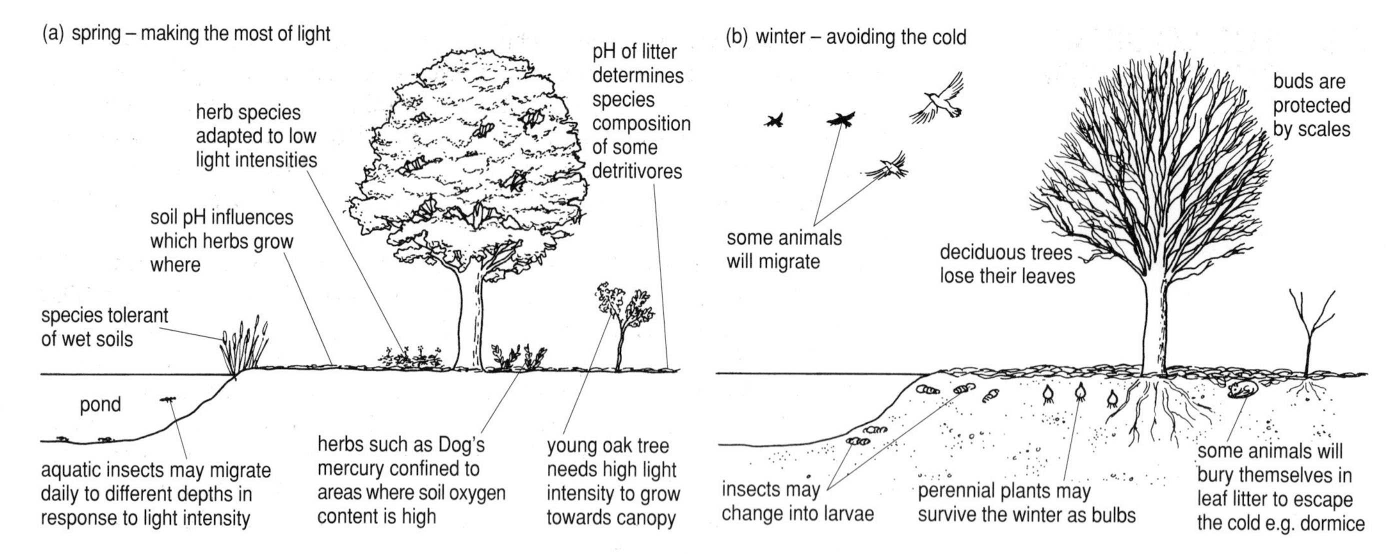

Fig 7.9 Abiotic factors in an oak woodland.

Sometimes abiotic factors may effectively prevent the establishment of a particular species in an area. Many plants, for example, cannot grow in acid soils. In other cases, seasonally changing factors, such as temperature and frost incidence, may mean that some species are only present in an ecosystem at certain times of the year. Bluebells, for example, may appear in deciduous woodland in May and may last through until June, but are not seen again until the following May. Many of the physiological processes of plants are closely tied to the seasons, and because of this an oak-hornbeam woodland, for example, may look very different at different times of the year (Figure 7.10).

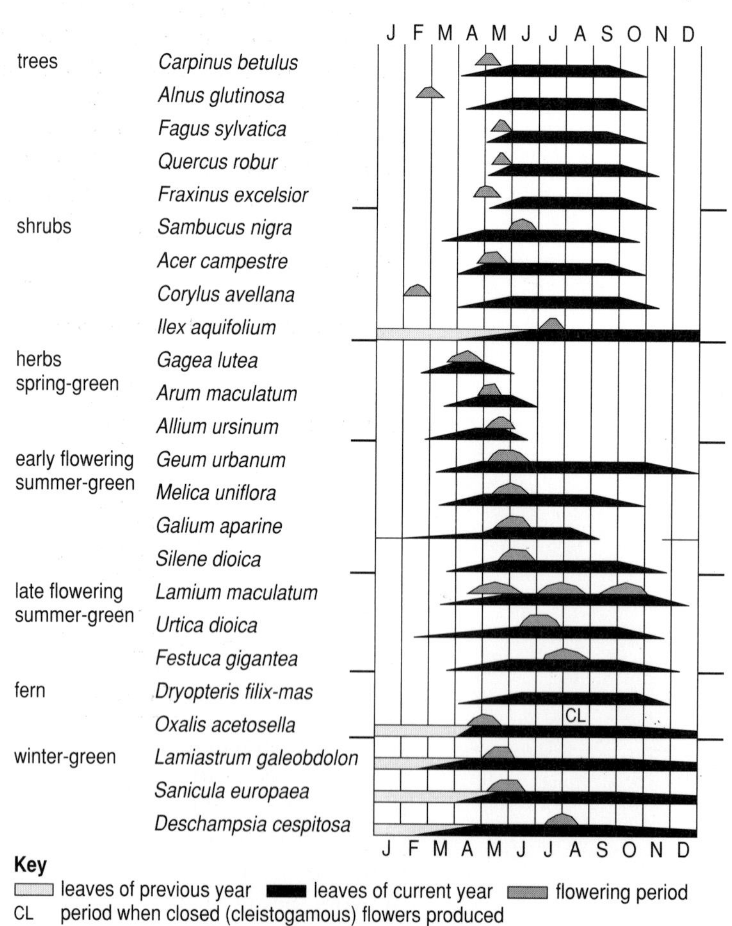

Fig 7.10 Abiotic influence on the appearance of trees and woodland plants.

7.3 ECOLOGICAL SUCCESSION

It is clear then that the physical characteristics (for example light availability or temperature) or chemical characteristics (for example soil pH or salinity) of a habitat influence the species of plant or animal that live there. However, the species that live in a particular habitat also change that habitat. For example, a canopy of beech trees may drastically reduce the amount of light reaching the soil surface, and this may, in turn, affect the water content and temperature of the soil.

By absorbing particular nutrients from the soil or by adding leaf litter, plants may change soil pH. Soils beneath conifers such as Norway and Sitka spruce may become ten times more acidic over a period of a hundred years. Such changes may allow species that were previously excluded from the habitat, that is, those that could not tolerate the conditions, to invade the habitat. These new species may then change the habitat, making conditions less favourable for the species already there, the populations of which might then decline. In this way, there may be a gradual progression from one community to another, and this progression is called **natural succession** (Figure 7.11).

Primary succession involves the introduction of species into an area that has never previously been colonised; for example, bare rock formed from cooling lava or a manufactured pond. The first species to enter an area are called **primary colonisers** and form a **pioneer community**.

Secondary succession involves the re-establishment of species into an area that once contained populations of plants and animals. Thus the gradual transition of an abandoned field back into deciduous woodland would be an example of secondary succession.

As succession continues, the number of different species in the ecosystem increases, and the food webs become more complex. Eventually a stable ecosystem develops which is in equilibrium with its environment and which normally undergoes little further change. This is called the **climax community**. Over most of Britain the climax community is mixed deciduous woodland (Figure 7.11), whereas in equatorial regions the climax community is tropical rainforest. The stable ecosystem is therefore broadly determined by climate and is known as a **climatic climax community**. Within these climate regions variations in the stable ecosystem may also occur. For example, differences in soil types produce different climax communities, known as an **edaphic climax**.

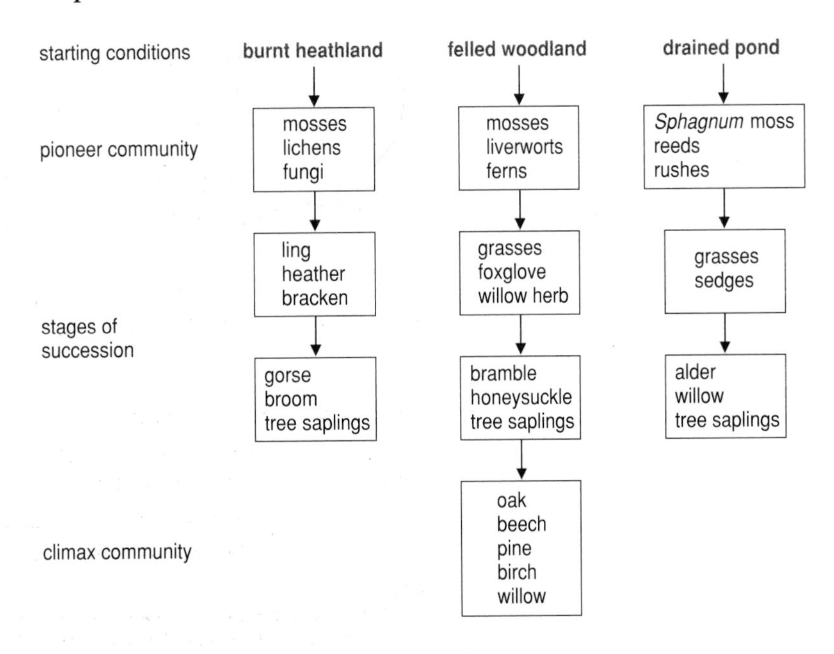

Fig 7.11 Succession leading to mixed woodland in the UK.

Climax communities may, of course, be subject to change, and this is often caused by humans. The introduction of non-native species, that is, those species that do not normally exist in this country, or the effects of acid rain or global warming, may alter what was previously a balanced community, as does deforestation, the planting of crops and the grazing of cattle. A climax community produced by the action of humans is known as a **plagioclimax**. In fact, there is no such thing as a totally natural ecosytem; humans have already affected every ecosystem on earth. Our aim, however, should be to use our understanding of ecosystems to try to minimise our harmful effects. Looking at the points that we have covered in this chapter, this might involve the following:

- maintaining complexity in natural and manufactured ecosystems
- avoiding the introduction of non-native species
- maintaining a diversity of plants and animals, to allow for succession
- avoiding drastic or rapid changes in abiotic factors, for example the pH of rain.

It seems evident that the main threat to the biosphere, the giant ecosystem, is humans. The human population has rocketed in the last two centuries, increasing the potential and actual damage humans can cause to the biosphere. Population and its regulation are discussed in Chapter 9.

SUMMARY ASSIGNMENT

1. Define the terms:
   (a) ecosystem
   (b) habitat
   (c) community
   (d) population
   (e) symbiosis.

2. The figure shows a transect across an area of coastal sand-dunes.

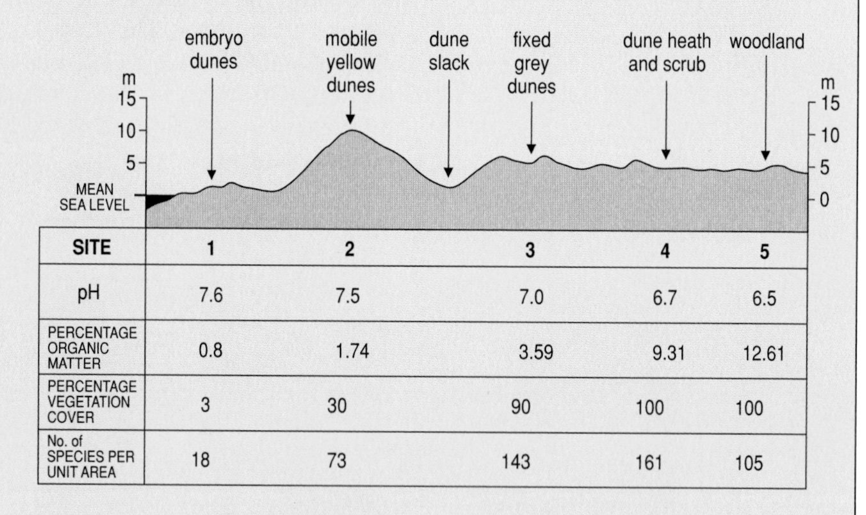

SITE	1	2	3	4	5
pH	7.6	7.5	7.0	6.7	6.5
PERCENTAGE ORGANIC MATTER	0.8	1.74	3.59	9.31	12.61
PERCENTAGE VEGETATION COVER	3	30	90	100	100
No. of SPECIES PER UNIT AREA	18	73	143	161	105

(a) Explain the conditions which:
  (i) inhibit the number of plant species at site 1.
  (ii) encourage a variety of plant species at site 4.

(b) How would you account for the decline in the number of plant species from site 4 to site 5?

(c) Suggest the effects that recreational activity might have on the succession shown.

3. The diagram shows a food web for an aquatic ecosystem.

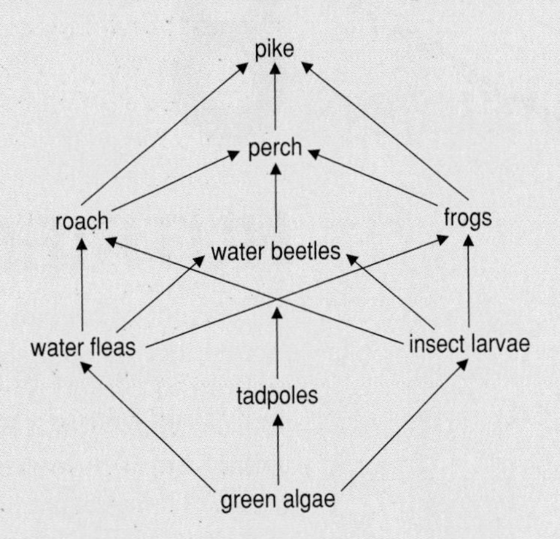

(a) From the information in the diagram, name:

(i) a primary consumer

(ii) a tertiary consumer.

(b) Suggest how this community might be altered if the population of water beetles died out.

(c) Only a small percentage of the energy absorbed by green algae is incorporated into the tissues of pike. Give three reasons why this should be so.

ULSEB January 1991 (Biology)

Chapter 8

NUTRIENT CYCLES

All living organisms need elements such as carbon and nitrogen. The availability of such elements is, of course, finite; we cannot increase the total amount of carbon which is available even if, as we saw in Chapter 1, we can alter its distribution between sinks. It is essential therefore that the elements locked up in living organisms are released when they die, that is, they are recycled. Nature is a superb recycler and this chapter will consider how these nutrient cycles work.

LEARNING OBJECTIVES

After completing the work in this chapter you will be able to:

1. outline the general principles of nutrient cycling
2. describe, in detail, the carbon, nitrogen and phosphorus cycles
3. explain how humans are influencing these natural cycles.

8.1 THE SIMPLE CYCLE

All plants and animals need a source of elements such as carbon, nitrogen, phosphorus and sulphur. Carbon is the basis of all organic life, and green plants are able to convert carbon dioxide gas from the atmosphere into carbohydrates (Chapter 5). Animals can then eat the plants to obtain their carbon (Chapter 6).

Table 8.1 Plant macronutrients and micronutrients

Element	Cations	Anions
Macronutrients		
nitrogen	NH_4^+ (ammonium)	NO_3^- (nitrate)
calcium	Ca^{2+}	HPO_4^{2-}
magnesium	Mg^{2+}	
potassium	K^+	
phosphorus		$H_2PO_4^-$
sulphur		SO_4^{2-} (sulphate)
Micronutrients		
copper	Cu^{2+}	
iron	Fe^{2+}	
manganese	Mn^{2+}	
zinc	Zn^{2+}	
boron		BO_3^{3-}
molybdenum		MoO_4^{2-}
chlorine		Cl^-

However, neither plants nor animals can survive on carbohydrates alone. Plants need fourteen elements in varying quantities in order to make proteins, nucleic acids and the hundreds of other substances that they require for healthy growth. Some of the elements are needed in relatively large quantities and are known as **macronutrients**, whereas **micronutrients** or **trace elements** are only needed in minute quantities. The macronutrients and micronutrients required by plants and the form in which they are usually absorbed are shown in Table 8.1.

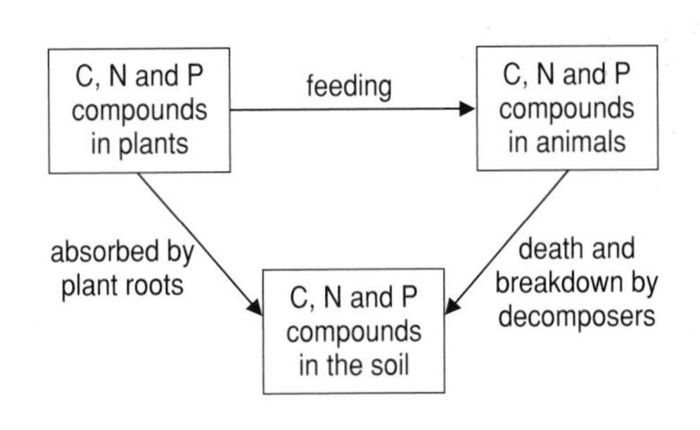

Fig 8.1 The nutrient cycle.

Terrestial plants absorb these elements from the soil, itself created by the weathering of rocks (Chapter 3). When plants or animals die, their bodies are broken down by invertebrates and micro-organisms such as bacteria and fungi (Chapter 7). This process releases nutrients back into the soil, from where plant roots can absorb them once again. Alternatively, a plant may be eaten by an animal, but this too will return nutrients to the soil through urine and faeces and ultimately through the death of the animal. This rather simplified **cycle** of events, which nevertheless applies to all nutrients, is shown in Figure 8.1, and is the basis of the cycles. However, Figure 8.1 misses out the role of the atmosphere and the rock cycle. As we shall see, these are crucially important in the cycles of particular nutrients.

8.2 THE CARBON CYCLE

The carbon cycle is shown in Figure 8.2.

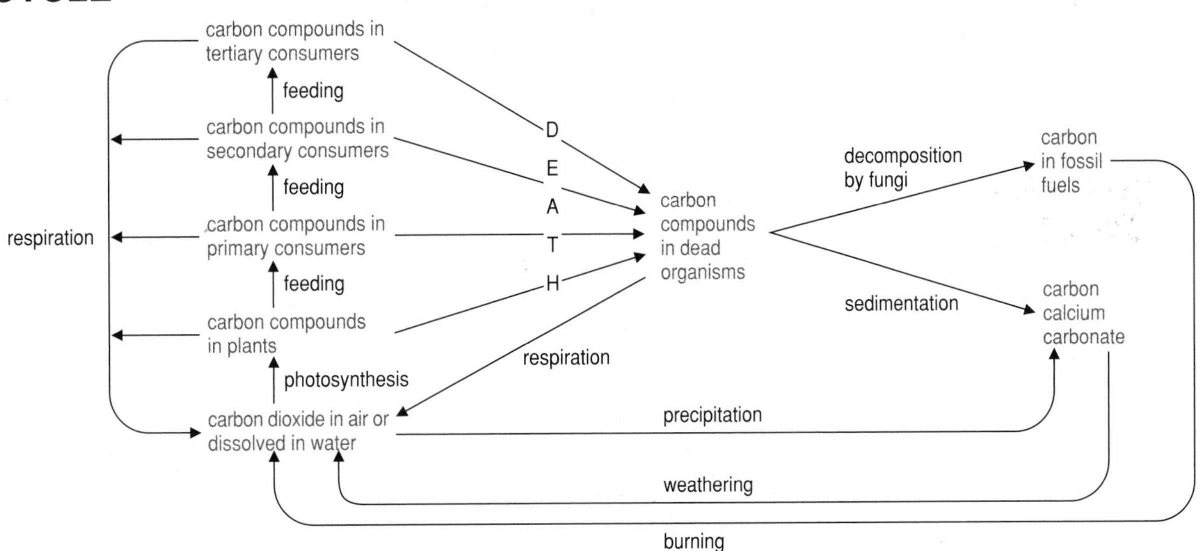

Fig 8.2 The carbon cycle.

QUESTION

8.1 Why is the sun essential to the carbon cycle?

Atmospheric carbon

The troposphere contains 0.035% gaseous carbon dioxide (Chapter 1). Carbon dioxide is removed from the atmosphere by photosynthesis (Chapter 5) and is added when plants, animals and micro-organisms respire or when carbon-containing material such as coal or wood is burned.

Organic compounds in plants

Plants use the carbon in carbon dioxide to produce organic substances such as carbohydrates, fats and proteins. Cellulose, a key component of cell walls, is a complex carbohydrate made from glucose. The pigment chlorophyll, for example, is also an organic substance and has the formula $C_{55}H_{72}N_4O_5Mg$, showing 55 carbons in each molecule.

Organic compounds in consumers

Through eating plants, primary consumers obtain their organic compounds, the energy from the release of which may be passed on through the trophic levels.

Carbon in freshwater and marine ecosystems

Gaseous carbon dioxide dissolves in water to form hydrogen carbonate ions (HCO_3^-). Large quantities of carbon are also trapped in the form of carbonates in sea-bed sediments, which over millions of years will form sedimentary deposits (Chapter 3). In some coastal waters in the tropics, coral reefs lock up carbon as calcium carbonate. Areas where carbon is locked up, for example in the oceans or in the stems of trees, are known as **carbon sinks**.

Human effect on the carbon cycle

You are also a small part of the carbon cycle. The carbon atoms in your left hand have been around since the beginning of the earth. Since then, they have formed part of the body of millions of different plants and animals; your thumb may contain a carbon that once belonged to a dinosaur!

In a few weeks, as the surface layer of your skin flakes off and falls into your bed as you sleep, bacteria (the ones you sleep with every night), will oxidise the carbon to carbon dioxide, which may diffuse through the gaps in your bedroom window. It may be absorbed by a tree in your garden or by a tree in Scandinavia. The winter robin that you see next Christmas might have that same carbon atom in its red breast. All living organisms are connected via the carbon cycle.

QUESTIONS

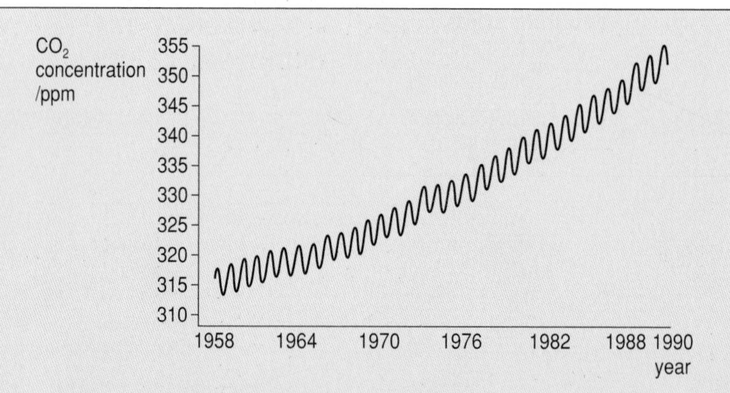

Fig 8.3 The changing atmospheric levels of carbon dioxide.

8.2 Figure 8.3 shows atmospheric carbon dioxide levels over the last 40 years.

(a) Describe three ways in which human activities have led to this increase.

(b) How do you account for the fluctuations in the graph?

8.3 A farmer concerned that rising CO_2 levels would accelerate the enhanced greenhouse effect has decided to 'do his bit' by planting a large broadleaved woodland. He has estimated that his family produce, through heating, cooking and driving, 15 tonnes of carbon dioxide annually. The farmer intends to retire in 20 years' time and wants to plant an area of woodland which, by 20 years of age, will have removed exactly the same amount of carbon dioxide that his family have produced over that time. This, he argues, will mean that *his* family, at least, will not have had any effect on so-called global warming.

The cumulative carbon fixation of a broadleaved woodland is shown in Figure 8.4.

(a) Calculate the amount of carbon dioxide the farmer's family will produce in 20 years.

(b) Calculate the number of hectares that the farmer should plant so that in 20 years' time the woodland will have removed the same amount of CO_2 as that produced by his family.

(c) Is the farmer justified in thinking that his family will have had no effect on global warming?

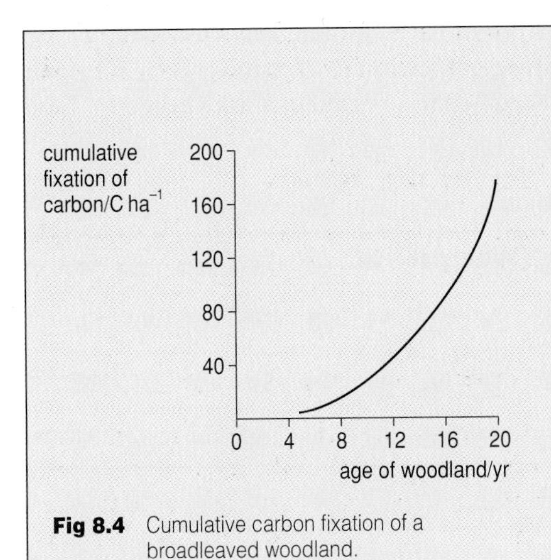

Fig 8.4 Cumulative carbon fixation of a broadleaved woodland.

8.3 THE NITROGEN CYCLE

All living organisms need a source of nitrogen in order to make molecules such as protein and DNA. Although nitrogen is by far the most abundant gas in the troposphere, atmospheric nitrogen is unavailable to plants and animals. Only some specialised micro-organisms are able to use this huge potential source. The nitrogen cycle is shown in Figure 8.5. At first sight it might look complicated but it becomes much simpler if we look at the three basic stages:

- Nitrogen in the atmosphere is made available to plants by the process of **nitrogen fixation**. This is the conversion of nitrogen gas into ammonia by nitrogen-fixing bacteria, which then dissolves in soil moisture to form ammonium ions.
- Ammonium ions may be absorbed by plant roots or may be converted into nitrite ions and then nitrate ions by **nitrifying** bacteria.
- Nitrate ions may be absorbed by plant roots, may be leached from the soil into groundwater supplies, or may be converted into nitrogen gas by **denitrifying** bacteria.

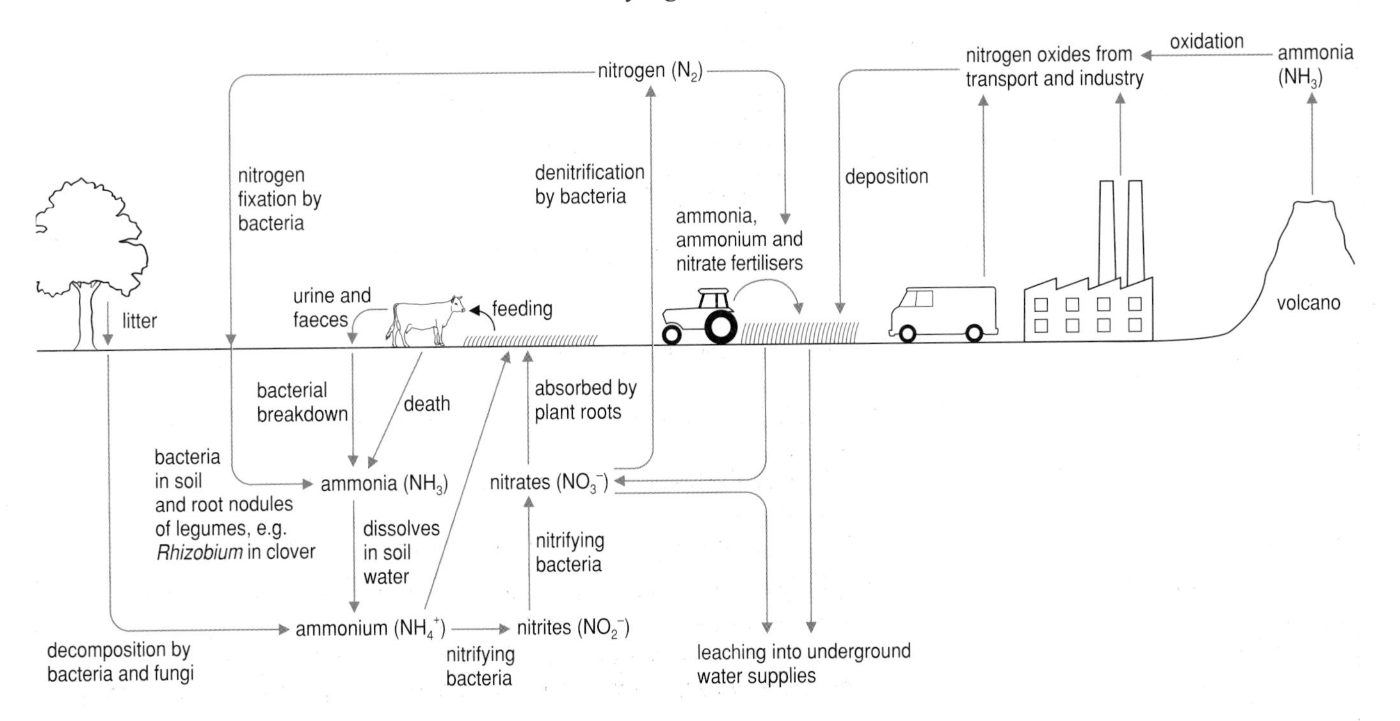

Fig 8.5 The nitrogen cycle.

Only some specialised bacteria and some cyanobacteria (formerly called blue-green algae) can carry out **nitrogen fixation** to fix atmospheric nitrogen. The most important types of nitrogen-fixing organisms are shown in Table 8.2.

Table 8.2 Nitrogen-fixing organisms

Name of organism	Where it lives
Rhizobium	root nodules of leguminous plants
Azotobacter	aerated soil
Clostridium	anaerobic soils
cyanobacteria, e.g. *Nostoc*	wetlands

Rhizobium deserves special mention because its relationship with the leguminous plant is a good example of **symbiosis**, where two organisms live intimately together to the benefit of both. *Rhizobium*, which is present in most soils, invades the root hairs of the legume which responds by forming root nodules. The bacteria carry out nitrogen fixation in the nodules, supplying the plant with ammonia. In return, the bacteria receive sugars from the phloem of the plant.

Besides nitrogen fixation, nitrogen compounds can also be added to the soil by lightning, by aerial deposition in acid rain (Chapter 10), from the decomposition of dead plants and animals and their wastes, and from manufactured fertilisers.

Dead plants and animals, broken down by bacteria and fungi, release ammonia into the soil. The ammonia is converted to ammonium ions (NH_4^+). Plant roots can absorb these ions and use them to make proteins. Animals obtain their nitrogen, in complex organic form, by eating the plants. However, not all ammonium ions are absorbed directly by plants. In aerobic conditions, specialised **nitrifying** bacteria can convert the ammonium ions into nitrate ions, as follows:

$$\underset{\text{ammonium ions}}{NH_4^+} \xrightarrow{\textit{Nitrosomonas}} \underset{\text{nitrite ions}}{NO_2^-} \xrightarrow{\textit{Nitrobacter}} \underset{\text{nitrate ions}}{NO_3^-}$$

This process is called **nitrification**. These nitrate ions are also absorbed by plant roots.

Denitrification is the conversion of nitrates in the soil back into nitrogen gas, which may then escape into the atmosphere. This is carried out by bacteria such as *Pseudomonas denitrificans* and occurs under anaerobic conditions.

QUESTION

8.4 Under what circumstances might soils become anaerobic? What could farmers do about it?

Human effect on the nitrogen cycle

Human effects on the nitrogen cycle includes the following:

- The large-scale use of ammonia-based fertilisers requires the industrial conversion of N_2 gas into ammonia (NH_3) in what is known as the Haber Process. A modern factory produces around 7000 tonnes per day, which requires 4.84×10^6 m^3 of nitrogen gas.
- Inappropriate application of nitrogen fertilisers has led to large leaching losses, causing contamination of underground water supplies and, along with phosphorus, accelerating eutrophication (Chapter 10).

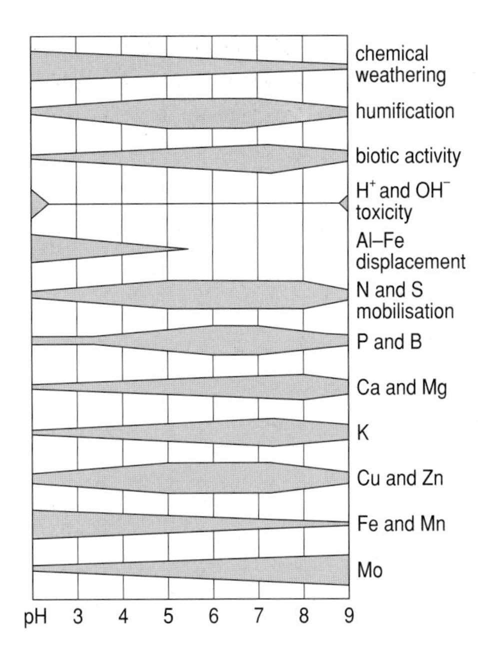

Fig 8.6 The effect of pH on soil processes and mineral availability.

- Ammonia from animal urine is released.
- Huge volumes of nitrogen oxides from vehicle exhausts are emitted. This releases the nitrogen which has been locked up in fossil fuels such as oil for millions of years.
- Nitrogen oxides, for example from car exhausts, react in the atmosphere, producing nitrous and nitric acids which dissolve in water droplets to produce acid rain. The acids are then deposited on the soil when rain falls, which may have serious effects on ecosystems. For example, as soils acidify, the types and numbers of soil organisms will change, and this will affect food chains and the rate of recycling of other minerals. The solubility and availability of minerals, and the many factors which affect the rate of soil formation, are influenced by soil pH (Figure 8.6).
- High inputs of nitrogen into the soil may initially increase productivity – it is, after all, essential for making proteins – but if soils acidify, other minerals may be lost, reducing plant growth. In fact, many areas which are of high conservation importance (for example heaths, dunes or calcareous grasslands) depend upon soils which have a low nitrogen content. Increasing the nitrogen content of such soils changes the balance of species, as it allows aggressive species, which were not previously important components of the ecosystem, to take over (Chapter 7). Diversity decreases and rare plants may be lost.

QUESTIONS

8.5 Study Figure 8.6 and answer the following questions.

(a) Under what soil pH conditions might aluminium reach toxic concentrations?

(b) State the range of pH over which nitrogen and sulphur are most readily available to plants.

(c) Decomposers break down materials such as leaf litter, twigs and fallen fruits and release nutrients back into the soil. Under which pH conditions might this process be drastically slowed? What effect would this have on the amount of litter on a forest floor?

8.6 Recent scientific studies have investigated the possible effects of increasing nitrogen deposition on herbivorous insects. In particular, scientists were keen to investigate possible links between nitrogen deposition and the defoliation (leaf loss) of herbaceous plants on motorway verges. Some of the possible connections are shown in Figure 8.7.

(a) Suggest why aerial deposition of nitrogen might be high near motorways.

(b) Suggest two reasons why herbaceous insects might spend less time chewing leaves which had a high nitrogen content.

(c) Suggest two reasons why the population of herbivorous insects might increase.

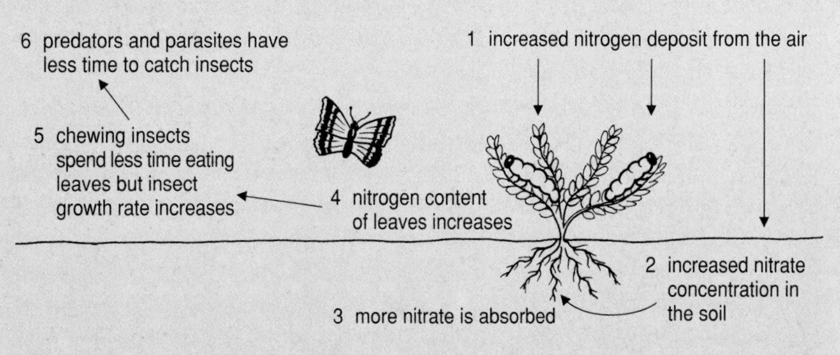

Fig 8.7 Nitrogen deposition and herbivorous insects.

8.4 THE PHOSPHORUS CYCLE

All living organisms need a source of phosphorus to make molecules such as protein, DNA and ATP. The amount of available phosphorus can therefore have a dramatic effect upon the productivity of an ecosystem.

As Figure 8.8 shows, inorganic phosphorus-containing compounds are leached from rocks and minerals. Plants absorb these compounds to make organic compounds. Animals then obtain organic phosphate by consuming plants. Inorganic phosphate is lost from animals through urine or other wastes.

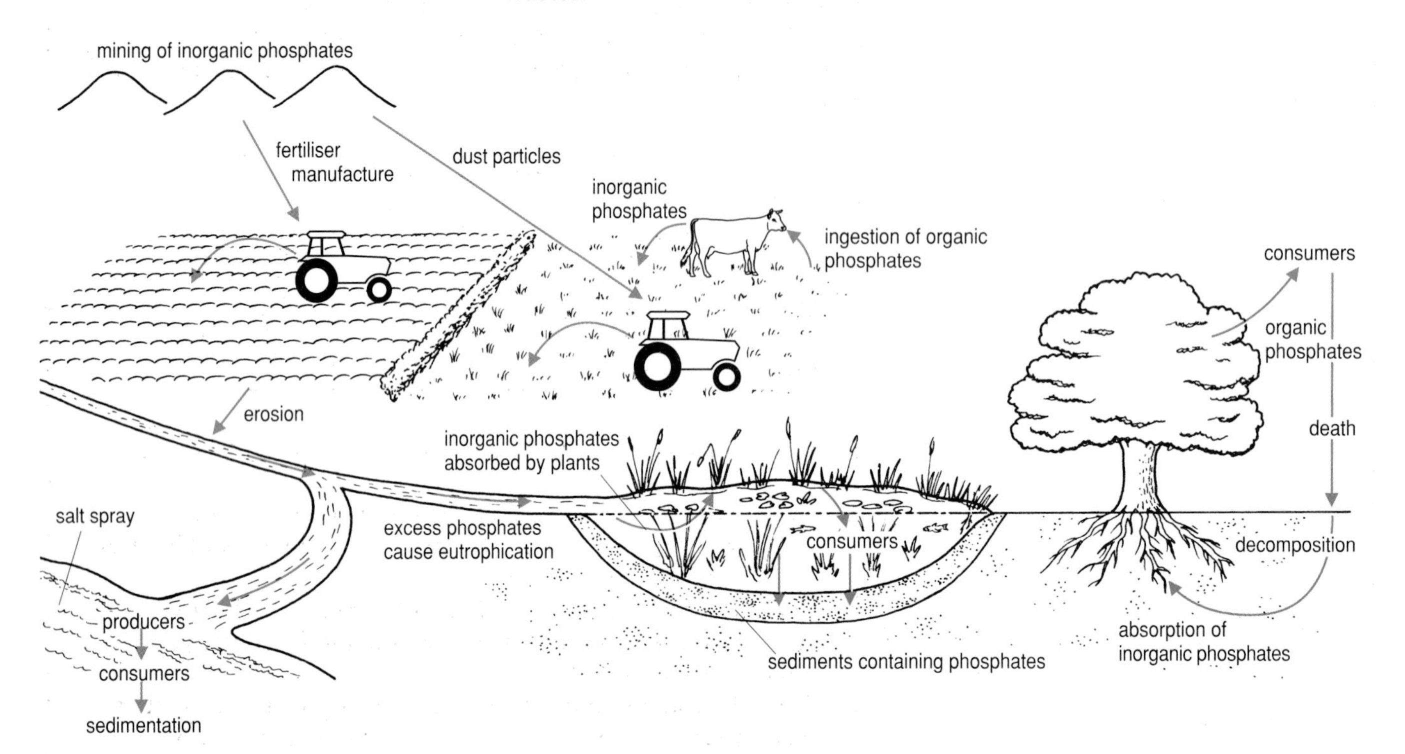

Fig 8.8 The phosphorus cycle.

A major difference between the carbon cycle and the phosphorus cycle is that phosphorus does not have a gaseous state. Plants can therefore only obtain their phosphate from the soil in which they are growing. If plants are harvested and removed, then all of the phosphorus they contain is lost from that area, and replenishment through weathering and leaching would take thousands of years.

The major phosphorus deposits exist at the bottom of the oceans. The deposits that are mined now in order to make inorganic phosphate fertiliser are simply the exposed ocean deposits which were formed millions of years ago.

The lack of a gas phase and the fact that replenishment through weathering is very slow means that phosphates, like fossil fuels, have to be regarded as a finite resource (Chapter 12).

Human effects on the phosphorus cycle

- Sedimentary phosphate deposits are rapidly being mined to make fertiliser.
- Inappropriate use of phosphate fertiliser or inappropriate farming practices have led to large losses from agricultural land. In addition, huge amounts of polyphosphates contained in detergents have entered freshwater ecosystems through sewage. Phosphate is frequently the limiting factor in freshwater ecosystems, and small extra inputs can rapidly accelerate eutrophication (Chapter 10).

SUMMARY ASSIGNMENT

1. Explain how a carbon atom, which was part of the eye of a rabbit, may now be in your eye.
2. Explain the effect of each of the following on the carbon cycle:
   (a) burning peat
   (b) the enhanced greenhouse effect.
3. It may be possible to increase the use of plant biomass as a source of fuel. Scientists have therefore investigated the production and recycling of plant biomass in three forest ecosystems by determining the dry mass of certain components. The results are given in the table below.

type of forest	total dry mass in kg m^{-2}			
	living plant biomass per year	new plant material per year	plant litter production	humus content of soil
coniferous	26.6	0.7	0.5	4.5
deciduous	40.7	0.9	0.7	1.5
tropical rain	52.5	3.3	2.5	0.2

(a) (i) Comment on the differences in total living plant biomass between the three ecosystems.
(ii) Compare and comment on the production of new plant material per year in relation to the total living plant biomass in the three ecosystems.
(b) In each ecosystem the mass of plant litter produced per year is less than the mass of new plant material produced per year. Suggest two reasons for this.
(c) From the data for tropical rainforest, what predictions might be made concerning the annual production of new herbivore biomass in this ecosystem? Explain your answer.
(d) The conversion of plant litter into humus is an important stage in the recycling of minerals. What do the ratios of litter production to humus content suggest about the relative rates of mineral recycling in the three ecosystems?
(e) Using the information in the table, compare these three forest ecosystems in terms of their suitability as sources of biomass for fuel.

ULSEB June 1990 (Biology)

Theme **2**

HUMAN IMPACT

Over the last two hundred years, the human population has grown exponentially. In the process of trying to satisfy the needs of these growing numbers, we have changed our planet; for example, millions of hectares of forests have been cleared to supply timber and land for homes and agriculture, deserts have grown, some rivers have dried up, and the air and oceans have been polluted.

Over the last twenty-five years, we have learned a great deal about the earth's life support systems and how our actions have threatened them. On a planet with finite resources, we must try to preserve the quality of our soil, air and water. In this section we will review the damage that has been done and look at ways in which humans can reduce their harmful impact.

Chapter 9

POPULATION

It has been said that population growth is the most important environmental issue. A rapid expansion in human numbers over the last century has resulted in an imbalance between resources and consumption. This paved the way for many of the current environmental problems: the over-use of fossil fuels has led to the enhanced greenhouse effect; the ever-growing demand for food has resulted in intensive farming and the widespread environmental problems associated with it; and levels of pollutants have risen beyond the ability of ecosystems to deal with them. This chapter will consider the factors which affect the size of populations, the consequences of unrestrained human population growth, and the ways through which population growth may be controlled.

LEARNING OBJECTIVES

After completing this chapter you will be able to:

1. explain the factors that affect the size of a population
2. choose and implement methods of measuring population size
3. describe and explain human population growth in different periods and in different countries
4. interpret population profiles.

9.1 POPULATION DYNAMICS

A population is the number of individuals of one species which occupies a particular geographical area at a given time. Population sizes are affected by birth, death and migration, which are in turn influenced by food supply, fertility, predation and disease. An apparently static population occurs when these factors are balanced. If an imbalance occurs, there may be a **population explosion** (a very rapid growth) or a **population crash** (a very rapid decline). To compare populations, the number of births, deaths and individuals joining or leaving the population in question need to be measured in a standard way.

Reproduction

The birth rate is obtained by dividing the number of live births in a given time, such as a year, by the initial population. Like the other rates, it is often expressed in the form of **X births per 1000** or as a percentage. For example, a birth rate of 0.017 would be expressed as 17 births per 1000 or as 1.7%.

The birth rate is dependent on the maximum possible rate of reproduction for the species under consideration, the number of individuals in the population physically able to reproduce, and the extent to which these individuals have the opportunity to breed.

Mortality

The death rate is obtained by dividing the number of deaths in a given interval by the initial population. It is dependent on the natural lifespan of the species, the age profile of the population (a population with a large number of old individuals will have a relatively high death rate), disease, predation, accidents, fighting, and environmental factors such as food supply.

Migration

The net migration rate is given by the difference between the number **immigrating** (moving in) and the number **emigrating** (moving out) in a given time period, divided by the initial population. Migration includes insects being carried from place to place by winds, or animals travelling, such as herds of cattle.

Population growth

The rate of increase of a population is given by:

rate of increase = birth rate – death rate + net (im)migration

This can be expressed in the same form as birth rates and death rates; however, an alternative is to use doubling time, the time required for a population to double its numbers, assuming the rate of increase remains constant. Table 9.1 shows doubling times which correspond to various rates of increase.

Table 9.1 Doubling times

Rate of increase /per annum	Doubling time /years
0.5%	139
1%	70
2%	35
5%	14
8%	9

If the rate of increase remains constant, the population grows exponentially; it doubles each year. Figure 9.1 illustrates such a population growth over two years. This is a model for rabbit populations, which assumes that:

- females become fertile at three months
- rabbits live for two years
- on average, each female rabbit produces a litter of eight
- litters are produced every quarter
- half of each litter is female.

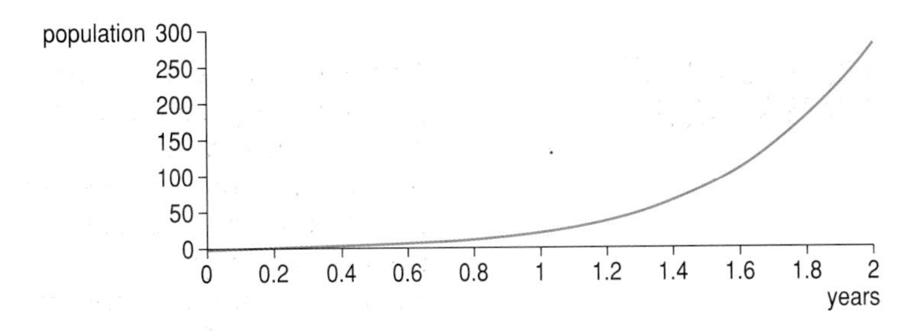

Fig 9.1 Exponential growth of a rabbit population.

In fact, this is likely to be an underestimate; the gestation period (the time taken for an embryo to develop before birth) for rabbits is only five weeks! As can be seen, the population increases from a single breeding pair to over 250 within two years.

This curve is known as a **J-curve**. It is characterised by its slow start followed by a very fast rise. If the population continued to increase in this way, it would reach over 300 million in 50 years.

Why do populations stop growing?

There is a limit to the size of population that can be supported by any environment. The maximum number of individuals of a particular species that can be supported by a particular ecosystem in the long term is called the **carrying capacity**. The rate of increase is only likely to remain constant when numbers are small compared to the carrying capacity. If the population exceeds the carrying capacity, death rates begin to exceed birth rates, and a catastrophic decline or population crash may result.

Some populations go through a periodic population explosion–crash cycle. This pattern is characteristic of fast-reproducing, short-lived species at the lower trophic levels. Examples include rodents, insects and annual plants. Figure 9.2 a illustrates typical behaviour for such populations. The numbers for these species are mainly regulated by external causes, such as predation, nutrition or climatic conditions.

Longer-lived species with relatively few predators, such as primates, follow a different pattern of growth, shown in Figure 9.2 b. This is known as a **sigmoidal curve**. The initial growth is still exponential, but as the population approaches the carrying capacity, growth slows. The factors slowing the growth rates of such species are known as **environmental resistance**.

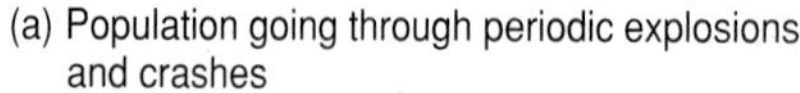

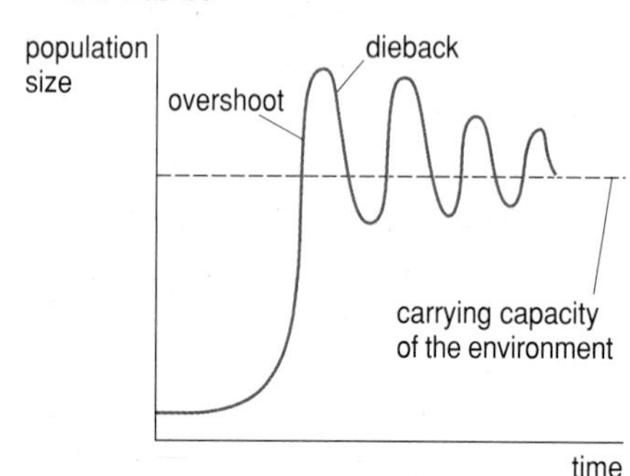

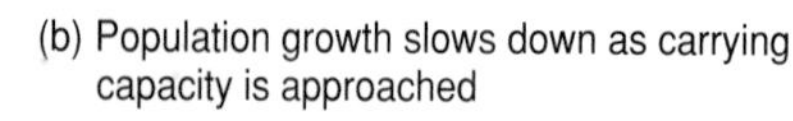

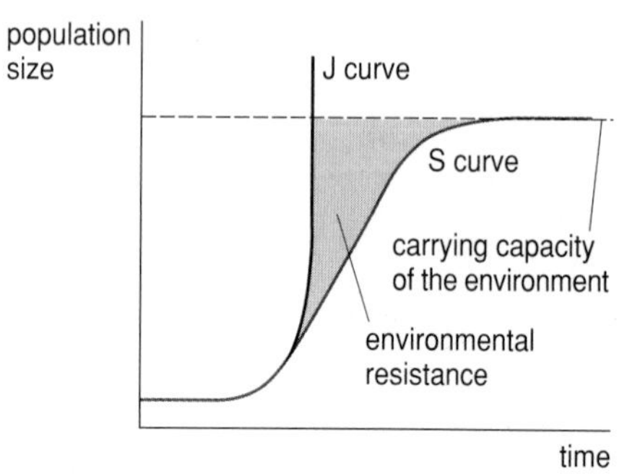

Fig 9.2 Population patterns.

QUESTION

9.1 The diagram shows a model for population regulation of a commercially important North Sea fish.

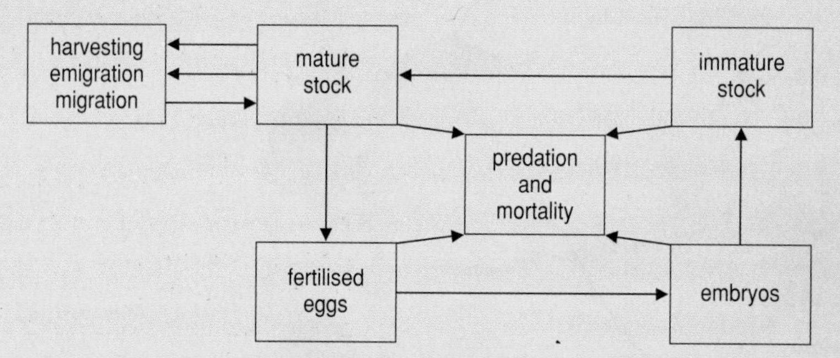

(a) Under what circumstances will the population remain stable?

(b) Explain why, in a managed population, it is desirable to keep the population in the log-phase of growth.

(c) Fishing nets may catch fish of all sizes. How might this make it difficult to maintain a stable population when harvesting marine fish?

AEB June 1992

Factors that limit population growth

The factors that limit population growth affect the fertility or mortality rates. These factors may be classified as:

- **density-dependent** (effects are stronger as population density increases) or **density-independent** (the effect is the same whatever the population density)
- **biotic** (caused by living organisms) or **abiotic** (caused by non-living components) of the environment.

Density-independent factors

Density-independent factors are usually abiotic. The most important ones are weather (conditions at a given time) and climate (long-term conditions). Extreme or unseasonal heat or cold, drought, rain, floods and fire can all have a dramatic impact. These impacts need not be negative; a change in climate, for example, will usually favour some species while harming others. Thus, as the last Ice Age ended in Britain and the land warmed, species of plants adapted to more arctic conditions, for example dwarf willows were replaced by species adapted to more temperate climates, such as silver birch, beech and oak.

Density-dependent factors

A density-dependent factor acts to decrease fertility or increase mortality as the population grows. The larger the population density (number of organisms in a given space), the greater the effect the density-dependent factor will have. For example, the amount of fighting amongst prairie voles increases as the population density increases. Such fighting leads to an increase in the rate of emigration of these rodents from the area of prairie which they are inhabiting.

Predator–prey relationships

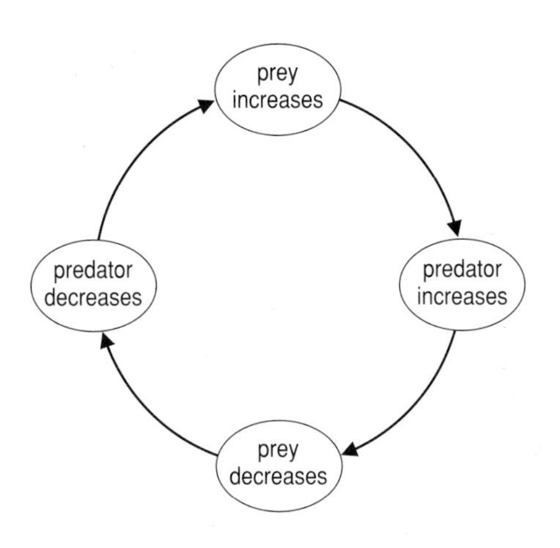

Fig 9.3 The predator–prey relationship.

This refers to the relationship that exists between the size of the population of a prey species and the size of population of its predator species. The populations of predator and prey are inextricably linked. Consider, for example, a simplified system consisting of fish and gulls. If the fish population increases, then more food is available for gulls. After a time lag, the gull numbers will therefore increase. The increased gull population increases the mortality rate of the fish, due to predation, causing a decline in fish stocks. The decreasing availability of food then causes a reduction in gull numbers, allowing fish to increase again. This relationship is illustrated in Figure 9.3.

If a graph of population against time were to be plotted for both species, it would show a similar shape for both, but with the predator lagging behind the prey, and the size of the oscillations smaller for the predator. Generally, this shows a regular cyclic pattern; the numbers of both species oscillate within a given range. This cycle may, however, be disrupted by external factors; in the fish/gull example, the introduction of unlimited fishing by fishermen could cause the fish numbers to decline to a level too low for recovery, resulting in starvation for the gulls.

Predation may be of benefit to the prey population as a whole, since the predator selectively removes old or sick individuals (since they are easier to catch) and helps to prevent the population from exceeding the carrying capacity and the ensuing population crash.

QUESTION

9.2 Why is there a time-lag between a change in prey numbers and a change in predator numbers?

Parasitism

Parasitism can be considered as a form of predation; the main difference is that the parasite usually co-exists with the host rather than killing it. Some parasites such as honey fungus, which may attack deciduous or coniferous trees, may exist first as a parasite on the tree but then as a saprophyte after it has killed the tree.

Interspecific competition

This occurs when two species are competing for the same resources within an ecosystem. It may result in a balance in which the two share the resources, or one species may take over to the exclusion of the other. A domestic garden being overrun with weeds is an example in which a number of weed species co-exists, but the original domestic plants may well be totally excluded.

Intraspecific competition

When species numbers are small, there will be little effective competition between individuals for resources, and provided numbers are not too small for individuals to find mates readily, growth will be high. As the population approaches the carrying capacity, this intraspecific competition becomes significant. The stronger individuals claim a larger share of the resources.

Many species deal with intraspecific competition by being **territorial**. An area is held by an individual or pair, who fend off rivals by display or, if necessary, fighting. The most reproductively successful individuals will be those that hold the largest territory.

Disease

High population density results in the ready transmission of infectious diseases, the effects of which will be made worse if sufferers are already malnourished (lacking nutrients essential for health), due to the scarcity of resources.

Stress and crowding

Very high population densities, even in the presence of adequate resources, seem to result in a variety of physical and behavioural symptoms, including aggression, hyperactivity, reduced fertility, lack of parental instinct and low resistance to disease. This has been seen in laboratory conditions among rats.

QUESTIONS

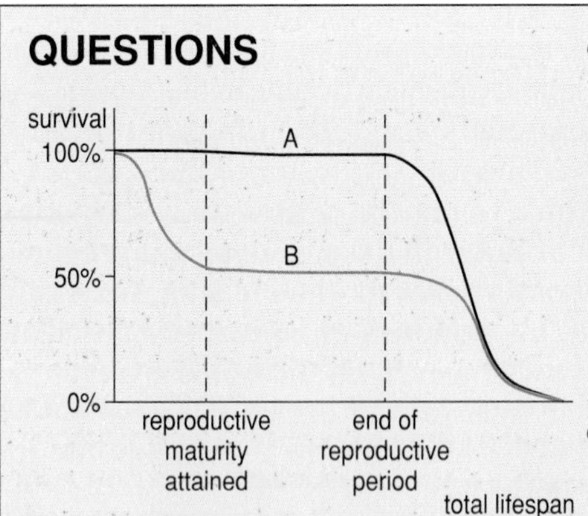

Fig 9.4 Survival curves.

9.3 Figure 9.4 shows survival curves for two species, A and B. These illustrate the percentage of individuals who will survive to a given age.

(a) Approximately what percentage of species B survives to maturity?

(b) What can be deduced from the curves about the mortality rates of each species throughout its lifespan?

(c) What factors are likely to be the important limiting factors to growth for each of A and B? Explain your reasoning.

9.4 The reintroduction of wolves into the UK has been suggested.

(a) Sketch curves to show:

(i) how the wolf population would grow after the introduction of just a few breeding pairs into an area

(ii) how you would expect the population numbers to behave in the long term.

(b) One of the areas suggested for this reintroduction is the highlands of Scotland. It has been suggested that introducing wolves would avoid the need for deer-culling in this area.

(i) Why are deer culled?

(ii) What are the likely consequences for the deer population from this strategy?

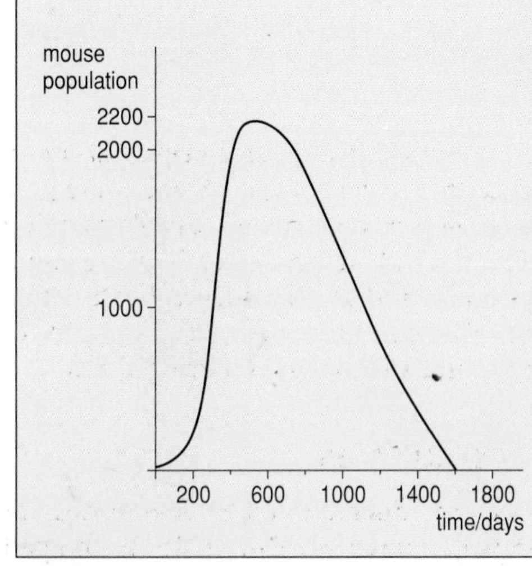

9.5 The graph shows the growth curve for a population of mice reared in an enclosed environment.

(a) Define the term *population*.

(b) Explain the shape of the curve between day 200 and day 600.

(c) State two density-dependent factors which may have caused the population to crash.

(d) State **two** environmental factors controlling wild population levels which do not operate in this case.

AEB June 1990

9.2 WORLD POPULATION – DISTRIBUTION, GROWTH AND IMPACT

The world's human population has increased dramatically in recent years (Figure 9.5), and its doubling time is now approximately 40 years. Currently world population increases at about 92 million people per year.

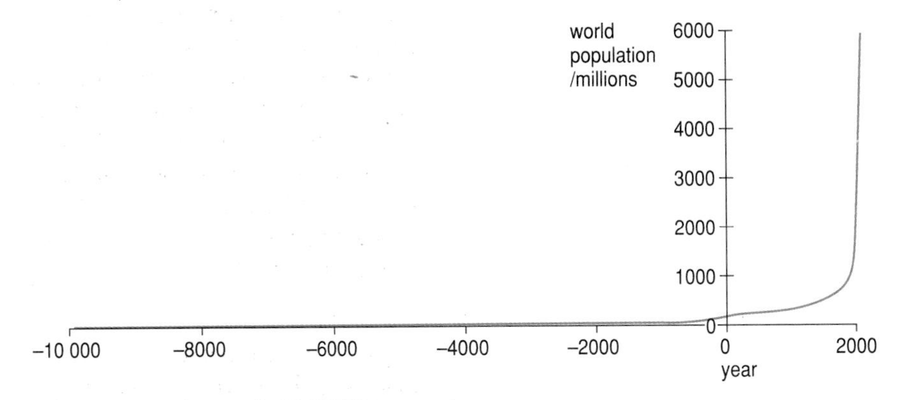

Fig 9.5 Population growth, 10 000 BC to present.

Why has the population increased dramatically?

Until recently, human populations were controlled by the same limiting factors as the rest of the animal kingdom – food supply and disease. Other factors were war and cultural influences. Although the development of agriculture had some impact, the major change came with the Industrial Revolution. The increased availability of power, enhanced technology for food production, and improved standards of hygiene and medical care allowed people to survive longer and hence to reproduce more.

The reduction in infant mortality was particularly significant. In the eighteenth century, half of all newborn babies died in infancy. In these circumstances, the optimum strategy for parents was to produce a large number of children in the hope of some children surviving. Children were needed both as a source of labour and as effective 'insurance' in case the parents survived past the age when they were able to work. When infant mortality began to fall as a result of medical advances, the birth rate did not, and so the population rose rapidly.

In the twentieth century, life expectancy in many countries has increased, many previously fatal diseases have been almost eliminated, and food production has continued to rise. Although the birth rate has decreased in some countries, in others it is still high, and globally it is far outstripping the death rate. The reproductive life has increased and far more people are also surviving into their seventies and eighties. World growth is exponential. If the current pattern continues, we can anticipate a world population of above eight billion by the year 2020.

What next?

The key issue is whether the population will stabilise (to give a sigmoid curve) or whether it will increase above the earth's carrying capacity and undergo a catastrophic dieback. There is little agreement on what the earth's carrying capacity is. Optimists believe that continuously improved technology and a more even distribution of resources would allow the survival of a much higher population than at present, whereas others suggest that we have already exceeded the carrying capacity and are 'living on borrowed time'.

Population differences of countries

The world may broadly be classified into **developed** and **developing** (or **more-developed** and **less-developed**) countries. People in developed countries have, broadly speaking, higher per capita income, better access to contraception, a higher average level of education and greater access to technology.

The developed countries have, for the most part, stabilised their own populations. The social pressure to have children has been all but eliminated, and the availability of contraception, allied with the increasing independence of women, has reduced the birth rate so that it is near or below the level for replacement.

In the developing countries, improved health care has reduced infant mortality and prolonged life, but the birth rate has remained high. For people in a low-technology, labour-intensive society, reducing the number of children hardly seems a wise option, and the threat of over-population is not as high a priority as compared to their immediate needs. The birth rates have fallen to a certain extent, but are still way above replacement levels.

Figure 9.6 illustrates the contribution of the more-developed and less-developed countries to the global population total. Figure 9.7 compares fertility rates, showing both the drop over the last thirty years and the comparison between countries.

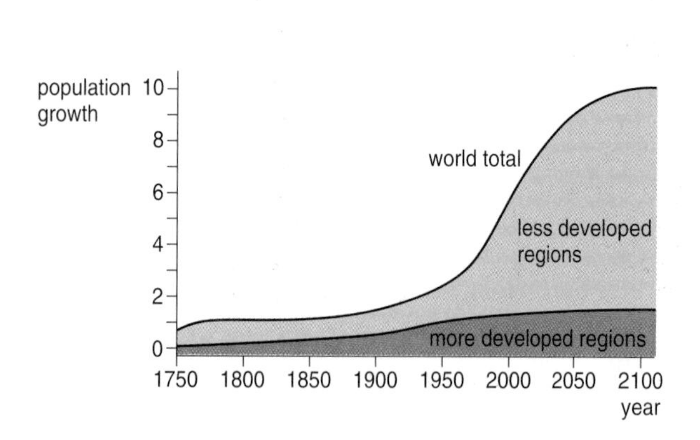

Fig 9.6 Human population growth in more and less developed regions.

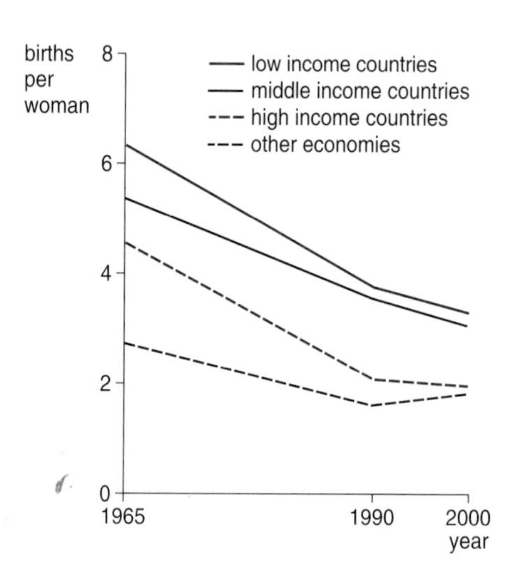

Fig 9.7 Total fertility rates.

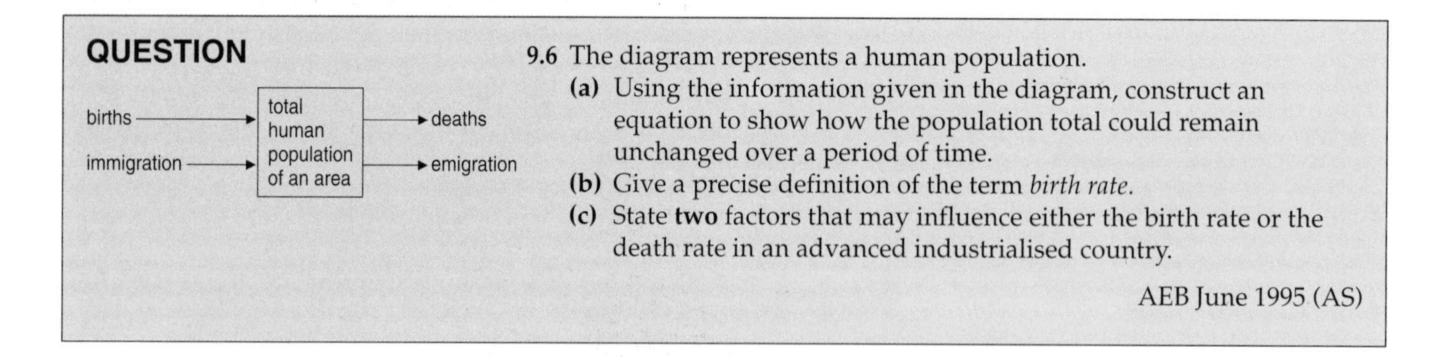

QUESTION

9.6 The diagram represents a human population.

(a) Using the information given in the diagram, construct an equation to show how the population total could remain unchanged over a period of time.

(b) Give a precise definition of the term *birth rate*.

(c) State **two** factors that may influence either the birth rate or the death rate in an advanced industrialised country.

AEB June 1995 (AS)

9.3 POPULATION PROFILES

In order to make predictions and analyse the factors determining the size of a particular population, it is necessary to consider more than just the number of individuals in that population. Their age distribution is also vitally important. **Population profiles** show the distribution of individuals by age and sex in a population. This makes it possible to see immediately:

- the proportion and number of the population able to reproduce currently

- the number of elderly people compared to the rest of the population, which is important for welfare and healthcare assessments, and is an indication of how the population has changed over time
- the number of children in the population, which is useful for healthcare assessments and is an indication of the size of the reproductively active population of the future.

Figure 9.8 shows population profiles for a developed and a developing country in 1990. The profile for the developing country has a characteristically pyramidal shape – numbers of people decrease sharply with age. This means that a high proportion of the population is young, which is associated with both a low life-expectancy (few people survive to a great age) and a high population growth.

The profile for the developed country, on the other hand, has a profile of more or less uniform thickness until late middle-age, when natural ageing occurs. This shows a high life-expectancy and a negligible population growth.

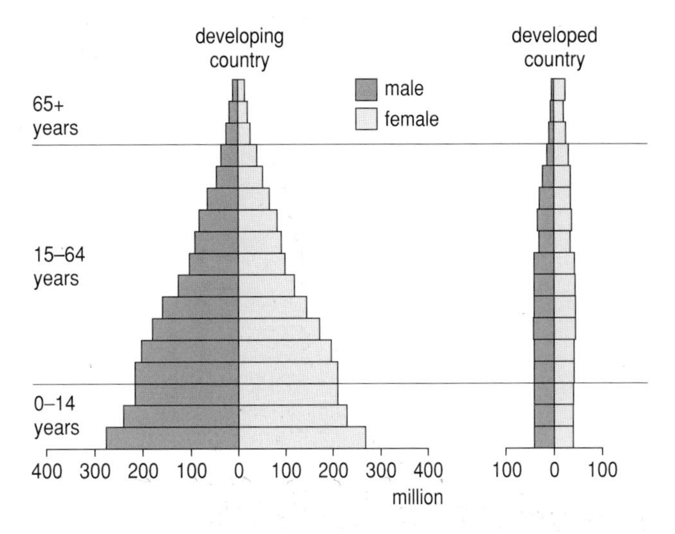

Fig 9.8 Population profiles for developed and developing countries, 1990.

Given a knowledge of the current population profile and current birth rates and life expectancies, the population profile for the future may be predicted. Figure 9.9 shows the predicted structure for a developing country in the year 2020.

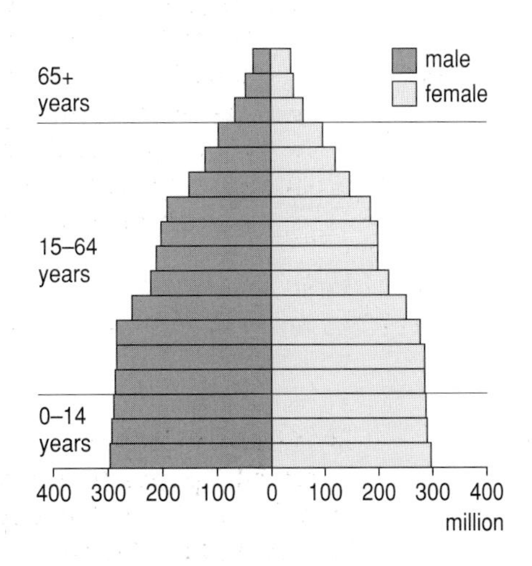

Fig 9.9 Population profile for a developing country, 2020.

QUESTIONS

9.7 Describe and account for **three** important differences between the 1990 and 2020 profiles of a developing country (Figures 9.8 and 9.9).

9.8 Figure 9.10 shows the population profiles of Denmark and Côte d'Ivoire.

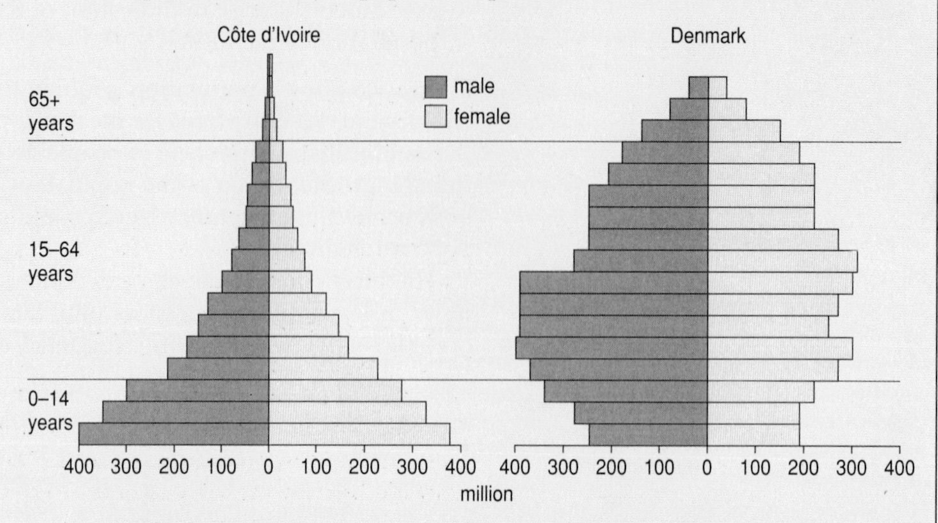

Fig 9.10 Population profiles for Denmark and Côte d'Ivoire.

Comment on the population growth of each country.

9.9 The diagram shows three types of age pyramid.

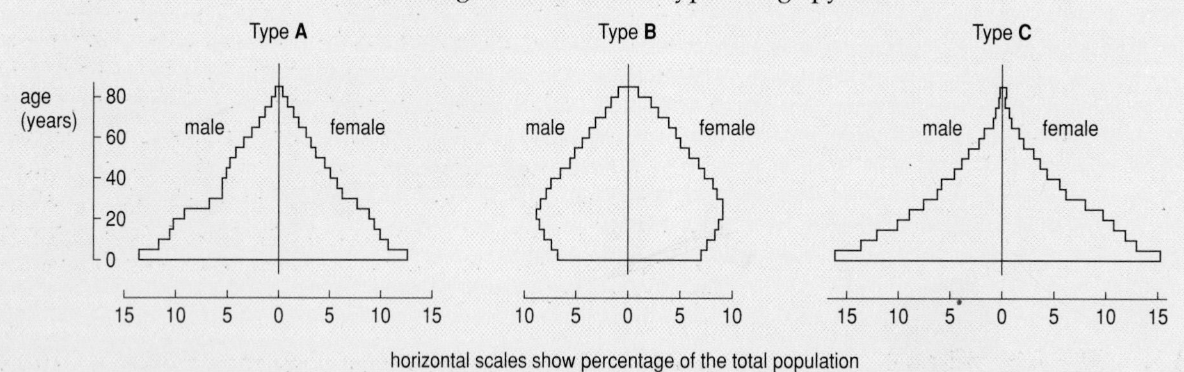

(a) Assign each pyramid to the appropriate description.
- **(i)** Rapidly expanding population
- **(ii)** Declining population
- **(iii)** Stable population

(b) **(i)** Which pyramid would most closely resemble that applicable to the present population of the UK?
- **(ii)** Explain the shape of this pyramid.

AEB June 1988

9.4 REGULATION OF POPULATION

From an environmental point of view, it seems inevitable that actions should be taken to reduce population growth drastically. However, the objections which are raised include:

- religious objections to the use of contraception
- objections that regulation is not necessary because the world can accommodate many more people
- objections that regulation is not necessary because populations will stabilise of their own accord, following the pattern of the developed countries
- objections about maintaining the human 'right' to reproduce freely

- objections to restrictions from, and on behalf of, developing countries; these objections are from groups who argue that the developed countries have had their opportunity to breed freely and use resources as they will, and that it is unreasonable and hypocritical of them to expect the developing countries not to do the same; it seems that the developing countries are effectively paying for the developed countries' luxuries.

Methods of regulation

Natural regulation

This includes disease, starvation and fighting, which may all be due to overcrowding – the factors are discussed in Chapter 9, Section 9.1. They are not the desired methods of regulation for society, but they will have an effect where the technological means of averting them are not available.

Social and political regulation

Some countries have tackled the population problem by officially limiting reproduction. China, for example, adopted a 'one child per family' policy in the late sixties, and provided free contraception. There was also strong pressure for those who infringed upon the rules to have an abortion. Couples who pledged to limit themselves to one child received better pay, better housing and an improved education for the child. Although this certainly had an effect – the birth rate halved – it was not without its problems. There is now a deficit of girls in China; many families were keen to ensure that if they were to have only one child, it should be a boy, and methods, from selective abortion to infanticide, were used to guarantee this.

Other countries have used similar campaigns involving free contraception and sterilisation, as well as programmes of education. These campaigns often conflict with religious and cultural traditions. The effect of such campaigns is often determined by the level of women's education in a particular country.

Even education, however, will have little effect if the individuals concerned have good reason to believe they will be disadvantaged by having fewer children. In many developing countries, labour is the substitute for wealth, and while infant mortality is high, families will wish to ensure they are left with sufficient children.

9.5 MEASURING POPULATIONS

Measuring every individual of a population may be either physically impossible or extremely expensive, therefore a **sample** is taken, from which the experimenter aims to infer information about a population as a whole. The sample must be, in some sense, representative of its parent population. The main methods used to ensure this are **simple random sampling** (see the case study for details), and **stratified random sampling**. In general, the more detailed and specific the information required, the larger the sample must be. It is also necessary to bear in mind the requirements of any statistical tests that are to be carried out on the data.

CASE STUDY

Using random sampling to estimate the number of bluebells in a wood

1. Place tape measures along two sides of the woodland. Poles may be inserted at 1 m or 5 m intervals.
2. Often a 50 cm × 50 cm quadrat is used. If possible, try to sample 1% of the total area. For example, if the woodland is

$$20 \text{ m} \times 10 \text{ m} = 200 \text{ m}^2, \text{ a } 1\% \text{ sample would be } \frac{1}{100} \times 200 \text{ m}^2 = 2 \text{ m}^2$$

   If each quadrat was 0.25 m², this would involve sampling in

$$\frac{2}{0.25} = 8 \text{ quadrats.}$$

3. Use random number tables to identify co-ordinates at which you will place the quadrat. Co-ordinates which fall outside the woodland are ignored.
4. Count the total number of bluebells inside each quadrat.
5. Repeat for all the co-ordinates.

NB This assumes the distribution of bluebells is random. It might, however, be related to soil conditions which could change within the woodland.

Another major problem in random sampling is that it may mean that large areas of a habitat are missed out. There is no guarantee that random numbers will cover all areas of a habitat equally.

To overcome this, stratified random sampling may be used. Here, an area is divided up into more homogenous units, for example in the case study example, according to soil pH. Samples can then be selected at random from each of these units. The number of quadrats to be placed in each unit is calculated by using the formula:

$$\text{number of quadrats in unit} = \frac{\text{area of unit} \times \text{total no. of quadrats}}{\text{total area}}$$

This ensures that important areas do not get missed out.

The method of collecting the sample will vary according to the organisms concerned; the methods will be considered under the headings of non-moving organisms and moving organisms.

Sampling non-moving organisms

Non-moving organisms include plants in general and animals such as barnacles. Methods may be **subjective** or **quantitative**.

Subjective

The ACFOR scale records the presence of each species at a given site. The numbers of each species are put into one of the following categories:

- A = Abundant
- C = Common
- F = Frequent
- O = Occasional
- R = Rare

This method is liable to errors. Conspicuous organisms tend to be over-estimated and inconspicuous ones under-estimated. Some of the subjectivity may be removed by quantifying categories, for example deciding that 'abundant' refers to at least 75% of ground cover. This method, however, has the advantage of being quick and easy to implement.

Quantitative

Quantitative methods actually produce a figure for the frequency of species. They are carried out using **quadrats**. Two types of quadrat will be considered: the frame quadrat and the point quadrat (see Figure 9.11).

Frame quadrat

The quadrat is placed on the ground in several locations, whose co-ordinates are selected by using random number tables. Although, as an alternative, many students attempt to throw a quadrat 'at random', this will not result in a truly random sample because of the fact that the handedness of the thrower

(a) Frame quadrat
e.g. 0.5 m × 0.5 m = 0.25 m^2
Frame size will vary depending on species density of diversity within an area

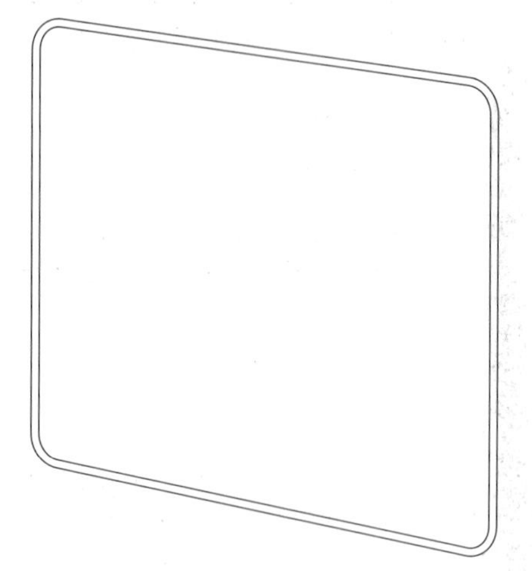

(b) Using a point quadrat

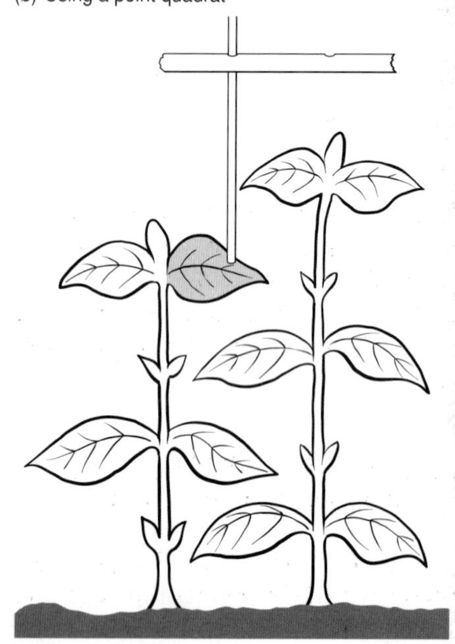

Fig 9.11 Types of quadrat.

– not to mention their likely preference for avoiding mud or nettles – will affect the outcome.

The incidence of the species under investigation within the quadrat is noted in each case. Two figures may be calculated from this:

$$\%\text{ frequency} = \frac{\text{no. of quadrats in which species was present}}{\text{total number of quadrats}} \times 100\%$$

$$\text{species density} = \frac{\text{average no. of individuals per quadrat}}{\text{area of quadrat}}$$

Percentage frequency is the easier and quicker figure to find, as recording the presence or absence of a species is simpler than ascertaining an exact number. However, it does not provide an effective comparison between two reasonably abundant species, as it does not take into account that higher numbers of one species may be present in a given place in the area under investigation.

Point quadrat

Random number tables are used to select sampling positions. The point quadrat is suspended above the area to be sampled, and the pointer is passed through each of the holes in turn. A record is kept of the species touched by the pointer on its way down. Percentage cover is measured, using the formula:

$$\%\text{ cover of species} = \frac{\text{no. of times point touches species}}{\text{no. of point quadrats taken}} \times 100\%$$

This method is difficult to use if the vegetation is tall, and if the leaves of plants have a vertical pattern of growth, their numbers may be under-estimated.

Sampling moving organisms

It is often very difficult to count directly the number of moving organisms within an area; they may not all be easily visible, and they may move during the sampling process. Instead, an indirect method – the **mark-release-recapture** method – is used. The procedure is as follows:

- capture a sample of the population
- mark each individual in a non-harmful way
- recapture organisms after an appropriate length of time
- note the number of marked organisms recaptured.

The aim is to give an indication of the total number of individuals in the population. To understand how this works, consider the following case: There are 100 organisms in a population. Ten of them are captured and marked, and then released. These mix back in with the rest of the population, so that 10% of the total population is marked. We would therefore expect 10% of the next sample to be already marked.

The formula used is:

$$\frac{\text{no. in population}}{\text{no. in first sample}} = \frac{\text{no. in recaptured sample}}{\text{no. already marked in recaptured sample}}$$

This requires that the time between marking and recapturing is long enough for the marked organisms to mingle with the rest of the population but short enough so that the population numbers have not yet changed significantly. It also requires that marking has no effect on the marked individual.

Capturing mobile organisms

Aquatic organisms

Aquatic organisms may be sampled by collecting them in a net. Three of these are illustrated in Figure 9.12 a, b and c – the **tow net**, the **drop net** and the **surber sampler**. The tow and drop nets are used for, respectively, collecting plankton and collecting bottom dwellers.

Other organisms are collected by **disturbance sampling**. Organisms are dislodged from rocks or within the substrate. They may be scraped into a submerged net or swept downstream into a net. **Kick-sampling** may be used to churn the substrate or, for greater accuracy, the surber sampler is used.

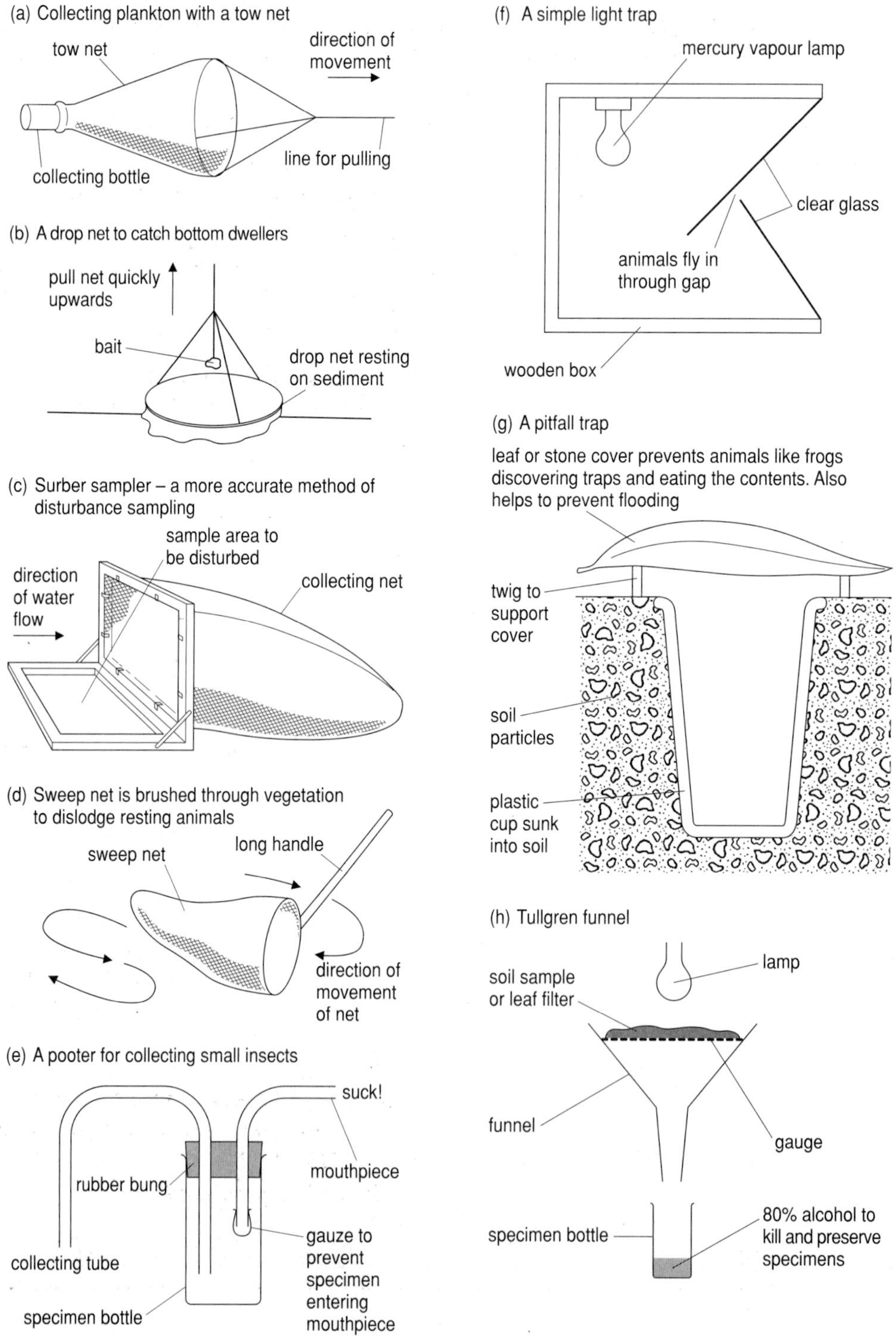

(a) Collecting plankton with a tow net

tow net
direction of movement
collecting bottle
line for pulling

(b) A drop net to catch bottom dwellers

pull net quickly upwards
bait
drop net resting on sediment

(c) Surber sampler – a more accurate method of disturbance sampling

sample area to be disturbed
direction of water flow
collecting net

(d) Sweep net is brushed through vegetation to dislodge resting animals

sweep net
long handle
direction of movement of net

(e) A pooter for collecting small insects

suck!
rubber bung
mouthpiece
collecting tube
gauze to prevent specimen entering mouthpiece
specimen bottle

(f) A simple light trap

mercury vapour lamp
clear glass
animals fly in through gap
wooden box

(g) A pitfall trap

leaf or stone cover prevents animals like frogs discovering traps and eating the contents. Also helps to prevent flooding
twig to support cover
soil particles
plastic cup sunk into soil

(h) Tullgren funnel

soil sample or leaf filter
lamp
funnel
gauge
specimen bottle
80% alcohol to kill and preserve specimens

Fig 9.12 Capturing mobile organisms.

Collecting organisms in air or vegetation

Various types of nets may be used to collect flies or animals resting on vegetation. A **pooter** (Figure 9.12 e) may be used to suck small animals into the collecting chamber. Various types of **trap** (Figures 9.12 f and 9.12 g) may be used to attract insects.

Collecting invertebrates in soil or leaf litter

Invertebrates that move across the surface of the ground may be trapped by a **pitfall trap** (Figure 9.12 g). A **tullgren funnel** (Figure 9.12 h) is used to extract small invertebrates from leaf litter or soil. The invertebrates in the soil sample move away from the drying effect of the bulb and fall through into the specimen bottle below. Another method for animals such as earthworms is to draw them to the surface by an irritant such as dilute potassium manganate (VII). They may then readily be sampled.

SUMMARY ASSIGNMENT

1. **(a)** Give two assumptions that must be made when using the mark-release-capture method to estimate population size.

   (b) In a survey of deer population, 80 deer were marked and released. Two weeks later a second sample was captured. Of these deer, 17 were seen to be marked and 3 were unmarked. Calculate the estimated population size.

   (c) The graph shows the effect of marking increasingly large samples of deer on the estimated population size.

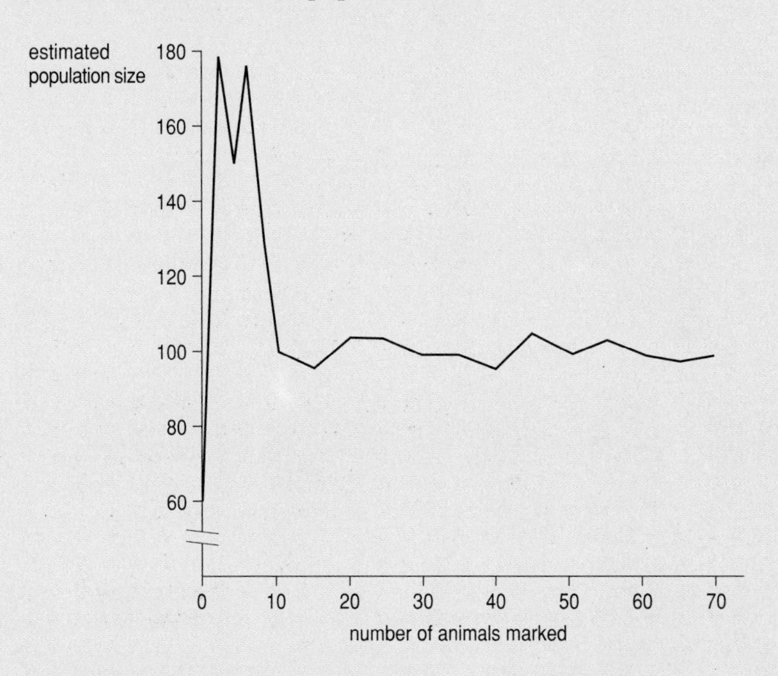

   (i) Suggest the minimum number of deer that should be marked in order to obtain a reliable estimate of the size of this population.

   (ii) Explain your answer.

2. The lemming is a small herbivore. The relationship between its population and that of its principal predator, the snowy owl, is shown on the graph.

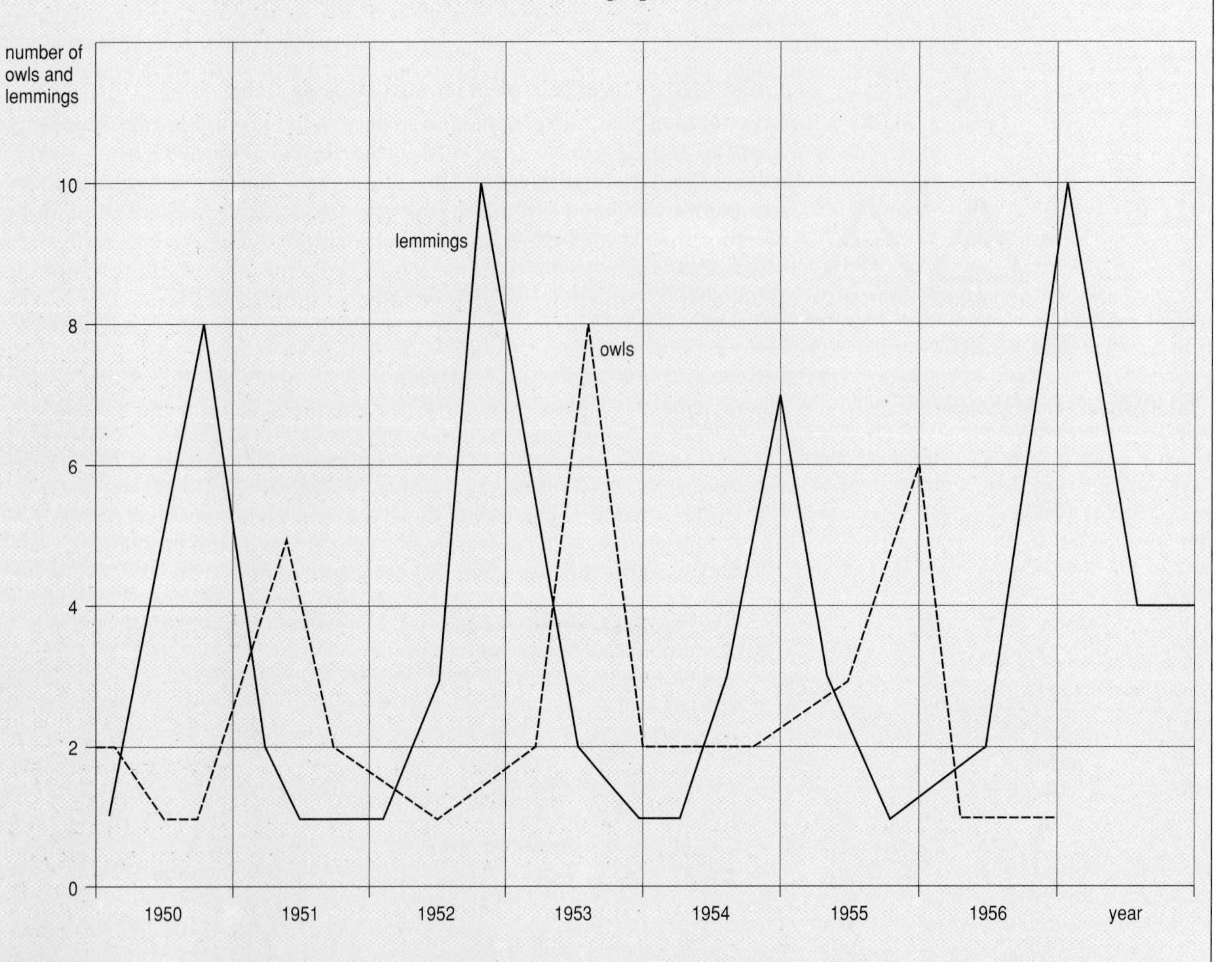

(a) Estimate the mean cycle time for:
 (i) the lemming population
 (ii) the owl population.

(b) Explain why the lemming and owl populations vary in an approximately regular manner.

(c) **(i)** Explain clearly the differences between regulatory factors which are density-dependent and density-independent.
 (ii) Into which category does the regulatory factor shown by the graph fall?

(d) Excluding predation and climate, name **two** other factors which might limit the lemming population.

(e) Locust populations are much affected both by drought and excessive rainfall, but they have few predators. Sketch a graph showing the possible changes in locust population over a 50-year span and explain its shape.

AEB June 1986

3. Why do many developing countries have high population birth rates?

4. The population (age/sex) pyramids below show the structure of the population of England and Wales in 1871 and 1971 and that of India in 1971.

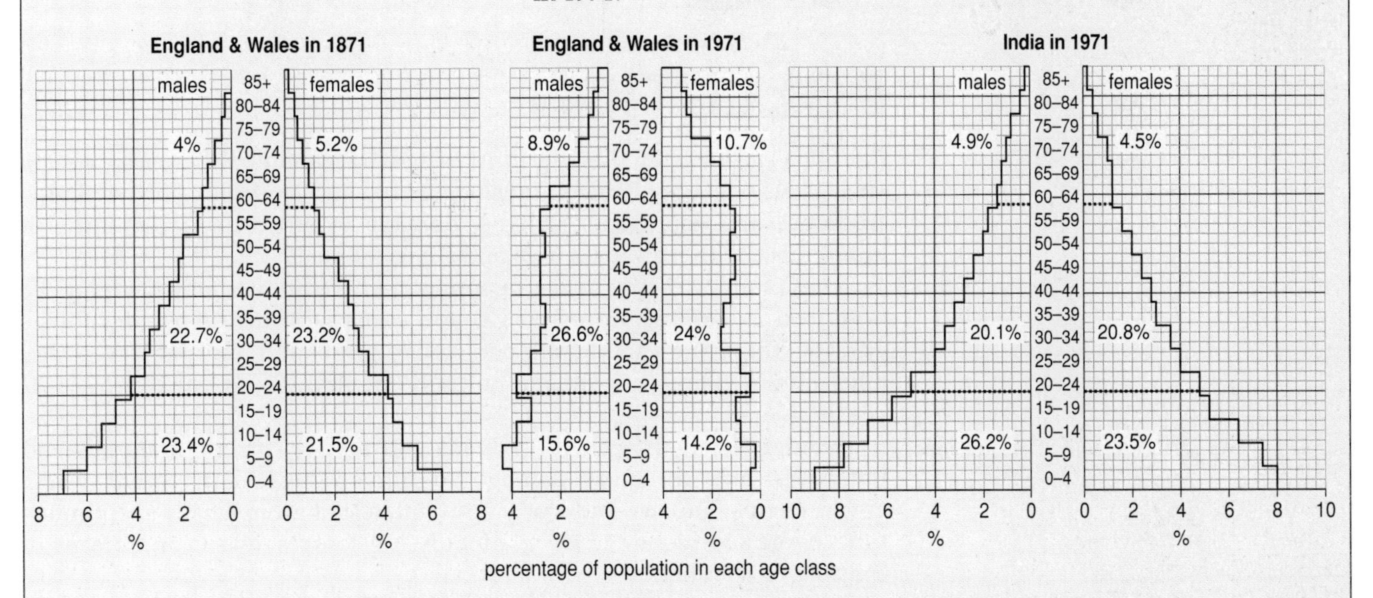

(a) In England and Wales, what was the percentage of females, under the age of ten, in the total population in the years:
  (i) 1871
  (ii) 1971?

(b) If the population of England and Wales in 1971 was 50 million, how many of these were females under 10 years of age? Show the stages of your calculation.

(c) State **one** difference between the population pyramids of England and Wales for 1871 and 1971 for each of the following age ranges:
  (i) 0–19
  (ii) 20–44
  (iii) 45–85+.

(d) The *dependency rate* (d_r) of a population can be calculated using the formula

$$d_r = \frac{(\text{total \% aged 0–19}) + (\text{total \% aged over 60})}{(\text{total \% aged 20–59})}$$

  (i) With reference to this formula, explain *dependency rate*.
  (ii) Using the information on the population pyramids, calculate the dependency rate for:
    A England and Wales in 1971
    B India in 1971.
  (iii) The *true* dependency rates were different from those that you have calculated. How, and why, were the *true* rates different from the calculated rates in:
    A England and Wales
    B India?
  (iv) What are the social/economic consequences of a high dependency rate?

(e) In 1971 the population structure of England and Wales was different from that of India and the shape of the population pyramids is different. Explain the reasons for these differences.

AEB June 1995 (AS)

Chapter 10

POLLUTION

Pollution can be defined as 'the introduction by man into the environment of substances or energy liable to cause hazard to human health, harm to living resources and ecological systems, damage to structure or amenity, or interference with legitimate uses of the environment'.

This definition distinguishes manufactured pollution from what might be described as natural processes. Do you agree with the definition? Silage effluents (liquids released from cut, rotting grass) and autumn leaves may have similar effects in a stream, but only the first would be recognised as pollution.

Pollutants may affect wild animals and plants, corrode materials and buildings, and do great damage to human health. In recent years, governments, industry and many other interested groups have spent much time, effort and money to try to identify the sources, effects and means of controlling pollution.

LEARNING OBJECTIVES

After completing the work in this chapter you will be able to:

1. describe the pathways of pollutants
2. explain the biological effects of common pollutants
3. describe the causes and effects of acid rain and eutrophication
4. explain the concept of cost-benefit analysis in terms of pollution control
5. describe the usual approaches to pollution control.

Before we try to define what we mean by the term 'pollution', consider the list below. Which of these do you think counts as 'pollution'?

- A volcanic eruption which releases millions of tonnes of ash into the atmosphere.
- An accidental release of radiation into the environment.
- Autumn leaves falling into a woodland stream.
- Cows releasing methane when they belch.
- A factory releasing a bright orange, odourless, non-toxic chemical into a nearby river.

Can accidental and natural events be counted as pollution? Given the definition of pollution provided in the introduction to this chapter, how many of the above statements do you now think count as pollution?

10.1 THE PATHWAYS OF POLLUTANTS

Identifying the source of pollutants may be extremely difficult. Some pollutants may have a dramatic effect only metres from their source, such as radioactivity and thermal pollution entering fresh water (Chapter 12), while others, such as acid rain, may travel thousands of miles before damaging ecosystems.

Air pollutants may be chemically and/or physically changed. For example, sulphur dioxide is converted into sulphurous acids before being deposited either in dry or wet form. Similarly, pollutants which enter freshwater or marine ecosystems may be biologically or chemically transformed. This can make precise identification of the source very difficult, since what has been released is not what is causing the damage. Furthermore, such changes may increase or decrease the harmful effect of the pollution.

Two or more pollutants may interact with each other, increasing the harmful effect of one of the substances (**synergism**). For example, the harmful effects of a mixture of SO_2 and NO_2 (which are both released from coal-burning power stations) may be much greater than the effects of the individual pollutants added together. Synergism is thought to be involved in the death of large areas of European forests, such as the Black Forest in Germany.

Pollutants may damage buildings and plant surfaces by simply coming into contact with them. Alternatively, they may enter plants in solution, through the leaf epidermis, through the stomata or via the roots. Animals, including humans, can also be affected. This can happen by pollutants entering the body by ingestion (for example pesticides on lettuce), by inhalation (for example benzene at a petrol station), or by absorption through the skin (for example mercury). Some of the possible pathways of pollutants are shown in Figure 10.1.

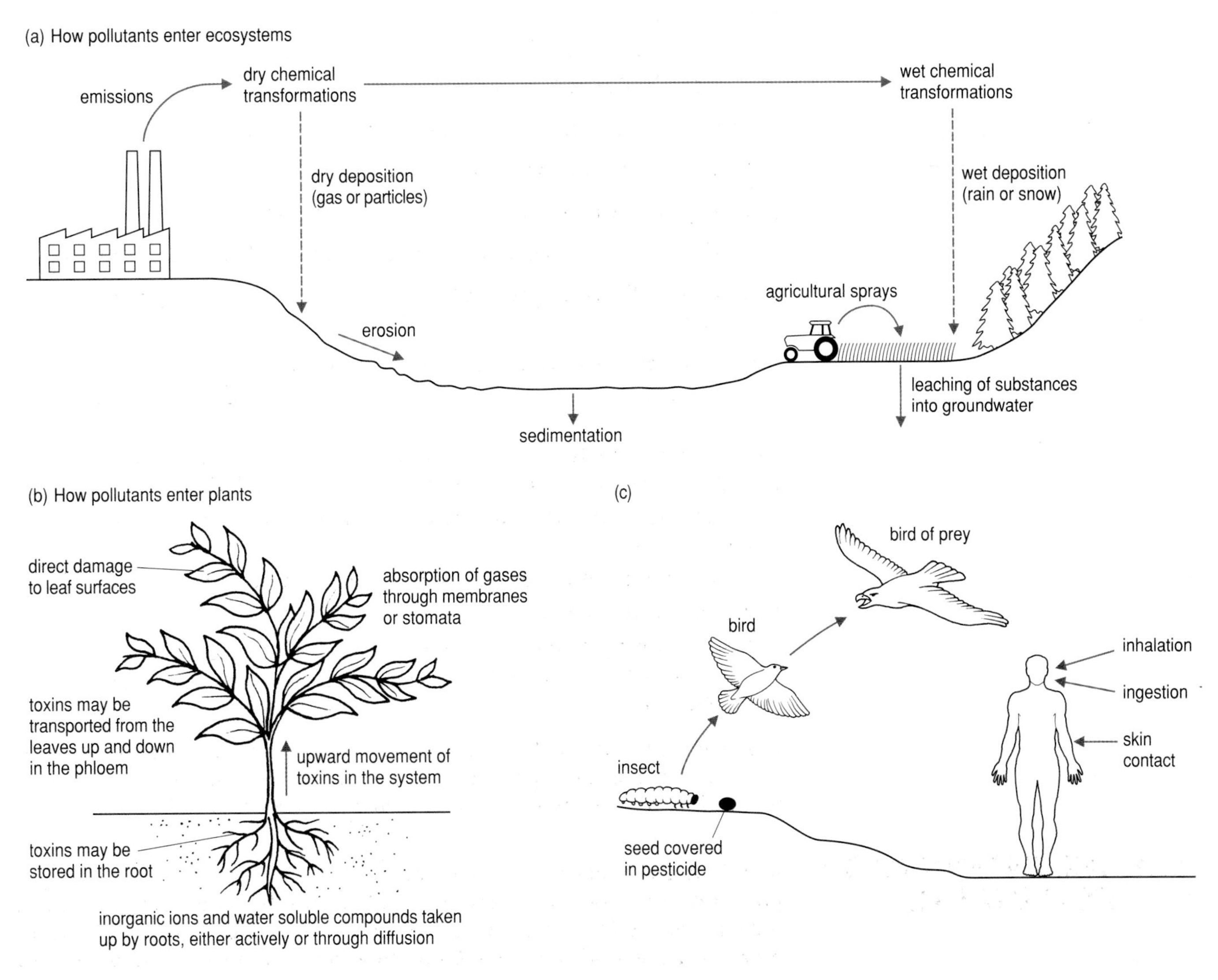

Fig 10.1 Pathways of pollutants.

10.2 GENERAL EFFECTS OF POLLUTANTS

Substances may be acutely toxic, causing rapid death, or they may exert their effects over long periods of time, weakening the individual so that it succumbs to another pollutant or disease. The majority of pollutants have sub-lethal effects, that is, they do not cause death but they may make existing problems worse. The biological effects of pollutants are shown in Table 10.1.

Table 10.1 Biological effects of pollutants

Biological effect	Outcome
toxic	interferes with physiological or neurological processes; may cause death
carcinogenic	causes cancer (uncontrolled cell division)
mutagenic	causes alterations to chromosomes
teratogenic	causes malformation of embryo

Pollutants do not last indefinitely, but some pollutants certainly last longer than others. Pollutants that enter fresh water may be so diluted as to become insignificant. However, freshwater pollution often becomes more of a problem in summer. Temperatures and evaporation rates increase, and this effectively increases the concentration of the chemical in the water. Suspended material will eventually sediment out and settle on the bed of a lake, estuary or a river. Some pollutants, such as heavy metals, may be only temporarily locked up in this way. If chemical conditions change on the bed, for example if oxygen becomes available as a result of mixing or through prolonged exposure to air at low water levels, then toxic metals such as lead or cadmium may be released.

Biodegradable organic pollutants will be broken down by micro-organisms such as bacteria and fungi, and although this will eventually eliminate the pollutant, it may cause serious ecological problems in the process.

Persistent or non-biodegradable substances cannot be broken down by living organisms and may therefore accumulate in organisms. In this way, small, sometimes apparently harmless, amounts of such chemicals may build up over a long period of time inside an organism until toxic concentrations are reached. Heavy metals such as lead, arsenic, mercury and cadmium may be ingested in water. Once inside the body, they may bind to enzymes, producing toxic effects.

Synthetic organic compounds, such as halogenated hydrocarbons, also accumulate in this way. Halogenated hydrocarbons are organic compounds in which a halogen (chlorine, bromine, fluorine or iodine) has replaced hydrogen. The most notorious class of halogenated hydrocarbons are the chlorinated hydrocarbons or organochlorines, which include DDT (now banned in Europe and North America), which are still widely used in plastics, solvents and in electrical insulation products. Examples of toxic synthetic organic compounds and their effects are shown in Table 10.2.

Table 10.2 The effects of synthetic organic compounds

Chemical	Mutagenic	Carcinogenic	Teratogenic
benzene	•	•	•
chloroform	•	•	•
carbon tetrachloride	•		•
polychlorinated biphenols	•	•	•
toluene	•		
xylene			•

Organisms at lower trophic levels of the food chain or web that have accumulated persistent toxins may then be eaten by members of the next trophic level. Because biomass decreases at each trophic level (Chapter 7), any accumulated pollutant is concentrated into the smaller and smaller biomass of organisms at the top of the food chains. Thus top carnivores may receive concentrations of pollutants that are millions of times greater than the concentrations originally released into the environment. This is termed **biomagnification** (Figure 10.2).

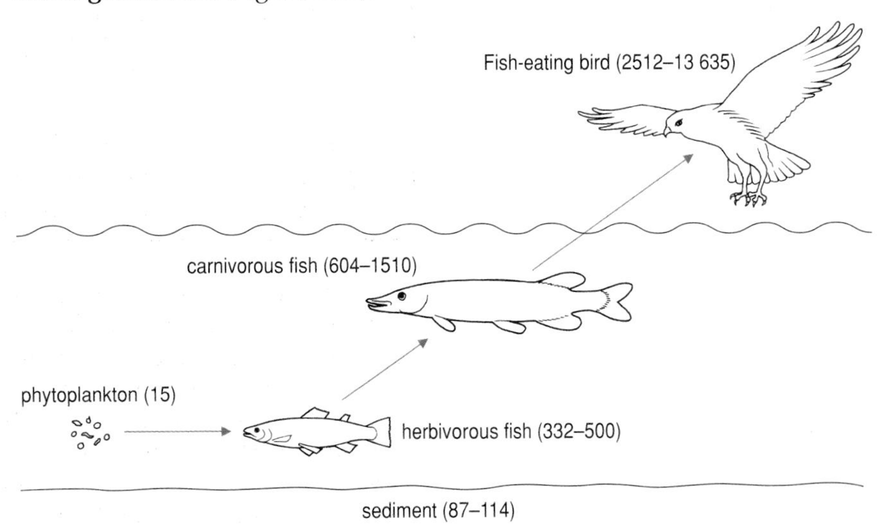

Fig 10.2 Biomagnification of mercury in Lake Paijanne, Finland (figures ppm).

QUESTIONS

10.1 Using Figure 10.2, explain the difference between bioaccumulation and biomagnification.

10.2 Calculate the concentration factor for the mercury pollutant in the food chain shown.

Biological indicators

Biological indicators are groups of species belonging to a particular community or taxonomic group, or sometimes a single species, sensitive to a particular type of pollution, and which disappear from a habitat when pollution levels increase above a certain level. Consequently, water quality can be monitored by measuring variations in communities of bottom dwelling (benthic) invertebrates in a river. The invertebrates studied have different oxygen demands, and the characteristics of communities can be used to compare streams and different sites on one stream for levels of pollution. Watercourse quality can be given scores according to the species present. There are three commonly used indices to asses water quality:

- Trent Biotic Index (TBI): the simplest index, based on presence or absence of invertebrate groups
- Chandler Score: a more accurate index, based on abundance of invertebrates
- Biological Monitoring Working Party (BMWP): a simplified version of the Chandler Score.

The use of benthic invertebrates has several advantages in water quality assessment because they are relatively large and easy to identify and count. In addition, many species have a very narrow pollution-tolerance range and provide a continuous monitor of water as opposed to an intermittent chemical analysis. They will also respond to a variety of, as well as combinations of, different water quality determinants and pollutants (synergistic effects).

The main disadvantage of using biological indicator species, apart from errors in sampling, is that it does not give information on the specific chemical pollutants or bacteria present. It is also important to realise that other factors determine the benthic community, specifically current velocity and nature of substrate, along with the life-cycle of species dependent on the time of year.

10.3 THE MAJOR SOURCES OF POLLUTION

Three major sources of air and water pollution are power stations, transport and modern agriculture.

Power stations

Power stations that burn fossil fuels generate most of the UK's electricity (Chapter 12). In doing so, they also produce significant quantities of air pollutants (Table 10.3). The combustion of coal in power stations is the largest source of carbon dioxide emissions in the UK (Figure 10.3). Although carbon dioxide is the least potent greenhouse gas (Chapter 1), it is the most important contributor to global warming simply because the volume of emissions is so great.

Table 10.3 Pollutants from power stations/thousand tonnes

Pollutant	Fossil fuel			
	coal	oil	gas	peat
SO_2	400–3000	500–3000	–	100–1000
NO	300–1300	250–1300	200–500	100–1000
NO_2	10–40	10–40	10–30	3–30
CO	–	–	100	2000–5000
UVHCs	10–60	100–2000	100–2000	30
HCl	80	–	–	–

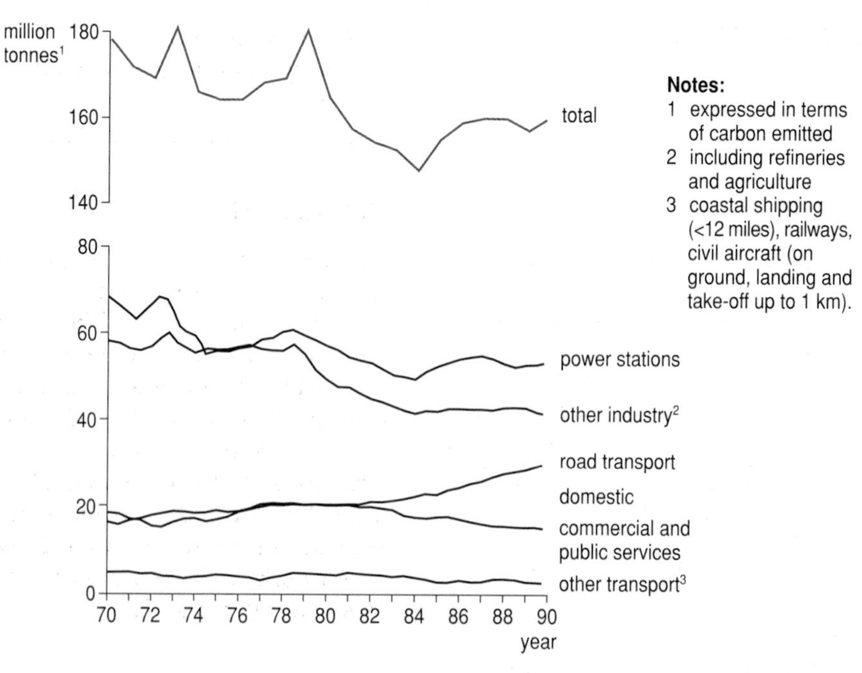

Fig 10.3 Carbon dioxide emissions by source, 1970 to 1990, UK.

Power stations are the most important source of sulphur dioxide (SO_2) which, along with nitrogen dioxide (NO_2), dissolves in atmospheric moisture, eventually forming sulphuric and nitric acids respectively. These are the major components of acid rain. Coal may also contain a small percentage of chlorides which, on combustion, produce hydrogen chlorides, thereby further increasing the acidity of precipitation.

Total sulphur dioxide emissions have been declining in the UK for several decades. This reflects the increasing use of sulphur-free fuels, such as natural gas, reduced industrial use of energy, and improved energy conservation measures. The European Union sets air quality standards for SO_2 and smoke, based on World Health Organisation (WHO) criteria. Member states must ensure that average daily concentrations do not exceed these limits (Table 10.4).

Table 10.4 EU limits for smoke and sulphur dioxide
(micrograms per cubic metre) (OECD units)

	Smoke	Sulphur dioxide	
Year (median of daily mean values)	80	If smoke less than or equal to If smoke more than	40: 120 40: 80
Winter (median of daily mean values October to March)	130	If smoke less than or equal to If smoke more than	60: 180 60: 130
Year (peak – 98 percentile daily mean concentration, i.e. the level exceeded by the highest 2% of daily mean values during the year)	250	If smoke less than or equal to If smoke more than	150: 350 150: 250

Note: The units are equivalent to British Standard units.

Member states must ensure that the annual peak values are not exceeded for more than three consecutive days and must try to prevent and reduce such instances where the values have been exceeded.

QUESTION

10.3 Study Table 10.4. Why do you think the maximum allowable daily SO_2 concentration depends upon how much smoke there is at the time?

Transport

Emissions from petrol engines include carbon monoxide (CO), carbon dioxide, volatile organic compounds, lead, nitrogen oxides and a wide variety of particulates. The contribution which road transport makes to emissions of major pollutants is shown in Table 10.5.

The number of vehicles on the road has increased by 500% since the 1950s (Figure 10.4), and in the 1980s road transport superceded power stations as the major source of nitrogen oxides (Figure 10.5). In contrast to declining emissions of SO_2, emissions of nitrogen dioxide (NO_2) have risen sharply, almost entirely due to increasing emissions from road transport. High concentrations are common in winter, particularly in calm, cold weather when pollutants are trapped close to the ground because of temperature inversions (Chapter 2).

Table 10.5 Emissions caused by road transport

Name of pollutant	% of total emissions caused by road transport
hydrocarbons	36
nitrogen oxides	51
carbon monoxide	89
lead	70
black smoke	42
carbon dioxide	19

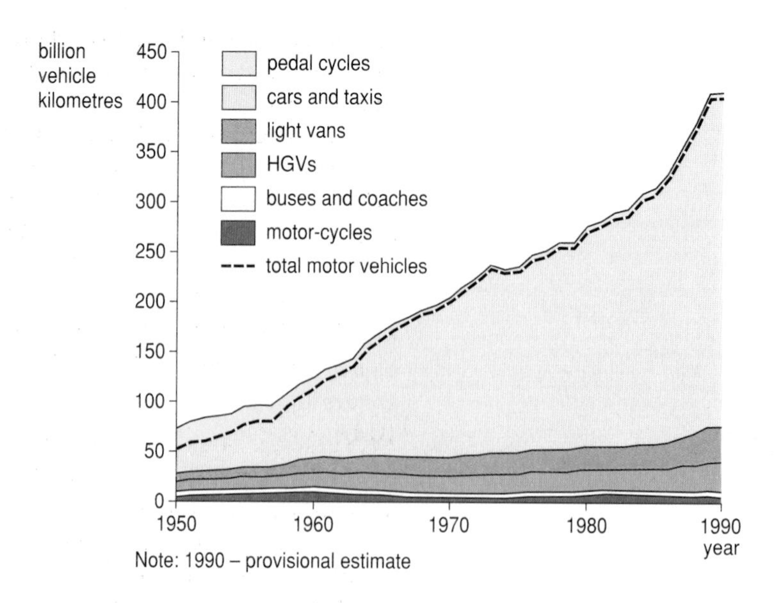

Fig 10.4 Growth of road transport.

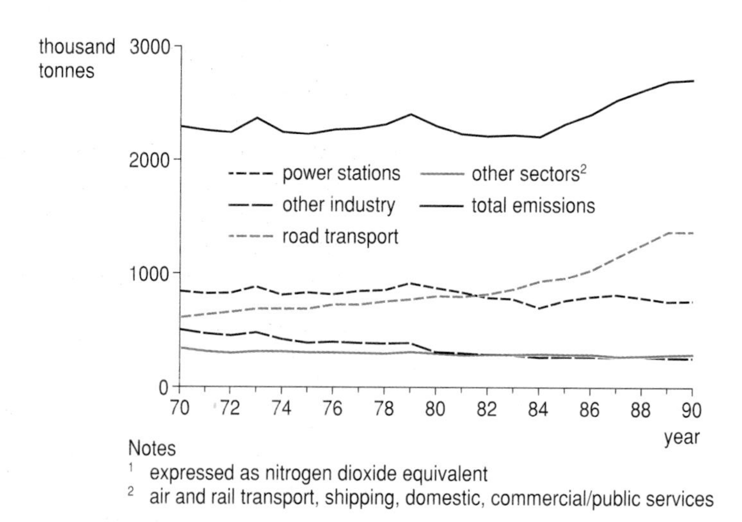

Fig 10.5 Sources of nitrogen oxide emissions[1], by sector.

In the presence of sunlight, nitrogen dioxide may react with hydrocarbons to form tropospheric ozone. This is an example of a **secondary pollutant**, that is, one that is not released directly but which forms through the chemical reactions of other pollutants. Tropospheric ozone is an eye irritant, and may have similar effects on people's lungs, nose and throat, but it has recently been clearly implicated as one of the causes of forest damage throughout Europe. Nitrogen dioxide is also involved in the atmospheric synthesis of peroxyacylnitrates (PANs), which cause eye irritation and damage plant tissues.

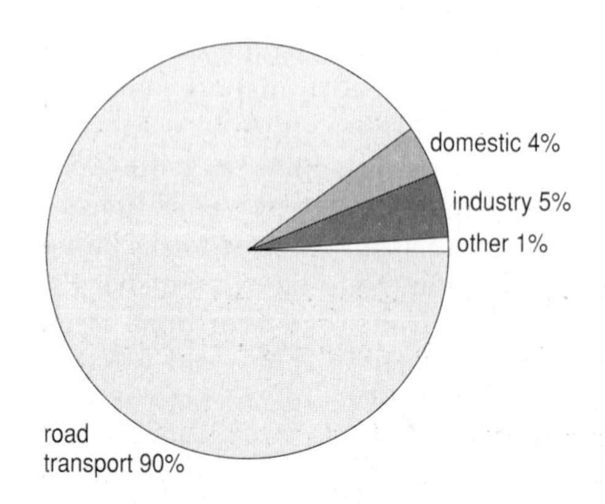

Fig 10.6 Sources of carbon monoxide.

The incomplete combustion of petrol accounts for 90% of carbon monoxide emissions (Figure 10.6). Carbon monoxide accelerates the breakdown of nitrogen dioxide in light (hence it increases ozone production) and is rapidly oxidised to carbon dioxide. Carbon monoxide is another important greenhouse gas and a serious health hazard since it combines strongly with haemoglobin to produce carboxyhaemoglobin, decreasing the amount of oxygen which the blood can carry.

Volatile hydrocarbons are released when fossil fuels are incompletely burnt. Several hundred types of hydrocarbon, including benzene, ethene and butene, are released. Hydrocarbons accelerate the production of tropospheric ozone while some, such as benzene, are known **carcinogens** (cancer-inducing substances).

The concentration of lead in the air seems to have been on the decline since the 1970s, as successive legislation reduced the permitted concentration in petrol. This eventually led to the introduction of unleaded petrol. Diesel engines emit less carbon monoxide and hydrocarbons but similar amounts of nitrogen oxides and many more particulates. Sulphate-containing particulates cause severe irritation of the eye and nasal passages and induce uncontrollable coughing. They may also be involved in the development of respiratory diseases and cancer.

ANALYSIS

Vehicle exhaust emissions

Study the statements below and Figure 10.7 and then answer the questions which follow.

- Vehicle exhaust emissions include NO, NO_2, CO, hydrocarbons, aldehydes and ketones.
- Within the troposphere the main source of ozone is the photolysis (breakdown caused by light) of NO_2 to form NO and O. An oxygen atom (O) then combines with an oxygen molecule (O_2) to form ozone (O_3).
- NO is rapidly oxidised to NO_2. This oxidation is accelerated by hydrocarbons and CO.
- In sunlight, aldehydes and ketones may react to form radicals.
- Radicals react with NO and NO_2 to form PANs.
- PANs and radicals cause severe eye irritation. High concentrations of NO_2 may cause breathing difficulties in people who suffer from asthma. High ozone concentrations may damage lung tissue and the effects of ozone are made worse by exercise.

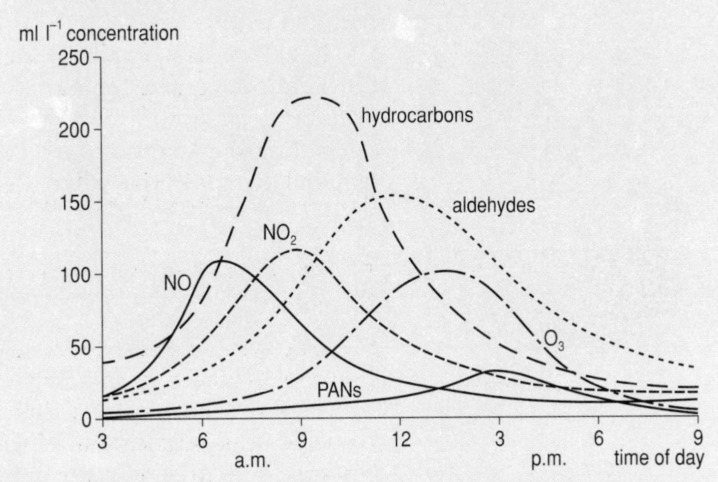

Fig 10.7 Components of photochemical smog on a sunny day in an urban area.

1. Why do the concentrations of NO, NO_2 and hydrocarbons increase in the morning?
2. Suggest an explanation for the timing of the peak concentrations of NO and NO_2.
3. Explain how hydrocarbons accelerate the formation of ozone.
4. Suggest an explanation for the increase in concentration of PANs in mid-afternoon.
5. Why might cyclists be advised not to leave work early in order to avoid rush-hour traffic?

Modern agriculture

With its reliance on high inputs of artificial chemicals such as pesticides and fertilisers, agriculture is the source of several kinds of environmental pollution. Leaching of nitrates and erosion of phosphates into streams, lakes and coastal waters has accelerated the natural process of eutrophication. Nitrate levels in drinking water supplies in several parts of England now exceed the EU limit of 50 mg l^{-1}. High levels of nitrates in drinking water may pose two health problems. Firstly, very young children may be at risk of developing methaemoglobinaemia, a condition in which the oxygen-carrying capacity of the blood is reduced. In severe cases this would prove fatal. Secondly, concentrations greater than 100 mg l^{-1} of nitrate could theoretically result in the formation of nitrosamines – known carcinogens – in the stomach. Groundwater supplies which contain more than the EU limit have to be blended with cleaner supplies before being allowed to enter the domestic supply.

The two major types of agricultural organic waste are animal slurry (urine and faeces) and silage effluent. **Silage** is grass which has been anaerobically digested by bacteria and which can then be used to feed livestock. The volume of both of these wastes has increased as the practice of intensive, indoor livestock rearing has become more common.

Slurry and silage are rich in organic compounds, such as proteins, lipids and carbohydrates. If such wastes enter streams, rivers or lakes, aerobic decomposers, such as bacteria and fungi, will begin to break down the wastes, rapidly consuming the oxygen in the water to do so and producing ammonia in the process. The resulting deoxygenation of the water kills most or all of the aerobic fauna. As turbidity (cloudiness) increases, light penetration to aquatic plants will decrease and these may also die. Anaerobic decomposition will result in the release of hydrogen sulphide, which is responsible for the characteristic foul odour of organically polluted water.

The ability of an organic effluent to deoxygenate the water is known as the **biochemical oxygen demand** (BOD). This is measured in the laboratory by examining a sample of the polluted water for five days at 20°C and calculating the amount of oxygen which has been used. The more organic material there is in the water, the higher the BOD.

QUESTION

10.4 Why is temperature kept constant when measuring BOD?

Pesticide usage continues to increase and there are now over 450 different pesticides authorised for use in the UK. For humans, most pesticide intake is through food consumption. The Ministry of Agriculture, Fisheries and Food (MAFF) undertakes routine tests to determine whether pesticide residues exceed maximum recommended levels. Although most residues which are detected are at a low level, the maximum residue levels are sometimes exceeded, particularly as a consequence of post-harvest treatment, especially fumigation to extend storage life (Chapter 14).

It is likely that the majority of people in the UK already have residues of pesticides, such as organochlorides, stored in their fatty tissue. We do not know what effect, if any, they have on our health, but it is perhaps reassuring that since the use of such pesticides has been prohibited or slowed, residues are falling.

ANALYSIS

The effect of organic pollution

The general effects of organic pollutants on the oxygen content of a river are shown in Figure 10.8.

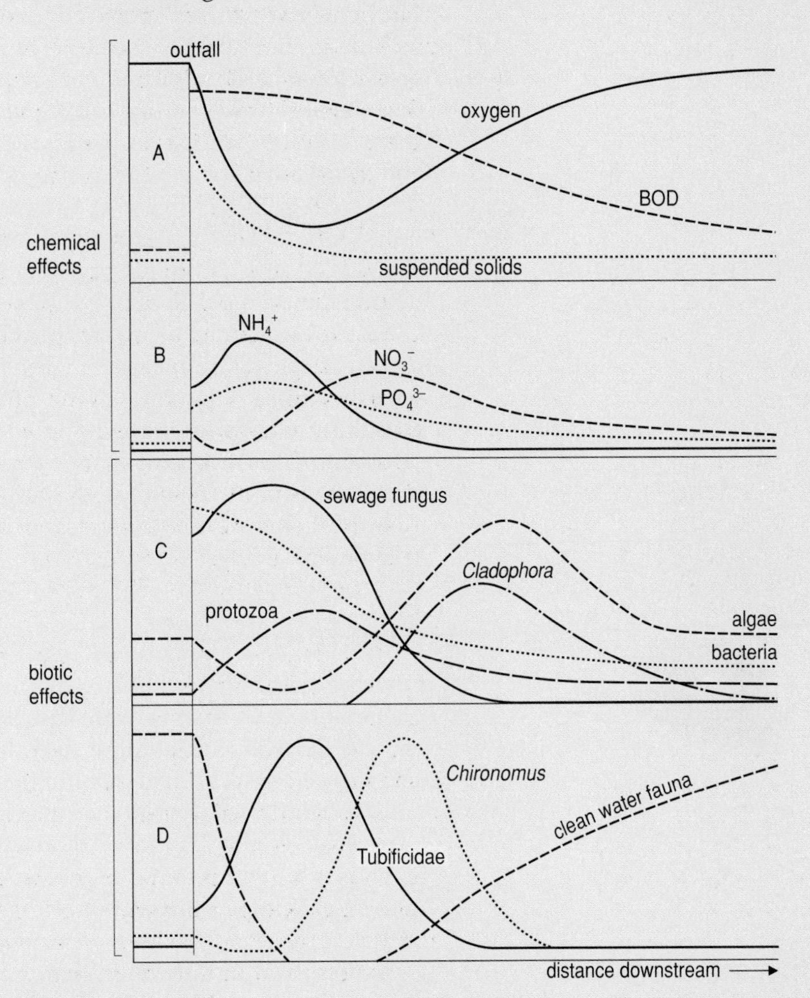

Fig 10.8 Effect of organic pollution

1. Suggest explanations for the change in the numbers of:
   (a) bacteria
   (b) algae
   (c) protozoa.
2. Suggest an explanation for the changes in the concentration of:
   (a) ammonium (NH_4^+)
   (b) nitrate (NO_3^-).
3. Explain why Tubifidicae may be used as a warning sign of organic pollution.
4. Suggest what effect, if any, the following would have on the recovery time of organically polluted water:
   (a) water temperature
   (b) water turbulence.

Pesticides and wildlife

Pesticides have been implicated in the decline of many species of insects, birds and mammals. Most environmental problems have been caused by the effects of insecticides, in particular by a class of insecticide known as **chlorinated hydrocarbons** or **organochlorines**. Organochlorines include DDT, aldrin, dieldrin, lindane, chlordane and methoxychlor.

Organochlorines posed such a great problem for a number of reasons. They covered a broad spectrum, that is, they killed a wide range of insects, including some useful species. They were persistent, in that they did not break down in the environment very quickly, thereby increasing the chances that they would harm useful populations. Because they were insoluble in water but very soluble in lipids, they tended to accumulate in fatty tissue, and bioaccumulation became a major problem. Scientists were first alerted to this problem in the 1950s by the rapidly declining populations of birds of prey. Once inside the tissues of the bird, organochlorines were very slowly chemically changed, releasing new substances which seemed to cause the thinning and breakage of egg shells.

At the same time, using aldrin and dieldrin as sheep dip and as dressing for cereal seeds poisoned several top predators, such as peregrine falcons and sparrowhawks (Chapter 7). The phasing out of organochlorines has seen significant recoveries in the populations of these birds.

Organochlorine pesticides may also be implicated in declining populations of otters, although it seems likely that habitat loss has also had a harmful effect. Useful insects, such as butterflies and bees, are also often killed by pesticides.

Acid rain

Like the greenhouse effect (Chapter 1), acid rain is an entirely natural phenomenon. Carbon dioxide dissolves in rainwater to form carbonic acid, and nitrous oxide and sulphur dioxide dissolve to form nitrous and sulphuric acid respectively. Oxidising pollutants such as ozone, hydrogen peroxide and peroxyacylnitrates (PANs) may accelerate the formation of sulphuric acid.

Such acids may be blown thousands of miles before falling as acid rain, and this is why acid rain is known as a **transboundary pollutant**. In addition to this long-range transport of acidity, dry deposition of suspended sulphur particles, for example, may occur near to the site of emission (Chapter 12). Once dissolved in rainwater, or in water on leaf surfaces, such particles will once again form a harmful acidic solution.

By releasing millions of tonnes of NO_2 and SO_2, humans have made acid rain an international problem which has damaged aquatic ecosystems, forests, crops and buildings.

Effect on aquatic ecosystems

In total, acid rain has destroyed or severely damaged aquatic invertebrates, fish and plant life in thousands of lakes and rivers. In Scandinavia alone, 9000 lakes now have severely depleted fish stocks. The major biological effects of acidification on aquatic life can be summarised as follows:

- Fish have difficulty in regulating their internal salt concentration. Sodium ions in particular may be lost through the gills. Acidicification causes a build-up of mucus in the gills which may lead to suffocation. The haemoglobin in the blood of fish becomes less efficient at picking up oxygen, again leading to suffocation.
- Soil acidification causes aluminium ions to become soluble in the soil. These may leach into lakes, where the ions may kill fish directly or they may increase the harmful effects of other pollutants (synergism). The toxic effect of aluminium is greatly increased if calcium levels in the lake are also low.
- In very acidic waters, crustaceans are unable to absorb the calcium which is essential for the formation of their exoskeletons.

- Populations of aquatic invertebrates, such as mayflies, decrease, as do the populations of birds, such as the dipper, which feed on them.
- The reproductive success of amphibians and fish decreases. The eggs of salmon and trout, for example, may fail to hatch.

In short, acidification may lead to a drastic reduction in both species number and diversity, thus causing the ecosystem to become unstable. Clear but almost lifeless lakes are the result.

Damage to forests

The precise contribution of acid rain to forest damage remains controversial. Some scientists believe that the forest dieback, which has occurred over huge areas of countries such as Germany, is due mainly or entirely to the effects of acid rain. Others believe that forests have been weakened or predisposed by pollutants such as ozone as well as by factors such as drought. There is evidence to suggest that pollutants, such as sulphur dioxide and ozone, may act synergistically.

Fig 10.9 Acid-damaged forests.

Forest canopies may act as efficient filters of particulate pollution. Because most coniferous species are evergreen, retaining their needles all year round, they are more effective at filtering pollution than deciduous species, and it is in coniferous forests that most damage has occurred. Such acid particulates may accumulate along leaf margins and veins, and may damage the waterproofing function of the cuticle or the waxy plugs which protect the stomata of conifers, thus allowing pollutants and pathogens entry to the leaf.

Sulphur dioxide may reduce tree growth by slowing down the movement of sugars from the leaves to areas of the plant which require them as an energy source. However, the concentration of SO_2 needed to cause this effect seems to vary with environmental factors such as temperature and light intensity, and there is evidence to show that prolonged exposure to SO_2 makes deciduous trees more prone to frost damage. Rainfall which trickles down leaves, petioles and stems will dissolve the acidic particles, and the final solution which reaches the ground may be many times more acidic than the original rainfall.

Such inputs have accelerated the natural process of soil acidification. As soils become more acidic, metals such as aluminium, manganese and copper become more readily available, and may reach levels which are toxic to plant roots. Soils which have low organic matter or low clay contents are most sensitive to acidification (Chapter 4).

The sources and effects of the most important air and water pollutants are summarised in Table 10.6.

Table 10.6 Air and water pollutants

Name/description	Source	Pathway	Biological effect
SPM or aerosols Solids particles or liquid droplets 0.1–25nm, e.g. dissolved sulphates, chlorides, smoke and soot, hydrocarbons	heating systems, power stations, diesel exhaust emissions, sea spray	suspended particles act as condensation nuclei during conditions of low temperature and high humidity, leading to fog formation; smoke and fog combine to form smog	loss of sunlight and increased cloud cover; reduced visibility; particles less than 500 nm in diameter may reach and stay in the alveoli for years, leading to lung diseases
sulphur dioxide – SO_2 acidic, colourless gas, pungent smell	combustion of sulphur-containing fuels; over 70% comes from power stations	it may remain in gaseous form or dissolve in rainwater or fog to form sulphurous and sulphuric acids	elderly and young and anyone with damaged respiratory pathways are at risk from bronchitis; the frequency and severity of asthmatic attacks increases; blocks stomata reducing CO_2 absorption, and can also cause chlorosis of the leaf (yellowing); erosion of limestone and sandstone buildings; synergistic effects with nitrogen oxides
nitrogen oxides – NO_x NO_2, N_2O and NO are the major pollutants	lightning, volcanic eruptions, bacteria in the soil breaking down organic matter, combustion of fossil fuels in heating systems, power stations and vehicles	components of photochemical smog; NO_2 is a component of acid rain; regional hazes; in the presence of sunlight, NO_2 and hydrocarbons produce ozone; N_2O is a strong greenhouse gas; it may also decrease stratospheric ozone levels	increases susceptibility to respiratory infections and asthma; NO_2 is a throat and eye irritant and may cause skin problems; decreases visibility; causes necrosis of leaves
photochemical oxidants (secondary pollutants)	action of sun on hydrocarbons and nitrogen oxides	ozone (O_3) and PAN (peroxyacetyl nitrate) are produced and contribute photochemical smog; the problem is greater in cities because of the large amount of traffic which emits both hydrocarbons and nitrogen oxides; may be carried very long distances by wind	both ozone and PAN damage plants; ozone causes leaf necrosis, reduced growth and growth abnormalities; PAN causes eye irritation and may cause respiratory problems; city-centre joggers may experience shortness of breath and throat irritation
ground-level ozone	found as a secondary pollutant when nitrogen oxides react with volatile hydrocarbons in the presence of sunlight	may be blown long distances – high concentrations are therefore often found in rural, traffic-free areas	may increase respiratory problems or cause eye irritation; if the concentration exceeds 200 microgrammes per m^3, many suffer eye, nose and throat irritation
dioxins	incinerators, vehicle exhausts, chlorine-bleached paper products	may travel through air or water	strongly absorbed via inhalation and skin contact; may cause skin problems
carbon monoxide – CO	incomplete combustion of fuels	released into atmosphere and rapidly oxidised into CO_2	forms carboxyhaemoglobin, thereby decreasing the oxygen content of blood; small amounts impair visual acuity and concentration

lead	tetraethyl lead is used as an anti-knock agent in petrol; many water pipes are still made of lead	released as very fine mist of inorganic lead	food may be contaminated from air, soil or water; causes disabilities and emotional disturbance in children
PCBs (polychlorinated biphenols)	found in protective sealants for wood and metal; also used as a coolant in transformers	PCBs are given off as a gas when matter is incinerated	may become concentrated in oily fish; large concentrations seem to be connected to bird deaths
thermal pollution	waste heat emitted from power stations, industrial plants and urban areas	may lead to increased precipitation and thunderstorms if released into air	direct toxic effects can result from heating of water – the solubility of gases decreases as the water temperature increases, so O_2 levels fall as the temperature rises; at the same time, the respiration rate of aerobic organisms increases, more organisms may therefore suffocate
VOCs (volatile organic compounds)	industrial solvents, cleaning materials, adhesives, airborne evaporation of petrol	may have local effect or travel long distances	some may cause drowsiness and eye irritation; some are carcinogenic, e.g. benzene; reactions with nitrogen oxides produce ground-level O_3
fluorides	brick factories, aluminium smelters	may enter atmosphere as hydrofluoric acid and cause effects similar to acid rain	plants in vicinity of source may accumulate fluoride, and if ingested by cattle, causes discolouring of their teeth – fluorosis
nitrates	nitrate fertilisers	highly soluble and therefore easily leached into groundwater	high nitrate levels lead to algal 'bloom' in contaminated waters; the bacteria which consume dead algae multiply, drawing greatly increased amounts of oxygen from the water, leaving the water anoxic; this process is eutrophication; this causes the death of bacteria, fish, etc.
organic waste	sewage, biodegradable industrial waste	often released from point sources such as effluent pipes	eutrophication, owing to high nitrate levels; light penetration reduced, affecting plant growth; waterborne diseases – dysentery, cholera, typhoid – in countries where drinking water is inadequately treated
pesticides e.g. chlorinated hydrocarbons such as aldrin	runoff from agricultural land, spraying	sprays may drift in the wind and be inhaled or ingested if they remain on food; pesticides may enter streams, rivers or groundwater supplies	even if present in very small concentrations in water, pesticides accumulate in body fat deposits, e.g. in fish; this results in higher concentrations of the pesticide being consumed by organisms higher in the food chain, e.g. birds; the pesticides may also become harmful to the organism concerned in times of stress, when the fat deposits are partially consumed
industrial waste e.g. iron salts, heavy metals, acids	direct discharge or leakage	may enter streams, rivers or groundwater supplies; may cause long-term soil contamination	depends on pollutant concerned, but may include deoxygenation of water, food contamination with heavy metals, and damage to fish

ANALYSIS

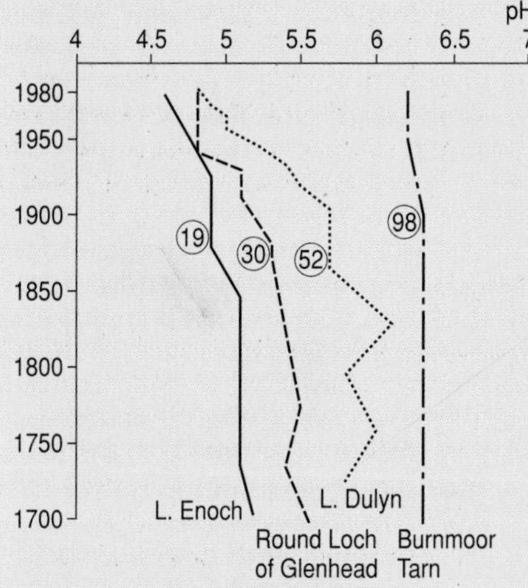

Fig 10.10 pH changes in Scottish lakes, 1700–1980.

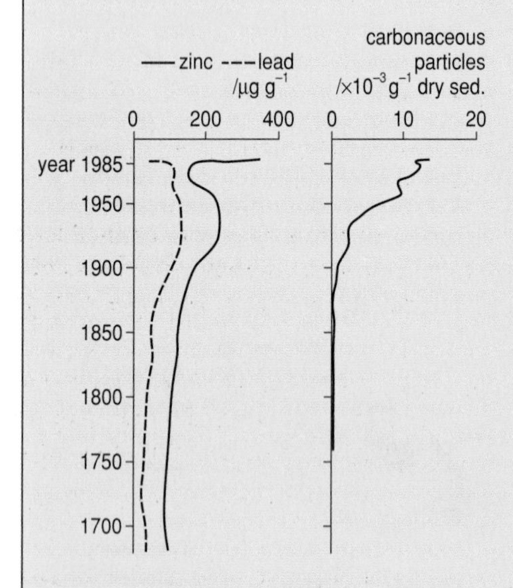

Fig 10.11 Lake bottom sediments of Lake Dulyn since 1700.

Ecological effects on lakes

1. Figure 10.10 shows pH trends in four upland lakes. The numbers in circles refer to water calcium concentrations.
   (a) Describe the trends shown.
   (b) Suggest an explanation for the trends shown.

2. Figure 10.11 shows trends in the concentration of lead in lake bottom sediments of Lake Dulyn since 1700.
   (a) Suggest an explanation for the change in concentration of lead in the sediments.
   (b) What factors might affect the rate of sediment accumulation?

3. Application of a herbicide to a young forest plantation resulted in contamination of an adjacent lake. A student intending to study the short-term effects of this pollution noticed that after only two days some of the large plants had begun to die. The student then drew up the list of ten criteria shown in Table 10.7.
   (a) For each criterion, indicate with a tick the expected effect of the herbicide pollution on the lake.
   (b) Use the completed list to draw a flowchart to show the expected effect of this pollution.

Table 10.7 Student's list

	Increase	Decrease
oxygen release to water		
food supply for herbivores		
population size of herbivores		
population size of secondary consumers		
overall population of aquatic organisms		
detritus levels in water		
population of bacteria		
light penetration into lake		
number of microhabitats		
turbidity of water		

Eutrophication

Eutrophication is the enrichment of fresh water by excess nutrients, usually in the form of nitrate and phosphate ions. It is a natural process which humans have greatly accelerated. The nutrient status of lakes increases naturally as sediment constantly reaches it in streams or through direct soil erosion. Thus an oligotrophic lake (low nutrient, low productivity) will inevitably change into a eutrophic one. Accelerated eutrophication has occurred as a result of the following:

- increased use of phosphate-containing detergents
- increased use of nitrate fertilisers which has resulted in increased leaching from soil into fresh water
- drainage or washings from intensive animal units
- bank erosion caused by the wash of boats
- increased soil erosion as a result of, for example, hedgerow removal.

Whereas nitrates are very soluble, phosphates are not, and so they usually enter the water as a result of erosion from land. Phosphorus is, however, a common limiting factor in fresh water and it is usually the extra phosphorus which results in excess growth of plants so characteristic of the early stages of eutrophication.

Eutrophication in lakes: the process

- Initially, extra nutrients may lead to increased abundance and diversity of plants and the animals that feed on them.
- Microscopic plants (algae) proliferate, causing algal blooms. Algae photosynthesise, releasing oxygen into the water, but they also block light to the lower depths, and this reduces the number of large plants (macrophytes).
- Zooplankton (microscopic animals) feed on algae but use macrophytes to escape predation by fish, so as macrophyte numbers decrease, more zooplankton are eaten by fish. The zooplankton population therefore falls.
- As zooplankton numbers decrease, less algae are eaten, and so their numbers increase further.
- Algae have a high turnover rate, that is, both the birth rate and the death rate are high. Dead algae are broken down by aerobic bacteria, which therefore begins to use up much of the oxygen in the water.
- Falling oxygen levels lead to the death of many aerobes – both plants and animals. Many food chains collapse.
- Oxygen levels continue to fall as the amount of dead material increases and is broken down by bacteria.
- The turbidity of the water increases, and detritus then sediments out on the bed of the lake.

QUESTION

10.5 The Norfolk Broads are formed from about forty shallow lakes in the low-lying marshes of eastern Norfolk. The Broads are under intense pressure from tourists on boating holidays and from farmers reclaiming the marshland to expand their intensive arable production. The data below describe some changes recorded in the Broads.

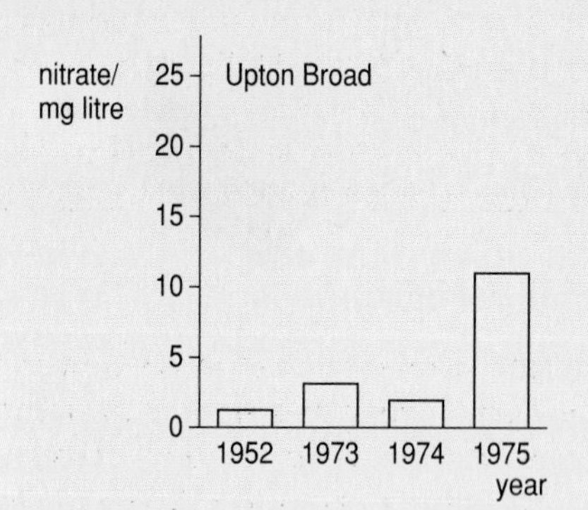

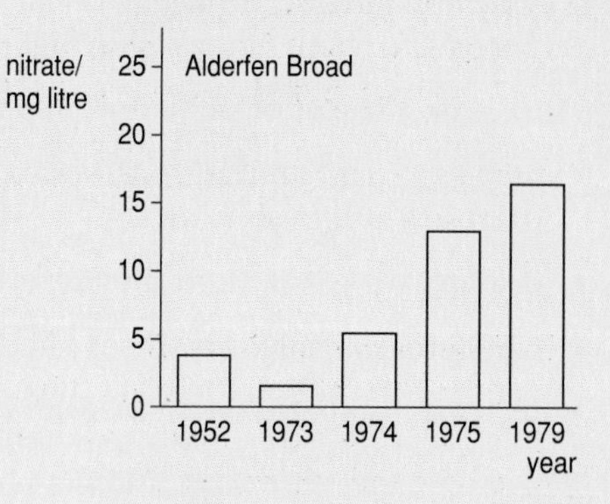

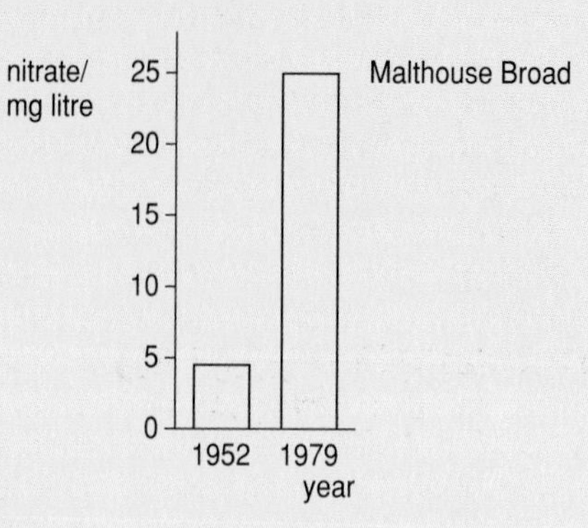

(a) Explain the changes in the maximum winter nitrate concentrations since 1952 as shown by the data.

(b) Suggest **two** reasons why changes in nitrate concentrations were measured in winter rather than at any other time of the year.

(c) The diagrams show phases of change in the Broads related to their phosphate concentration and rate of sedimentation.

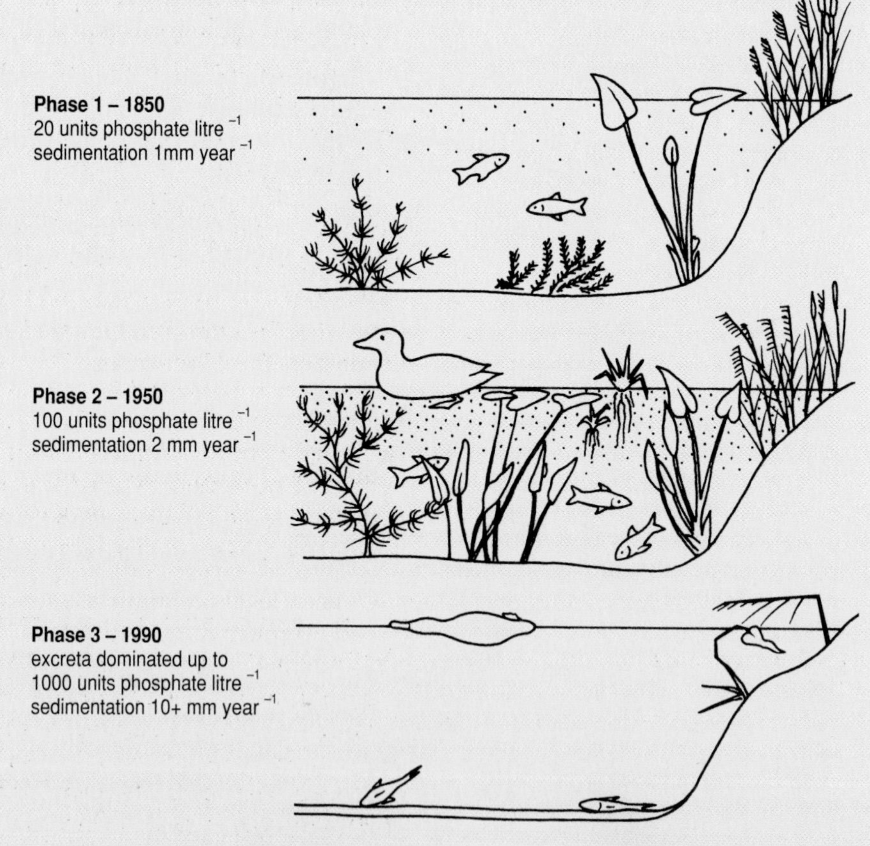

(i) What is the principal source of phosphate pollution and why has it increased?

(ii) Describe how changes in the phosphate concentration of the Broads have affected the ecology of their waters.

(d) Given the option of controlling further increases in the concentration of either phosphate or nitrate, state with a reason which would be the more realistic proposition.

(e) Describe **three** processes which could have accelerated the rate of sedimentation in the Broads.

(f) From the data, what evidence is there of oxygen depletion in the Broads' water?

(g) List **three** factors which are likely to determine the level of oxygen in the water of the Broads.

AEB June 1993

10.4 CONTROLLING POLLUTION

The Environmental Protection Act (EPA) 1990 firmly established the principle of Integrated Pollution Control (IPC). This principle regulates all of the major solid, gaseous and liquid pollutants which are released by industrial processes, and recognises that pollutants which are released into the air, for example, may well have an effect on aquatic ecosystems, hence the need for an integrated approach to their control.

It is important to note that total pollution control, that is, the complete removal of a pollutant, is rarely going to be possible or even desirable. Pollution control is an expensive business and becomes more expensive the greater the degree of control required. For example, it might be economically possible for a factory to reduce its gaseous emissions by 80%, but to reduce them by 90% might mean that the business would not be economic, causing it to become bankrupt. In any case, it is possible that the extra 10% reduction in emissions would have had little extra benefit to the environment. In deciding on the most appropriate level of pollution control, it is therefore necessary to compare the costs and benefits of the control measure. This is known as **cost-benefit analysis**.

Practically speaking, this means that industry is required to use the Best Available Techniques Not Entailing Excessive Cost (BATNEEC) to prevent or minimise the release of pollutants into the environment. Every factory, refinery and chemical works in the UK must obtain authorisation in order to operate. Public registers are kept which detail every application, along with details of whether or not authorisation was granted or refused. The registers, which can be inspected by any member of the public free of charge, also record details of any pollution incidents which have occurred and of any convictions which have been brought against the operator.

Since 1 April 1996, all waste regulatory functions of local authorities have been the responsibility of the Environment Agency. This has replaced separate organisations such as HMIP and the NRA.

CASE STUDY

A history of London air pollution

Air pollution has been a problem in London for a very long time. During medieval times, coal burning was considered such a problem that in 1285 a commission was set up to investigate what could be done. Despite this, the first moves for modern abatement of air pollution were not until 1820, although these early ideas were rejected by parliament.

In the past 'London smogs' have been a great cause of concern. The term 'smog' was coined in 1905 to describe how smoke combines with fog to form a thick dense smog. Smogs were such a problem that specific inquiries were set up.

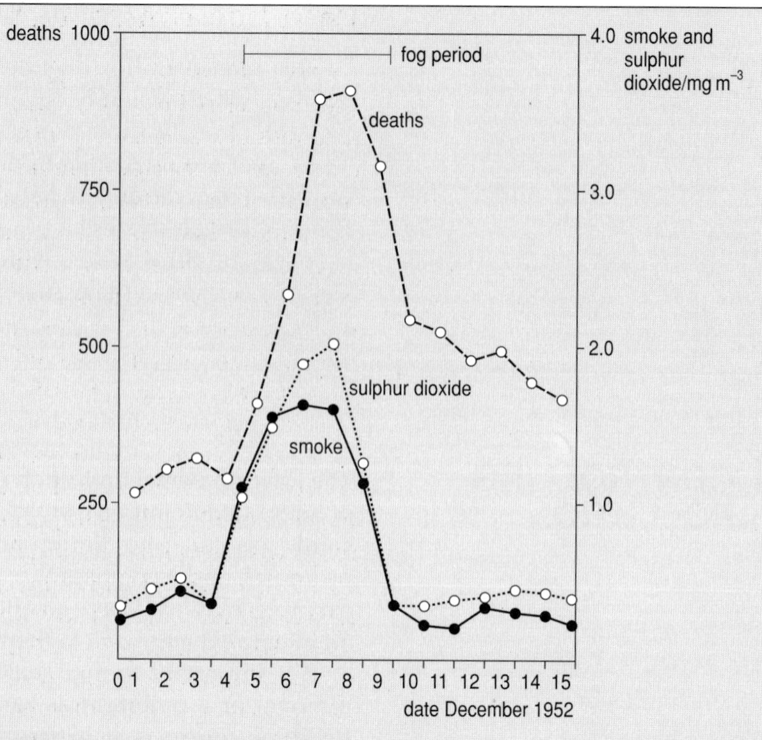

Fig 10.12 Deaths and pollution during 'The Great Smog'.

There was a decline in fog frequency in the early twentieth century, partly due to the work of the Coal Smoke Abatement Society (formed in 1899), which attempted to increase public education on the matter. There was also a series of Alkali Acts (for example 1862), which required companies to use the 'best practicable means' (BPM) to lower pollutant emissions. This was designed so that the companies would always have to improve their control technologies.

These early attempts did reduce air pollution, but not by as much as was hoped. Air pollution was still a major problem throughout much of the twentieth century, as highlighted by 'The Great Smog' which encased London from Thursday 4 to Tuesday 9 December 1952 (Figure 10.12). The smog was caused by a stable high-pressure air mass (Chapter 2), which was so still that it did not remove the pollution pouring from every chimney in the city. The result was a thick, pollutant-loaded air in which visibility was almost zero. The death rate increased from almost 250 to nearly 900 per day. The public outcry as a consequence of this disaster led to the 1956 Clean Air Act, which has been improved upon by subsequent acts.

Since 1952 there has been an 80% drop in the volume of smoke in the air. Thus, over the last 40 years, legislation to control smoke emissions can be seen to have been very successful. However, much remains to be done. In recent years, incidence of respiratory diseases has risen again in London.

The worst air pollution in London in 20 years took place during a week during the summer of 1991. A spell of unusually still weather trapped a range of pollutants from cars, lorries and industry to form a smog. Hourly average levels of NO_2 reached twice the World Health Organisation's recommended safe levels and, combined with black smoke and low-level ozone, caused severe difficulties for those with respiratory problems, even inducing asthma-type symptoms in those normally unaffected.

A study commissioned by the Department of Health reported that around 150 people died in this episode and accused government departments of failing to co-ordinate policies to monitor the health effects of air pollution or act to reduce emissions.

Pollution control technology

Power stations

The most effective way of cutting emissions from all kinds of power stations is to reduce domestic and industrial energy consumption. Domestic energy conservation measures are discussed in Chapter 12. Measures suggested to reduce business and industrial use include compulsory energy audits and stricter standards of building insulation.

Proposals have been recommended for a **carbon tax**, which would be based upon the energy content of a fuel and the amount of carbon contained in the fuel – the higher the carbon content, the higher the tax. This should reduce consumption of fossil fuels, such as coal, and stimulate increased use of lower carbon content fossil fuels, such as oil and gas, as well as nuclear and renewable forms of energy.

The UK has agreed to cut emissions of NO_2 by 30% by 1998 (taking 1980 as the baseline). It is hoped that this will be achieved by two measures: all 12 of the major coal-burning power stations in England and Wales are to be fitted with new burners which will significantly reduce the amount of NO_2 emitted; **catalytic converters**, which cut NO_2 emissions by 70%, have effectively been compulsory on all new cars since 1992.

In 1988 the UK agreed to the EU Large Combustion Plant Directive, which meant that SO_2 emissions from large power stations (50 Mw or more) must be reduced by 40% by 1998, using 1980 as the baseline. This is to be achieved by fitting **flue-gas desulphurisation** equipment to some of the largest emitters and by constructing new combined cycle gas turbine plants which use natural gas.

Transport

Catalytic converters reduce emissions of nitrogen dioxide, carbon monoxide and hydrocarbons, but increase fuel consumption and carbon dioxide emissions. They will, however, help to accelerate the reduction of lead in petrol, since lead rapidly poisons such catalysts. **Lean burn engines** increase the ratio of air to fuel (Figure 10.13) and, although emissions of nitrogen oxide decreases, those of hydrocarbons increase.

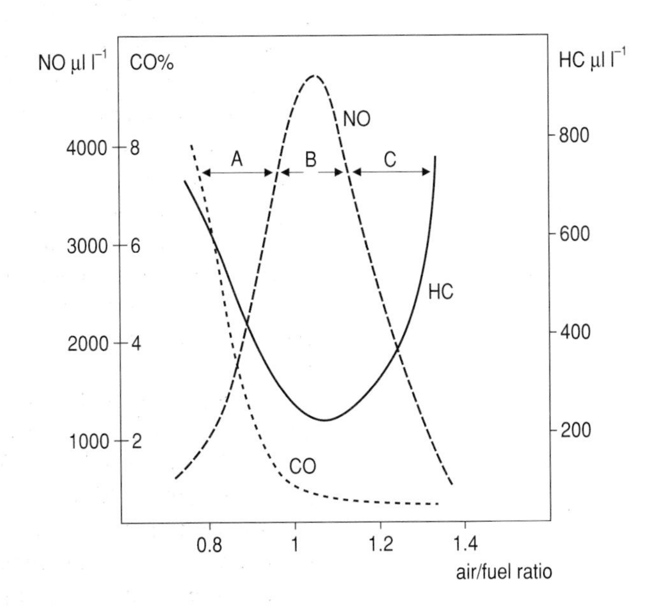

Fig 10.13 The effect of changing air/ fuel ratios on vehicle emissions.

The EU has set progressively tighter emission standards for all engine sizes and these now form part of MOT tests. Controls on volatile organic compounds which evaporate from the fuel tank will require fuel vapour collection canisters to be fitted.

Agriculture

The 1990 EPA required farmers to obtain waste management authorisations for the disposal of organic wastes on their land, and various regulations control the spraying of sewage sludge on land. This aims to prevent heavy-metal contamination of crops which are subsequently grown on the land. A number of EU directives relate to pesticides. Chlorinated hydrocarbons such as DDT, chlordane and heptachlor are banned, and food containing residues of these substances cannot be sold.

Over the last thirty years, environmental scientists have helped us to identify many local, international and global problems which humans have caused. As these effects have become better understood, governments have been forced to introduce pollution control legislation or issue tougher guidelines. Because some forms of pollution move from one country to another and because some threaten globally important processes, international legislation is becoming increasingly important. Some of the most important legislation affecting the atmosphere is summarised in Table 10.8.

Table 10.8 Protecting the atmosphere

Legislation	Aim
Clean Air Acts, 1956, 1968, 1993	reduce urban smog by banning the burning of coal in urban areas; replaced by the 1993 Clean Air Act which prohibited continuous emission of black smoke from any chimney
Environmental Protection Act, 1991	introduced Integrated Pollution Control and led to the development of BATNEEC (Best Available Technology Not Entailing Excessive Cost)
Montreal Protocol, 1987	international agreement to limit the production of substances which deplete the ozone layer; EU regulations phase out CFCs and CCl_4 ahead of the Protocol
Convention on Long Range Transport of Air Pollution	required countries to try to reduce air pollution, especially transboundary pollutants; acid rain was the major target
EU Large Combustion Plants Directive	required large but staggered cuts in emissions of SO_2 and NO_2 from major power stations

SUMMARY ASSIGNMENT

1. Explain how driving to college in Kent may affect:
   (a) the population of fish stocks in Sweden
   (b) the amount of shade available to picnickers in the Black Forest of Germany.
2. How can we use our knowledge of climatology (Chapter 2) to determine when and where urban air pollution is most likely to occur?
3. The graph compares the costs and benefits of a form of pollution control.

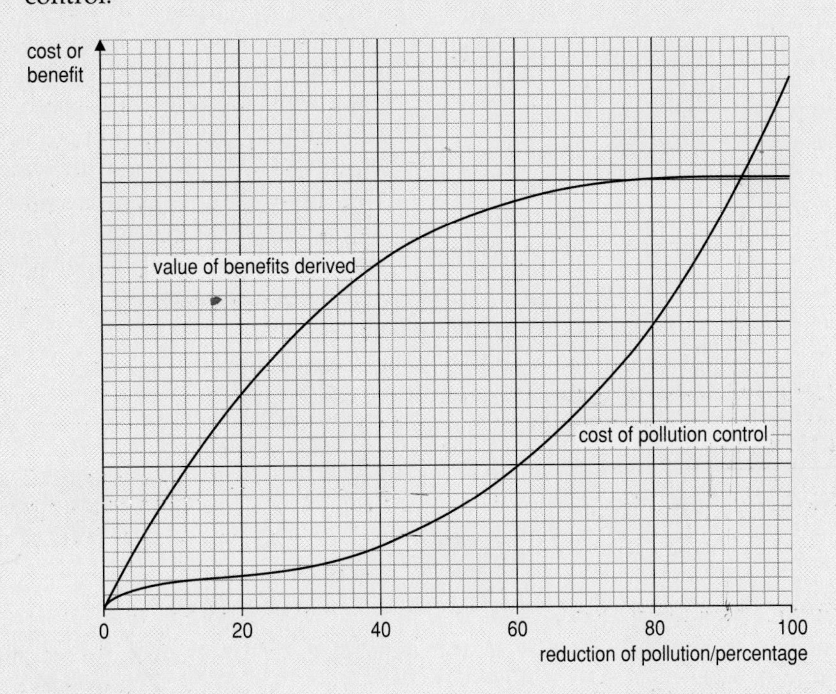

(a) From the graph:
   (i) state the level of pollution reduction which will give the greatest benefit:cost ratio
   (ii) explain why one hundred per cent control may not be the most economically desirable option.

(b) Catalytic converters (CATS) may be fitted to petrol-driven cars. CATS reduce emissions of carbon monoxide, hydrocarbons and nitrous oxides, but may increase emissions of carbon dioxide. Outline the possible environmental costs and benefits of fitting CATS.

AEB June 1995

Chapter 11

WATER RESOURCES

Water is probably the most basic of human requirements, upon which our society, economy and life itself depend. It is a renewable resource, in that it is constantly passing through the closed system of the hydrological cycle (Figure 11.1), but is distributed very unevenly in time and space (Figure 11.2). Worldwide, the demand for water is increasing dramatically. Although the world population has increased sevenfold over the last 300 years, water use has increased by 35 times. This chapter examines the value of water as a resource, the ways it is managed and the environmental consequences of water mismanagement.

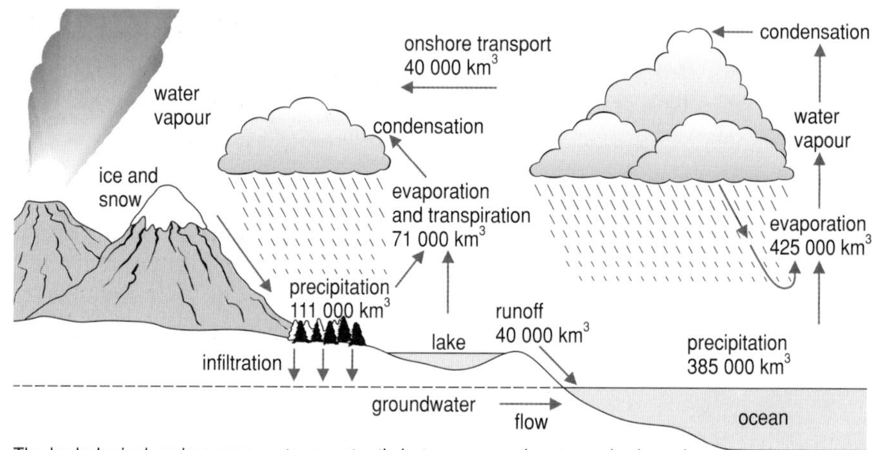

The hydrological cycle moves water constantly between aquatic, atmospheric and terrestial compartments, driven by solar energy and gravity. The total annual runoff from land to oceans is about 10.3×10^{15} gallons.

Fig 11.1 The hydrological cycle.

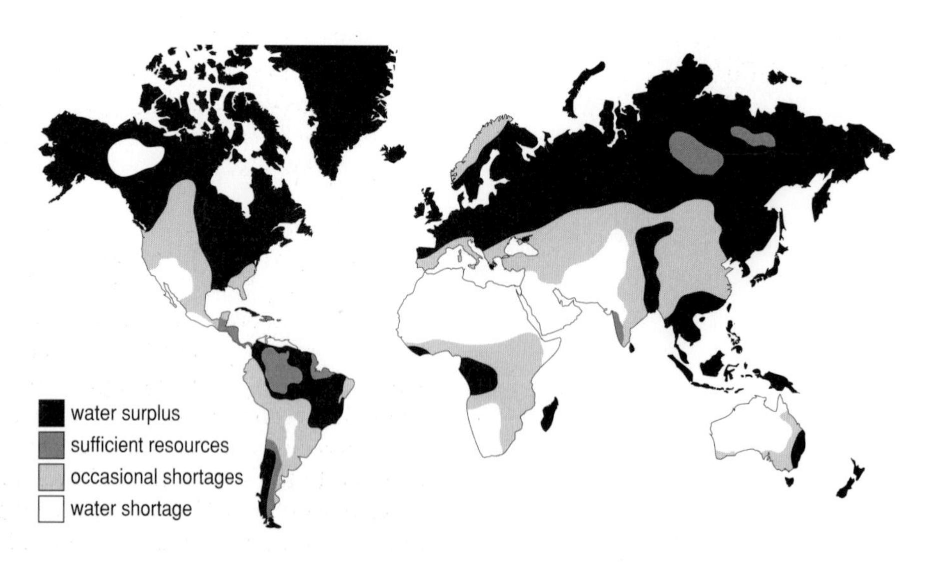

Fig 11.2 World water resources.

LEARNING OBJECTIVES

After completing the work in this chapter you will be able to:

1. describe the uses of water
2. describe the sources of water in the UK
3. explain how adequate supplies of clean water are maintained
4. describe strategies for conservation of supplies
5. explain the principles of river management and flood control.

11.1 THE MANY USES OF WATER

Of all our resources, water is probably the resource most often taken for granted. To many people in developed countries, this precious commodity is simply a transparent liquid available 'on tap' 24 hours a day, 365 days a year. In other parts of the world, simply obtaining enough water for drinking and cooking is a major part of life.

Domestic consumption

This includes water for drinking, washing and waste disposal. There are striking differences between domestic consumption in developed and developing regions of the world, and even within the developed world (Table 11.1).

Table. 11.1 Water consumption/m^3 $head^{-1}$ $year^{-1}$ by sector for selected countries

	Domestic	Agriculture	Industry
USA	211	796	951
Australia	209	1019	274
Japan	132	504	95
UK	122	2	64
Germany	80	3.6	526
India	18	569	25

QUESTION

11.1 How might you account for the differences recorded in Table 11.1?

Agriculture

Water availability is one of the most important physical influences on farming systems. The world's most successful and sustainable agricultural systems are those that have adapted to natural variations in water availability. A good example is South-East Asia's rice cultivation, with rice being grown in paddy fields which trap the summer monsoon rains, and with harvesting occuring in October, with the onset of the dry season. The nomadic habits of herders living in rainfall-marginal lands, such as the African Sahel, and the migration pattern of wild animals are similarly well-adapted to the semi-arid environment.

With the intensification of farming methods this century and expansion into more marginal areas, agricultural demands for water – through irrigation – have rapidly increased. Irrigated farming now accounts for 70% of water consumption worldwide, and over 90% in countries such as India.

Industry

The location of most manufacturing industry, especially the traditional 'heavy' type, has been strongly influenced by the availability of water. Water may be required in large volumes as a raw material, as in the case of textiles or paper-making, or it may be essential for the cooling of machinery or effluent disposal. The size of industrial demand for water obviously depends on how industrialised (that is, how 'developed') a country is, and the relative importance of the industrial and domestic demand in relation to the agricultural sector also increases with level of development (Table 11.1). However, in countries which are now experiencing a decline of heavy industries, such as the UK, the industrial demand is fairly static. It is the industrialising countries of the developing world whose demands are rising most rapidly.

Energy

Before the exploitation of fossil fuels, the power of flowing water was one of our most important sources of energy. In the late twentieth century, our reliance on clean, renewable water-power is again increasing. In the UK, less than 2% of our power requirement is generated by hydro-electric power, but there is considerable scope for expansion in developing countries, which have approximately 65% of global hydro-electric potential.

The 97% of the earth's water locked up in the oceans can also be used for electricity generation in the form of wave and tidal power. Underground water in contact with hot rocks is used to generate geothermal energy in volcanically-active countries, such as Iceland and New Zealand. Even 'conventional' thermal and nuclear power stations depend on large quantities of water for cooling. The world therefore depends on water to a large extent, both directly and indirectly, for its power supplies (Chapter 12).

Transport

Water transport is still the most economical means of conveying heavy and bulky goods over long distances. Before the development of railways and roads, **inland waterways** were the major mode of freight haulage, and are still used extensively in much of continental Europe. The European Union (EU) has introduced huge 4400-tonne barges and financed a programme of canal deepening and widening. It has also promoted the linking of the Mediterranean, North and Black Seas. **Ocean shipping** is still the obvious method of long-haul transport for very large loads of unperishable, low-value goods (perishable goods are transported by air).

Although both these types of water transport are economical in terms of fuel consumption, there is a serious potential environmental impact caused by leakage of oil, particularly from huge tankers, even though this pollutant does degrade with time.

Recreation and tourism

Water is intrinsically appealing. Lakes and coastlines appear to be particularly attractive to tourists of all ages. Water sports are increasing in popularity as people have more leisure time and disposable income. Water is undoubtedly a huge money-spinner, as tourism is now the world's most important industry.

There is a close link between water as a resource for transport and for recreation. The ocean cruise has for years been considered a highly desirable form of holiday-making. In Britain, the inland waterways, which by the mid-twentieth century had become sadly neglected, are now experiencing a resurgence of interest.

CASE STUDY

Birmingham's waterways

Fig 11.3 Birmingham's waterways.

Birmingham was once the hub of Britain's canal network. The waterways were open to commercial traffic up to the 1960s, but the movement of goods onto the growing motorway network signalled the apparent end of the canals' useful life. Once abandoned, the canals became neglected, polluted and unwholesome watercourses flanked by derelict land.

However, during the 1970s and 1980s, an interest blossomed in Birmingham's canals as fascinating pieces of industrial archaeology and as an integral part of the city's history. People realised that any schemes for redeveloping the city could focus on the 'canal image', which would not only improve the urban environment but also promote tourism and recreation.

In the late 1970s, the Birmingham City Council, the Department of the Environment and British Waterways started to implement the city's Canal Improvement Programme. In the early 1980s, the canals were dredged, the undergrowth was cleared, towpaths were repaved and access was encouraged through the provision of doorways and bridges. New moorings have been built, and around 50 permanent canal dwellers live on narrow-boats in areas such as Gas Street Basin and Granville Street Wharf. The towpaths are regularly used by cyclists and pedestrians, and fishing is on the increase. Canal holidays are also growing in popularity, and the Birmingham canal network, at one time avoided because of its barren, industrial surroundings, is now appreciated for its fascinating history. The city's economy is reaping the benefits.

The use of water in a recreational sense is not without its environmental problems. These range from the disturbance caused to wildlife by noisy activities, to problems of river bank erosion caused by the swash of pleasure-boats. The Norfolk Broads, for example, are visited by approximately 100 000 to 150 000 people each year in private boats, and bank erosion is a serious problem. The sails of boats terrify nesting birds, and species such as the goldeneye, which is a migrating duck, have been frightened away from the area. The Broads Authority is now promoting ecotourism, through activities such as 'whispering' electric boat trips to see wildlife.

QUESTION

11.2 The areas around many reservoirs have been provided with a variety of amenity facilities to encourage a wide range of recreational pursuits. Unfortunately, some of these pursuits may conflict with each other.

(a) Give two different examples of facilities provided by Water Companies to improve the amenity value of reservoirs.

(b) Explain how the use of time zoning and space zoning can help towards resolving recreational conflicts around reservoirs.

(c) Give two reasons why some reservoirs have very restricted access.

JMB June 1991

In considering the many different ways in which water can be regarded as a resource, it is clear that there is an implicit conflict between the various uses of water.

11.2 THE WATER SUPPLY CYCLE

Sources of water

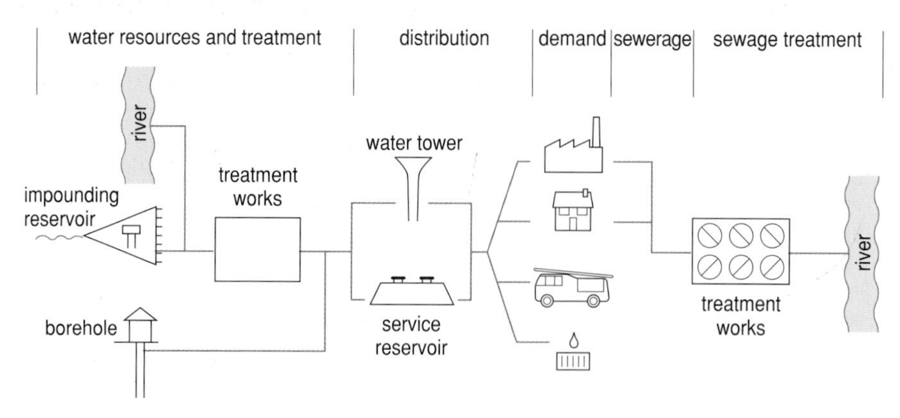

Fig 11.4 The water supply cycle.

The management of water resources for domestic consumption can be seen as a system, which involves taking water from the natural hydrological cycle and returning it after use (Figure 11.4). Water can leave the natural hydrological cycle and enter the supply system as surface water and as groundwater.

Surface water

Water which flow in rivers can be channelled off via an **intake** into a **storage reservoir**. The amount of water passing through the intake can be controlled; it is normal practice to fill the reservoir during high winter flows with a view to drawing it down during the summer. 'Raw' water in the reservoir is stored until it is required at the **water treatment works**, which is normally located alongside the reservoir.

Groundwater

Water-bearing rocks, or **aquifers**, act as important underground reservoirs. In areas of suitable geology, groundwater is normally preferable to surface water. It is cheaper and easier to abstract and treat, as it does not require reservoirs or extensive pipelines and is generally of higher quality than river water.

About one-quarter of the UK's potable (drinkable) water comes from underground, whereas other countries, such as the arid lands of the Middle East, are almost solely dependent on groundwater.

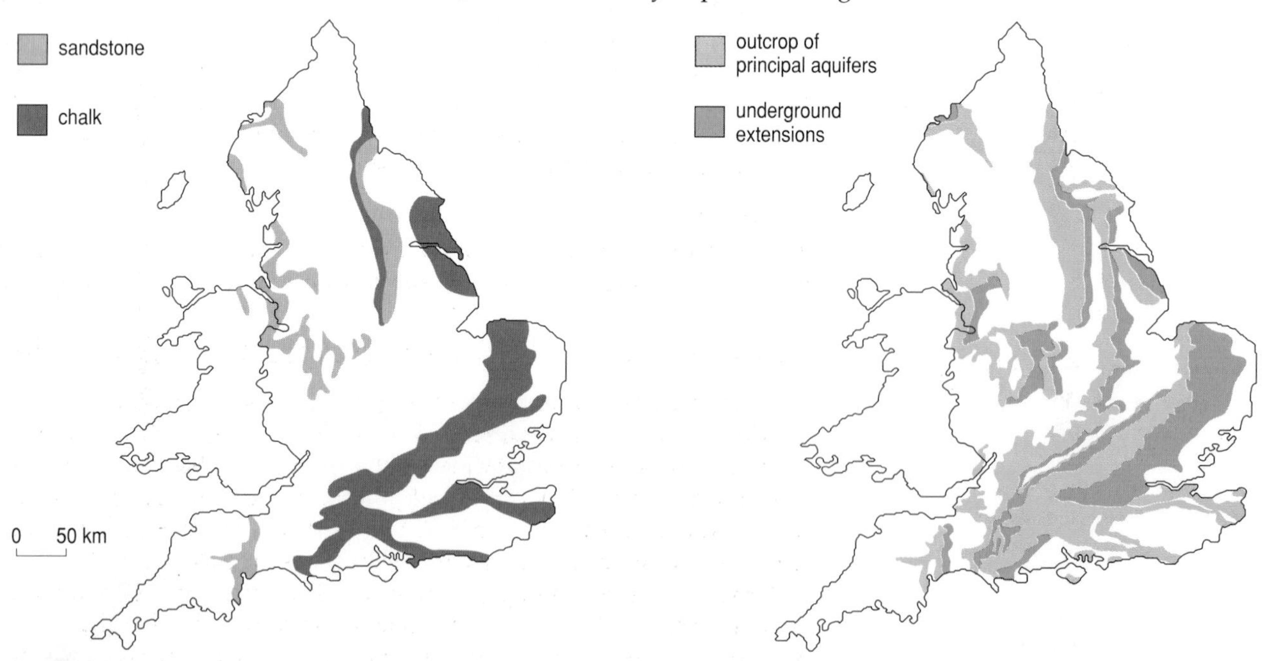

Fig 11.5 Rock types and principal aquifers of the UK.

QUESTION

11.3 Compare Figures 11.5 a and b. Which rock types are the UK's principal aquifers?

Groundwater abstraction

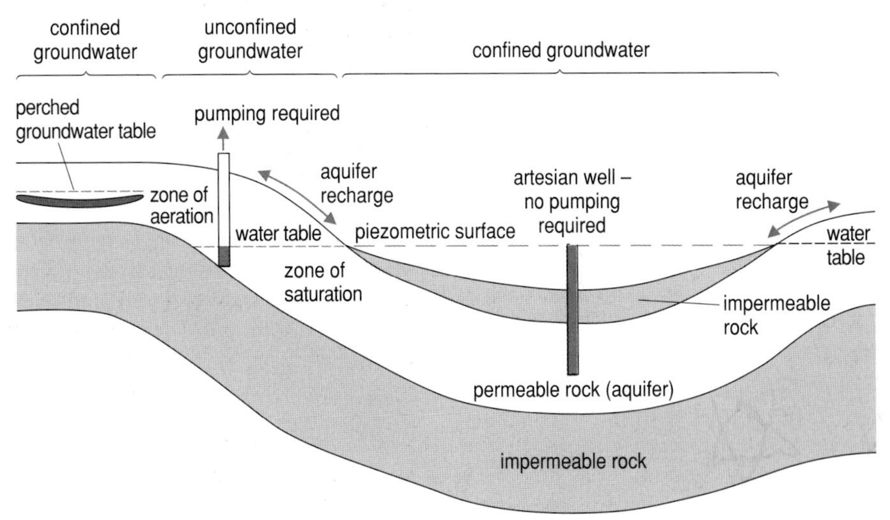

Fig 11.6 Types of groundwater.

Groundwater can occur in three different ways (Figure 11.6).

Unconfined

This is when the aquifer is exposed to the surface which recharges freely through percolation after rain has fallen. If a well is drilled into an unconfined aquifer, water will only rise to the level of the **water table** – the depth at which soil air spaces are saturated with water – and pumping will be necessary to draw water to the ground surface.

Confined

Confined groundwater occurs in a syncline, where the aquifer is 'sandwiched' between impermeable strata. The aquifer is therefore not exposed to the surface and is saturated with water under pressure. If a well is dug, the resultant release of pressure causes the water to rise to a level known as the **piezometric surface**, which is considerably higher than the top of the aquifer. Such a well is known as an **artesian well**, and requires little or no pumping to raise the water to the surface.

Perched

This type of groundwater is much more localised and occurs, for example, above a layer of impermeable clay within a bed of highly permeable gravels.

Problems associated with groundwater abstraction

Aquifers are a renewable resource which should be exploited on the basis of **maximum sustainable yield**. This would be the amount of water which could be extracted without causing any permanent decrease in the level of the aquifer. Abstraction should not proceed at a rate which exceeds the rate of recharge. Unfortunately, in too many parts of the world, over-abstraction has occurred, with serious environmental consequences.

Groundwater levels in the confined chalk aquifer of the London basin fell between 1875 and 1965. The piezometric surface has fallen by more than 60 m over an area of hundreds of square kilometres. The two most serious side-effects of over-abstraction are subsidence and salt-water incursions.

Subsidence

In Mexico City, one of the world's largest and fastest growing urban agglomerations, water use exceeds renewable supplies by 40%. The water table has been falling fast, and is now 3000 m below ground level. Widespread subsidence is occurring, and the cathedral in the main square is sinking into the ground.

Salt-water incursion of aquifers

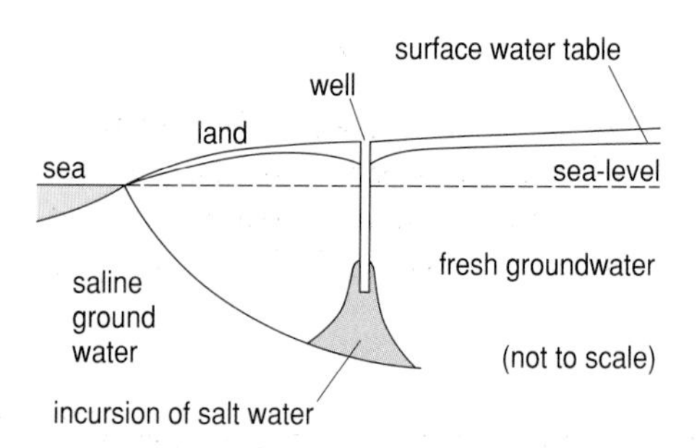

Fig 11.7 Salt water incursion of aquifers.

In coastal areas, over-pumping of aquifers can lead to the contamination of groundwater by salt water. An aquifer open to the sea will contain a layer of fresh water above the saline groundwater, as fresh water is less dense (Figure 11.7). Overpumping of a borehole will reduce the pressure of the fresh water acting down on the salt water (hydrostatic pressure) and induce the salt water to rise up the well.

Salt-water incursion is a particular problem in arid regions, such as the Middle East. In the Sultanate of Oman, for example, the country's oil wealth has in recent years led to a replacement of wooden buckets with diesel wells to draw water from aquifers. Along the fertile northern coast, salt-water incursion into the aquifer has killed date palms and lime groves. The increasing unpredictability of monsoon rain has made matters worse, by increasing demand for water from the aquifers.

Water treatment

Whether abstraction is from rivers or from the ground, a multi-stage treatment process is required to render the water fit to drink. Although treatment methods vary from place to place, the following sequence is typical of a 'traditional' works.

1. **Preliminary screening**
   Water enters a storage reservoir after passing through a series of coarse screens to remove gross solids such as twigs and leaves.

2. **Storage**
   Suspended solids settle out, and natural ultra-violet (UV) radiation kills some pathogens.

3. **Screening/microstraining**
   Water drawn from the reservoir is screened through fine mesh and microstrainers; rotating drums of fine mesh remove most particles and virtually all micro-organisms.

4. **Aeration**
   Water drawn from the ground or from the lower levels of a reservoir will contain little or no dissolved oxygen. Oxygen may be added by cascading the water down a tower structure.

5. **Coagulation**
   The very finest particles (**colloidal solids**), notably clay, metal oxides, large protein molecules and some micro-organisms, never settle out of suspension naturally. However, if a **coagulant** is added, colloidal solids may be induced to aggregate into larger particles called **flocs**. Aluminium and iron salts are commonly-used coagulants.

6. **Flocculation**
   Water is mixed gently so that flocs collide to form larger aggregates, which are more easily removed during clarification.

7. **Clarification**
   Here the flocs are removed by settlement. Water is forced to flow upwards from the base of a tank. The heavier flocs settle towards the

bottom of the tank. The clarified water at the surface overflows a weir to the next stage of the treatment process.

8. **Filtration**

   Some fine solids which have formed during coagulation are removed by sand filtration. Water is passed down through beds of fine sand underlain with coarse gravel. Pipes at the base of the bed drain the filtered water away.

9. **pH adjustment**

   The pH of the water will depend on the geology and soils of the area from which it was abstracted, and may need to be adjusted. If drinking water is too acidic, it may corrode distribution pipes and domestic plumbing, whereas alkaline water will cause deposition of salts within distribution pipes.

10. **Disinfection**

    Disinfection is required to remove any final pathogens and ensure that the water remains potable during its journey to the consumer. Three methods of disinfection are available:

    (a) Chlorination

    The addition of chlorine is the most traditional method of disinfection, but is falling out of favour due to its influence on taste.

    (b) Ozone

    This more expensive alternative to chlorine tends to be used where the natural water contains materials which would combine with chlorine to produce unpleasant tastes or odours. Ozone poses problems in that it does not provide **residual disinfection** (it ceases to work once water is in the distribution system). A low-level chlorine residual is thus usually required.

    (c) UV radiation

    The emission of UV radiation from special lamps can be very effective on a small scale.

11. **Softening**

    Water hardness results from dissolved calcium and magnesium salts and is a feature of chalk and limestone areas. Excessive hardness causes 'scaling' of household water-heating appliances, and therefore needs to be adjusted at the treatment works. The addition of lime (calcium hydroxide) or soda ash (sodium carbonate) converts soluble salts into insoluble ones which can be removed by sedimentation.

Developments in water treatment

Legislation controlling drinking water quality is now laid down under the EU Drinking Water Directive (1980) and the Water Industry Act (1991), and is becoming increasingly stringent. There is also growing consumer concern about drinking water, as testified by the mounting sales of domestic water filters and bottled water. Consequently, water supply companies are under constant pressure to refine treatment methods.

The introduction of ozone- and UV-disinfection as an alternative to chlorine-disinfection, which avoids some taste and odour problems, is one example of a modification to traditional methods. Another is the use of Granulated Activated Carbon (GAC) filters, which absorb organic molecules, such as pesticides and solvents, onto their porous surfaces. The use of these filters is now on the increase, often in conjunction with ozone disinfection.

CASE STUDY

Sewage treatment

Waste waters from houses and industry are conveyed by the sewerage system to sewage treatment works. Sewage treatment relies on a combination of physical separation of pollutants from water and biological removal of organic contamination and ammonia. Biodegradation, which would occur naturally in the environment, is accelerated by providing optimum conditions, such as good air supply to aerobic bacteria. Sewage treatment typically takes place in stages, as follows:

1. **Screening**
   Screens remove large debris such as paper, plastic, rags and bits of wood.

2. **Grit removal**
   The sewage flow rate is reduced to allow coarse grit to settle out.

3. **Primary settlement**
   The sewage is held in sedimentation tanks. Sludge settles to the bottom of the tanks and the liquid weirs over to secondary or biological treatment.

4. **Biological treatment**
   This usually takes one of two forms, biological filters or activated sludge.
   (a) Biological filters
   The sewage is sprayed over a bed of stones, providing a large surface area for the growth of the bacterial film which decompose the organic matter. Large pore spaces between the stones ensure that there is good contact between the air, the bacteria and the sewage.
   (b) Activated sludge
   This is an alternative biological treatment method where sewage is held in tanks into which air is blown. Aeration encourages rapid growth of bacteria which decompose the organic matter.

5. **Secondary settlement**
   Solids remaining from the bacterial treatment stage are known as humus solids. These are settled out in secondary settlement or humus tanks. After this stage of treatment the effluent may be clean enough to be discharged to a river.

6. **Tertiary treatment**
   A final treatment, by passing the effluent through grass plots, reed beds or sand filters, may be employed to remove residual solids and to further reduce the contamination by organic matter.
   Sludges collected from primary and secondary settlement are held in closed tanks for approximately two weeks at around 35°C. This process, known as sludge digestion, produces a product which is drier and less odiferous, and contains fewer pathogens. This digested sludge may be used as a fertiliser and soil conditioner. Methane is also produced by sludge digestion and may be collected and used to generate electricity.

QUESTION

11.4 Describe the role of micro-organisms in the treatment of sewage.

11.3 WATER RESOURCE MANAGEMENT IN ENGLAND AND WALES

This section examines how rising demands for water are being met in England and Wales.

Who manages England and Wales's water resources?

Before the water industry was privatised in 1989, ten Regional Water Authorities were responsible for water supply, sewerage and sewage treatment, and for the regulation of pollution, water conservation, land drainage, flood control and fisheries. In this sense they were self-regulating.

In September 1989, a new environmental regulator, the **National Rivers Authority** (**NRA**) was launched. The NRA has taken over the regulatory duties from the former water authorities and ensures that the new private water supply and sewerage companies (or 'utilities') conduct their activities within environmental legislation. In addition, the financial affairs of the water utilities are supervised by the watchdog **Office of Water Services** (**OFWAT**). Contrary to popular belief, the amount a water utility can charge for its services is strictly controlled by OFWAT, although significant regional variations exist. This is largely because some regions (notably the South West) must spend huge sums of money to comply with new European environmental legislation. The regions of the ten water utilities are illustrated in Figure 11.8. In Scotland and Northern Ireland, the water industry is still in state ownership.

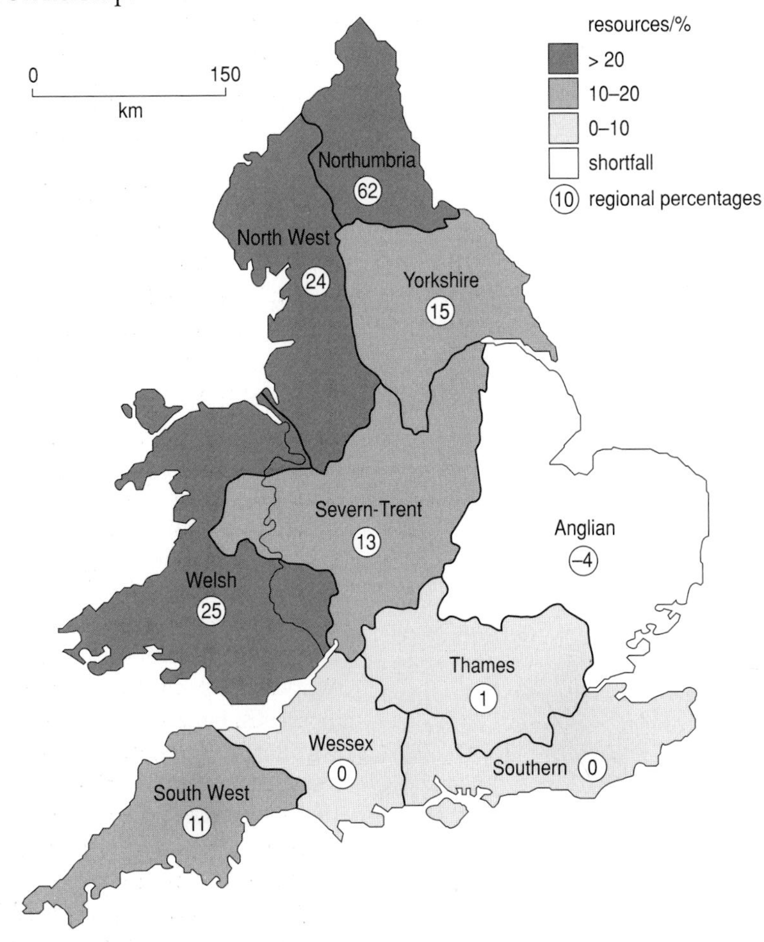

Fig 11.8 Regions of the ten water utilities.

Water demand

Demand for water is rising, especially in the domestic sector (Figure 11.9). Since the population of England and Wales is fairly static at present, this reflects an increase in usage per head. The lifestyle of the 1990s demands huge quantities of water: machines such as dish-washers are found in many homes, and people bath or shower more frequently than ever before. (Even a shallow bath uses around 50 litres of water.)

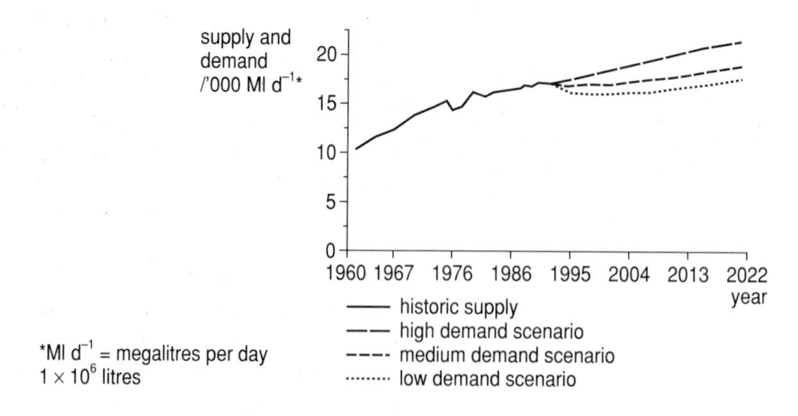

Fig 11.9 Water supply and demand, 1961 to 2021.

Forecasting future demands is not straightforward. Factors such as the future state of the economy, changes in domestic habits (which are likely to be affected by changes in the method of paying for water) and even climatic change need to be considered. It is therefore normal to examine a number of 'scenarios', as illustrated in Figure 11.9. This shows a maximum increase in demand of 23% for the period 1992 to 2022.

Assessing our resources

Is there enough water in England and Wales to meet this increasing requirement? This is a difficult question to answer because the water resources are unevenly distributed. Most precipitation falls in the north and west, since these areas bear the brunt of Atlantic depressions and their higher relief enhances rainfall.

The pattern of water demand is somewhat different. In the south, the concentration of affluent population and the irrigation requirements of agriculture (on average 9 years out of 10), coupled with a relatively low rainfall, produce a potential problem. Figure 11.8 shows that, according to projections, East Anglia will suffer a resource deficit by 2021 if no action is taken. Contrast this with a 62% surplus in rainy, sparsely populated Northumbria, with its huge Kielder Reservoir.

Water supply is further complicated by annual variations in resources and demand. It is when rainfall is lowest and water loss through evapotranspiration is highest (during the summer months) that demand increases. Farmers and gardeners require irrigation, swimming pools become popular, and baths and showers are taken more regularly. Long-term climatic change may exacerbate these problems. The droughts of 1976, 1988–92 and 1995 were maybe a foretaste of the future, and the water industry needs to prepare for this.

Meeting future demands

There is a wide range of options available to water resource planners, including continued provision of new water supplies and new reservoirs. (Those most likely to affect the UK's future are discussed below.) The most traditional reaction to the water supply dilemma is to increase storage capacity.

Fig 11.10 The Ataturk Dam in Turkey.

Dams and reservoirs

Around 700 new dams are constructed worldwide each year, and their dimensions are ever-increasing. The most ambitious scheme so far inaugurated (in December 1994) is the Three Gorges Dam on China's Yangtze river: a dam nearly 2 km long and 100 m high, impounding a 600 km-long reservoir.

Such schemes are normally associated with grandiose multi-purpose development projects. Three Gorges is typical in that its aims include flood control and hydro-electric power (HEP) generation, to reduce dependence on fossil fuels and stimulate industrial growth. Irrigation schemes are also often connected with dam-building. However, the human and environmental side-effects of major dams and reservoirs can be serious. These are summarised in Table 11.2.

Overall, the 'small is beautiful' approach to dams is the only way to minimise environmental and human loss. However, the international prestige to be won by developing countries for executing major feats of civil engineering and the vested financial interests of the developed world, which plays a major part in money-lending, are serious barriers to overcome.

Table 11.2 Human and environmental impact of dams and reservoirs

Effect	Example
Human	
eviction of people to make way for reservoir, with often inadequate provision for resettlement	1.2 million people are likely to be displaced by the Three Gorges Project
damming of reservoirs in tropical areas creates ideal breeding grounds for vectors of disease	Ghana's Lake Volta was associated with an 80% rise in malaria infection rates in local children within a year
people downstream of a flood-prevention dam may be lulled into a false sense of security and settle on flood-prone areas	this fear has been expressed by opponents of the Three Gorges Dam
Environmental	
dams trap sediment, as the velocity and capacity of rivers is much reduced; this lessens the storage capacity of the reservoir, starves downstream areas of fertile alluvium and increases the risk of downstream erosion	the capacity of Lake Nasser, Egypt, has been reduced and the Nile Delta starved of silt since the building of the Aswan High Dam; the river downstream of the dam now transports only 8% of its natural load and the delta is being eroded; Nile farmers are having to rely more heavily on chemical fertilisers
dams destroy habitats and disrupt ecosystems, for example preventing the breeding and migration of fish	endangered species whose habitats are under threat from the proposed Three Gorges Dam include the White Flag Dolphin and Siberian Crane
the huge weight of dams and reservoirs can lead to subsidence	Lake Mead, USA, produced a downwarping of up to 20 cm between 1950 and 1963
reservoirs provide a large surface area from which evaporation can take place; this leads to a concentration of dissolved salts in hot climates and increased soil salinity	Lake Nasser, Egypt
reservoirs change local groundwater conditions, which can reduce the stability of surrounding slopes and lead to landslides	a huge landslide, triggered by local changes in groundwater conditions, displaced a wall of water which overtopped Vaiont Dam, Italy, killing 2600 people in 1963
a reservoir can alter the local climate, notably by: – moderating temperatures due to the high specific heat capacity of water – increasing humidity, especially if slight winds favour the formation of a 'vapour blanket' over the water's surface – increasing wind velocities, as water is a low-friction surface	

Artificial recharge of aquifers

This involves artificially refilling an aquifer with surplus water at times of plenty for use during times of drought. Artificial recharge may be achieved either by inducing percolation from a water body, such as a lake, or by drilling a borehole and injecting water downwards. The very specific hydrogeological requirements means that this activity can only occur in a limited number of places, but it appears to be a relatively cheap option with limited environmental impact. So far, the main developments in the UK have been hydrating the chalk of north London.

Reallocation of supplies

One popular suggestion is that water supplies should be managed in the same way as electricity – via a National Grid. However, water is much more expensive to transport over long distances than electricity. Its flow is subject to gravity, and the cost of pipelines and pumps would be enormous.

Many 'inter-basin transfers' do exist, however. Birmingham has received its drinking water from the Elan Valley in central Wales for about 100 years. It has been suggested that further transfers could make use of the Midlands canal network.

CASE STUDY

Proposed River Severn to River Thames transfer

This scheme would involve abstracting water from the lower Severn in the vicinity of Deerhurst and discharging it near Lechlade in the upper Thames catchment (Figure 11.11). The River Thames would then be used to transfer these additional resources towards London, where the water would be stored in the existing reservoirs to the west of the city.

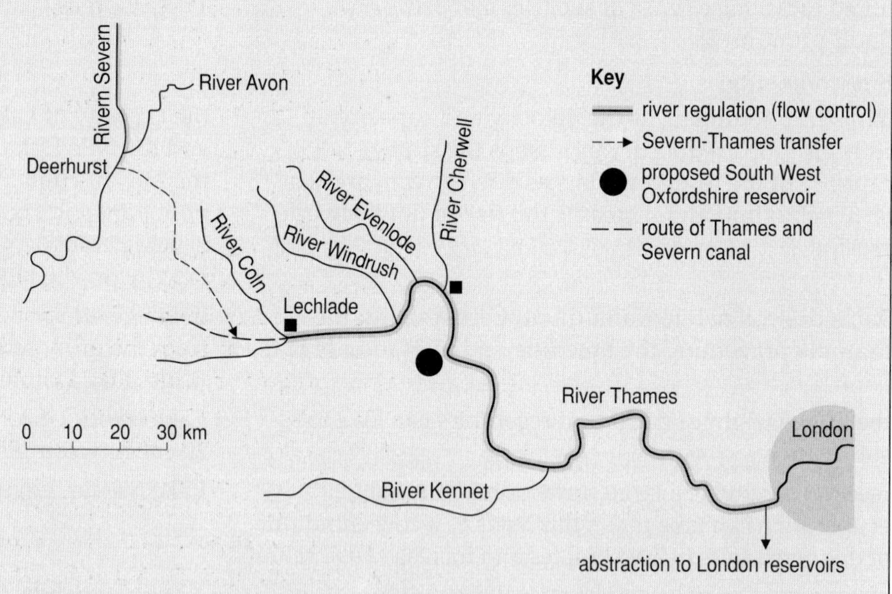

Fig 11.11 Water resource proposals in central England.

However, there would be environmental side-effects. One concern surrounds the effects of abstracting water from one area and discharging it into another at times of reduced flow when dilution will be low. According to an NRA report, 'in terms of water quality, there is a real risk of causing a significant change to the chemistry of the upper Thames' (NRA, 1994). It is not thought that the Severn would suffer major consequences, providing strict operating rules regarding appropriate flows are followed.

'Second-hand' and lower-quality supplies

The use of drinking-quality water for activities such as toilet-flushing and watering the garden does perhaps highlight a misguided use of resources. Although establishing a dual supply is unrealistic in Britain at present, in some arid regions, sterilised treated effluents are commonly used for 'non-consumable' activities.

In a sense, however, some effluent reuse already occurs in the UK. About 30% of raw water put into public supply has already been used further upstream, treated and returned to the environment in the form of **sewage effluent**. This figure is as high as 70% in south-east England. At times of low flow, the River Thames comprises around 95% treated effluent!

Effluent reuse is now a serious option for water conservation. Rather than returning effluent to a watercourse, from which it is likely to be 'lost' to the sea, it may be piped directly from a sewage treatment works and used for non-drinking purposes (for example golf-course irrigation).

Studies in the London area suggest that effluent reuse would delay the Severn-Thames transfer from the mid-1990s to 2006–2011, and the South-west Oxfordshire reservoir 'beyond the planning horizon' (NRA, 1994). However, winning public support for such a scheme, which has possible health implications, is by no means easy.

Reduce the need for additional water

Our final set of options tackles the problem at source, so should be given priority.

Leakage control

According to OFWAT (1995), the ten main water companies estimate that 25% of all water in the distribution system is lost through leakage. Although total elimination of leakage is impossible, leakage detection and control are being treated seriously by the NRA, who take distribution losses into account when considering water company requests for additional abstraction.

Demand management

Most domestic water users still pay for water regardless of how much they use; these 'unmeasured' bills are based on the old rateable property values. There is therefore little incentive to save water.

However, by the year 2000, an alternative method of payment will have to be introduced. Some members of the public favour universal metering on the grounds that it appears fairest to pay for what we use, and this will bring our water charges in line with the other utilities – gas and electricity. Most commercial and industrial premises are now metered, and most new homes automatically have meters installed.

Universal metering has not won the full support of the water industry. It is costly and often impractical to install, especially in the case of customers who share a supply pipe (for example in a block of flats). Metering also discriminates against large families who use water heavily. Any method of payment, however, has its winners and losers.

One real advantage of metering is that it does reduce water demand. The largest-scale metering trial to date has been run by Southern Water on the Isle of Wight. This was successful in reducing daily water consumption by 22% between 1988 and 1992.

QUESTION

11.5 What can we, as individuals, do to reduce our water usage? (You may recall the advice given out by water companies in times of drought.)

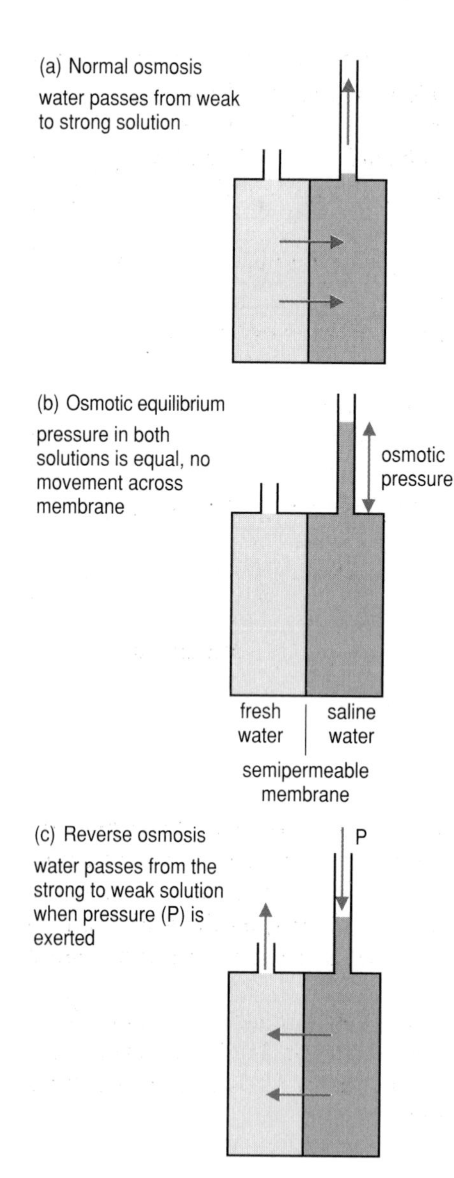

Fig 11.12 Principles of reverse osmosis.

Other methods of meeting water demand

On a global scale, the UK's problems of supply and demand are trivial. Figure 11.2 shows that there are large areas of the world where water is constantly in short supply and much more extreme measures need to be taken to meet demand.

Desalination

In many arid countries there are no surface supplies at all and, where groundwater is also scarce, the only option is to turn to the earth's largest reservoir, the sea. The main problem with desalination is that it is extremely demanding of energy and therefore very expensive. There are, however, about 7600 desalination plants worldwide, 60% of which are in the Middle East. The island of Jersey also has a desalination plant to meet peak summer tourist demands. Two major types of desalination process exist, multi-flash distillation and reverse osmosis.

Multi-flash distillation

This involves the evaporation of hot brine and condensation of fresh water in a series of chambers of increasing pressure. This occurs in Jersey.

Reverse osmosis

Osmosis is the net movement of water from a weak solution to a strong solution through a semi-permeable membrane (Figure 11.12 a). This movement occurs due to a difference in pressure on either side of the membrane (the **osmotic pressure**) and continues until the pressure difference is zero and the solution concentration on either side of the membrane is equal (Figure 11.12 b). By reversing this process, water is made to move from the more concentrated to the weaker solution. This is achieved by applying a pressure higher than the osmotic pressure on the concentrated solution (Figure 11.12 c). The fresh water collected is used for drinking.

Other options

Some arid countries import water from wetter parts of the world by tanker, but this renders them very vulnerable in the long term. Cloud-seeding, the release of silver iodide or ice crystals into clouds by aircraft to act as condensation nuclei, has been fairly successful in areas such as south-west Australia. The most outlandish option to date is to tow icebergs from high latitudes to the tropics. This has not yet been carried out, but may become increasingly attractive if the breakup of ice sheets due to global warming continues.

CASE STUDY

The 1995 drought

Water supply, or more precisely the lack of it, was a big issue in many parts of the UK in 1995. The hot, dry summer brought drought of varying severity across the country. The heat created a soil moisture deficit so that, even when rain did occur, there was little runoff to the rivers. Those areas of the country which relied on surface water sources to supply the reservoirs were particularly hard hit. Reservoirs fed by small catchment areas and rivers dependent on runoff fared very badly. This led to particularly acute problems in Yorkshire, which depended on the vulnerable South Pennine Reservoirs.

Winter rainfall had been sufficient to ensure that groundwater reserves were well recharged at the start of the summer. This contrasted with the recent droughts, such as that of 1989–90, when groundwater levels were low after inadequate winter rainfall. Rivers whose base flow was fed by groundwater were less badly affected, so water companies abstracting from such rivers, or directly from groundwater, were able to maintain supplies.

QUESTION

11.6 Any water-bearing rock that can yield its water in sufficient quantities to be exploited for public water supply is called an aquifer. The UK is well-endowed with such rocks which, through a combination of natural and artificial recharging, supply the country with considerable amounts of water.

(a) State the **two** main properties of any aquiferous rock which allow it to store and release water.

(b) Name the **two** main sedimentary water-bearing rock types in the UK.

(c) **(i)** State the main difference between a confined and an unconfined aquifer.
(ii) Name the other type of aquifer.

(d) **(i)** Outline briefly what is meant by 'the artificial recharging of aquifers'.
(ii) Give **one** advantage and **one** disadvantage of the artificial recharging of aquifers.

NEAB June 1995

11.4 IRRIGATION

Irrigation – the artificial distribution and application of water to arable land to enable crop growth – is by no means confined to arid and semi-arid areas. Anywhere which experiences prolonged periods of **water deficit** (when the soil cannot supply enough water to sustain plant growth) will need an artificial 'top-up' if crops are to be grown. Much of southern and eastern England falls into this category.

Irrigation is not without its problems, with even the most common form of irrigation, the traditional system of open channels, having a major effect on supplies. Two of the major environmental problems resulting from irrigation are waterlogging and salinisation.

Waterlogging

Up to 80% of irrigation water may never reach the crops at all. Most is lost from permeable channels and basins, with significant evaporation losses also occurring from sprinkler irrigation. Seepage of irrigation water into the soil will lead to a rise in the water table and, if this approaches the soil surface, waterlogging will occur in the root zone of crops. Waterlogging is detrimental to most major crops because it reduces soil aeration. Rice, however, thrives in such conditions. Therefore, many irrigated areas require effective drainage management.

Salinisation

Salinisation, or the concentration of soluble salts in the soil, often results from waterlogging. It is one of the major causes of land degradation and is threatening around one million hectares of farmland worldwide each year. Around 30% of the entire area of Egypt and 23% of Pakistan, for example, are now seriously waterlogged and/or saline.

The process of salinisation is illustrated in Figure 11.13 and is often exacerbated when slightly saline water, usually from artesian wells, is used for irrigation.

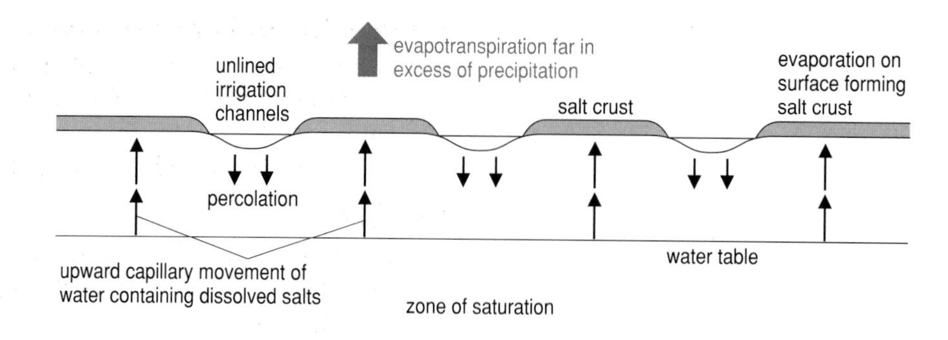

Fig 11.13 The process of soil salinisation.

CASE STUDY

Reclaiming saline lands: a case study of Pakistan

Approximately 13.4 million hectares of Pakistan are irrigated, and virtually all the country's food and cash crops are cultivated on this land. However, Pakistan is a poor country and irrigation methods are traditional: many channels are unlined and seepage accounts for 80% of aquifer recharge! Consequently, the water table in areas such as the Punjab in the south of the country has risen by about 16 m. Around 40 000 hectares of irrigated land are now thought to be lost each year due to waterlogging and salinisation.

Two methods have been employed by the Salinity Control and Reclamation Projects (SCARP):

- pumping water from freshwater aquifers to lower the water table
- developing a system of sub-surface drainage for saline water.

However, it is very difficult to evaluate the success of the scheme to date. Although some 15 000 tube wells and over 5000 km of surface drains had been installed by the late 1980s, there are fears that reclaimed land will only revert back to agriculture and the problems will begin again. In vulnerable areas, a change of land use, from wheat to woodlands for example, is the best solution.

Why is salinity a problem?

We noted that, through the process of osmosis, water will move from a weak to a strong solution across a semi-permeable membrane. In this way, plants which grow in a saline environment lose water to the soil, their cells collapse and they die. Plants vary in their degree of salt-tolerance; citrus trees, for example, are very sensitive, while crops such as wheat grass and Rhodes grass are **halophytic**, or highly tolerant to saline soils.

The death of crops due to soil salinity is not only an economic disaster for the farmers concerned, it may also lead to desertification, as soils become exposed to the agents of erosion, wind and water. The presence of the salts themselves may break down the structure of clay soils, rendering them vulnerable to erosion.

Other methods of reclaiming saline lands include:

Eradication

Salt can be washed out of the soil with fresh water. However, fresh water is normally scarce in areas where salinity is a problem, so this method is very expensive.

Conversion of harmful salts to harmless ones

A common example is the addition of **gypsum** (calcium sulphate) to salts such as sodium carbonate to form harmless calcium carbonate and leachable sodium sulphate:

$$CaSO_4 + Na_2CO_3 \longrightarrow CaCO_3 + Na_2SO_4 \text{ (leachable)}$$

Prevention

Once salts are removed, this is the surest way of ensuring that the problem does not return. Examples of good irrigation management include:

- lining channels
- switching to more efficient irrigation methods
- realigning channels so that water flows over less permeable soils.

11.5 RIVER MANAGEMENT AND FLOOD CONTROL

So far in this chapter we have been concerned with managing shortfalls of water. It is also important to consider how to deal with excesses. In this final section, we examine the causes of floods, both natural and human, and the ways drainage basins can be managed to minimise flood damage.

The storm hydrograph

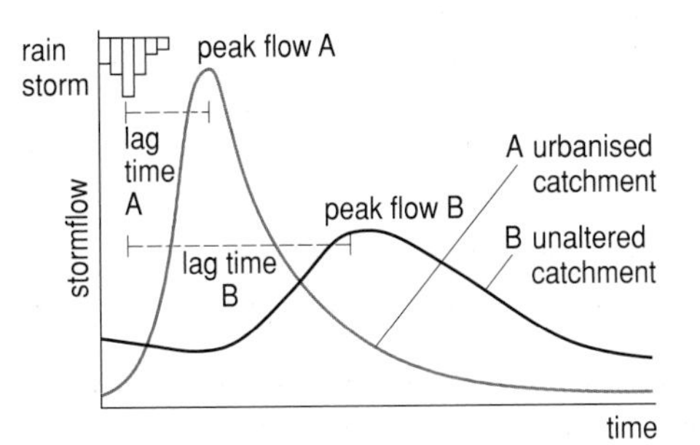

Fig 11.14 The effect of urbanisation on the storm hydrograph.

The **hydrograph** (Figure 11.14) is a useful means of recording a river's response to a rainfall event. Discharge – the volume of water passing a particular point in a unit time – is plotted against time. A bar chart of rainfall is superimposed on the hydrograph.

Before and during the storm, the river's discharge remains low, maintained by water flowing out of the ground (**baseflow**). The discharge starts to rise steeply to reach a peak. The delay between peak rainfall and peak discharge is called the **lag time**, and is the period during which surface runoff flows overland into the channel.

Once the peak discharge has occurred, the **falling limb** of the hydrograph indicates that the return to normal baseflow (dry weather) conditions is more steady than the rise to the peak had been. This is due to the steady input of water from the furthest extremities of the catchment and from out of the soil (**throughflow**).

The impact of urbanisation

Imagine heavy rain falling in a town. Most will fall onto pavements and roads, impermeable surfaces into which it cannot infiltrate and off which it runs into surface water drains, leading ultimately to a river. Some rain falls onto roofs, on which it may be trapped and re-evaporated back into the atmosphere (**interception loss**). However, much intercepted rain will flow down gutters and thence into surface water drains. Only in parks and gardens will rain be absorbed by the soil.

Downstream, much more water is available to boost the discharge, as little water infiltrates and is stored in the soil. This means that peak discharges may be very high. Because most of the stormflow is derived from surface runoff which is carried quickly by drains to the river, the peak also tends to occur relatively quickly and the ascending limb tends to be steep; the lag time is shorter in an urban catchment than in a rural one. The river will also return to baseflow levels quickly because throughflow is minimal. The impact of urbanisation on the storm hydrograph illustrates one of the most important ways in which humans have influenced rivers and increased the risk of flooding.

QUESTION

11.7 (a) Sketch the hydrographs likely to be obtained in the following catchments:

(i) a catchment before and after deforestation

(ii) a catchment in a granite area (impermeable rock) and one in a chalk area (permeable).

(b) What factors other than land use and geology are likely to affect the shape of a storm hydrograph?

Fig 11.15 Flooding in the UK in the early '90s.

Causes of floods

A **flood** can be defined as the inundation of land normally dry by excessive surface runoff or the sea. 'Excessive surface runoff' normally involves a river bursting its banks and inundating the land of its floodplain. Such flooding can occur on a grand scale in the world's largest river basins: the Amazon, for example, may spill its floodwaters for 20 miles on either side of its channel.

Any factor which increases the amount of runoff to enter a river channel will therefore increase the risk of flooding. The timing of the input into the drainage basin is also critical: a sudden increase in the runoff received by a channel is likely to produce a more dramatic flood than a steady input over a long period of time.

Coping with floods: protection versus abatement

There are two basic approaches to flood management, and considerable debate surrounds the relative emphasis which should be given to each.

Abatement

This involves tackling the problem at source by introducing measures to slow down and decrease runoff. These measures include afforestation, terracing steep slopes and, in urban areas, increasing the area of permeable surfaces. Flood abatement aims to increase a hydrograph's lag time and reduce the peak discharge.

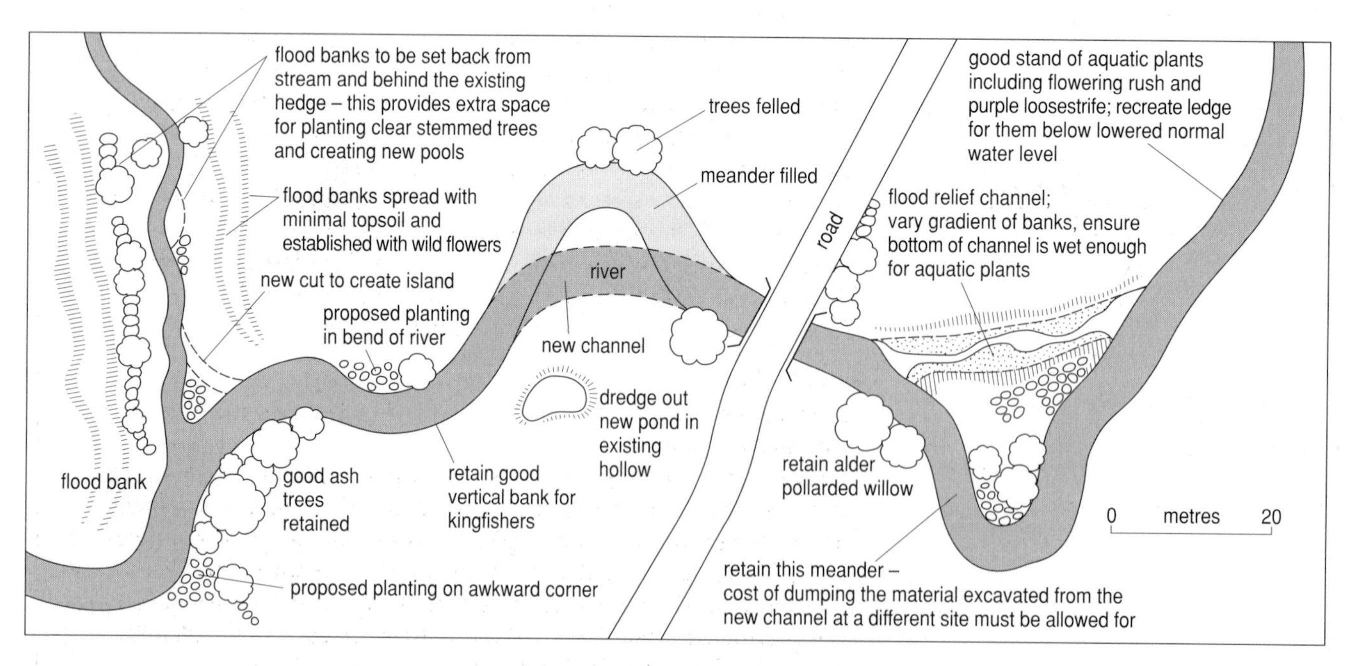

Fig 11.16 An 'environmentally sensitive' flood management strategy.

Protection

Flood protection comprises channel or bank modifications to enable a river to carry greater discharges more efficiently. **Levées**, or embankments, enable higher bankfull discharges. Dredging of river beds, in particular the removal of large boulders, reduces channel roughness and hence increases velocity. Flood relief has also commonly been a component of dam-building projects. Flood protection strategies in particular have a potentially major environmental impact. 'Heavy-handed' engineering solutions have been criticised for paying inadequate attention to aquatic and riparian ecology. It is now recognised that, prior to the implementation of any scheme, full biological surveys need to be conducted, and habitats worth conserving should be mapped. The use of heavy machinery should also be minimised.

QUESTION

11.8 Figure 11.16 shows an 'environmentally sensitive' flood management strategy for a stretch of river.

(a) Describe and explain the proposed flood relief measures.

(b) What measures have been taken to safeguard the ecology of the area?

In some parts of the world, another factor also enters the equation of flood management: the rights of native floodplain dwellers. Our final case study focuses on a country where flood protection is a very challenging and complex issue.

CASE STUDY

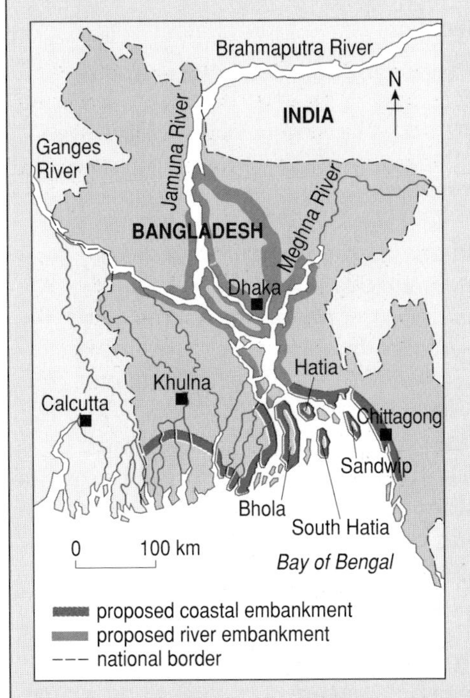

Fig 11.17 Bangladesh.

Bangladesh

Bangladesh is a very poor country on the Indian subcontinent (Figure 11.17). Much of the land consists of floodplain, which is frequently flooded by water from three great rivers: the Meghna, Brahmaputra (known as the Jamuna in Bangladesh) and the Ganges. These river systems comprise enormous catchment areas, the upper reaches draining the slopes of the Himalayas. By the time they reach Bangladesh, the rivers enter a low-lying delta system, a complex network of many channels and land built from the deposition of vast quantities of sediment. This area is a wetland of international importance.

Bangladesh's delta population is well adapted to living with floods and has long regarded the water as a resource for activities such as fishing and rice cultivation. However, with a natural growth rate of around 2.3% per annum, the population has been forced to inhabit increasingly hazardous areas. It is also possible, though not proven, that the incidences of 'abnormal' flooding are on the increase. There is no doubt that the risk to human life of flooding in Bangladesh is increasing and that a flood control strategy needs to be implemented.

Flooding occurs for a variety of reasons, notably:

Natural factors

- Seasonal snowmelt from the Himalayas is often coupled with intense summer monsoon rains.
- Monsoon rains falling on very steep slopes, especially in the north and east of the country, generate rapid and abundant surface runoff.
- **Storm surges** occur when tropical cyclones move up from the Bay of Bengal. The exceptionally low atmospheric pressure produces a local rise in sea level which, if coupled with a high tide and heavy rainfall, can produce devastating coastal flooding. An estimated 130 000 people died as a result of such an event in April 1991.

Human factors

- Clearance of forests by hill-dwellers in India, Nepal and Bhutan has traditionally been blamed for the floods suffered by Bangladesh. Deforestation is known to increase surface runoff and accelerate soil erosion, clogging river channels with sediment and reducing their efficiency. However, many now believe that events in the far-off hills have little impact on Bangladesh. There are even doubts that widespread forest clearance has occurred. This highlights the difficulty of fully understanding a problem in a region of the world where data are scarce.
- In many places, attempts to control flooding have actually worsened the problem. The construction of embankments to contain floodwaters, for example, may prevent water from draining back into river channels, resulting in channel congestion and flooding.
- Some coastal areas have been deliberately flooded with brackish water for commercial shrimp farming. This has reduced the area available for rice and livestock farming and encouraged groundwater salinisation.

Taming the flood

Devastating flooding in 1987 and 1988 provoked international interest in flood control in Bangladesh. By 1989, a five-year Flood Action Plan (FAP), co-ordinated by the World Bank, was approved by the Bangladesh government. The plan included the following components:

- strengthening the Brahmaputra right (west) embankment in the north of the country
- a cyclone protection study in coastal areas
- flood control and river bank protection for Dhaka city (the capital) and other urban areas
- flood forecasting and warning
- education in disaster preparedness.

The FAP has been criticised for not involving the participation of the local people. With any type of scheme involving social as well as environmental changes, it is essential that local people do not feel alienated from the decision-making process. In many ways, the native dwellers understand the natural processes of the delta better than any western planner, and they have adapted their lifestyles to their environment through many generations. Low-intensity flooding must be maintained, for without it, traditional production systems would be lost.

This last point raises the issue of adapting to, rather than controlling, floods. Although the FAP does include measures for forecasting, warning and education in coping with floods, many feel that there is an over-emphasis on large-scale engineering solutions. For a scheme to be truly successful, the floodplain must be viewed as a complex, dynamic environment, both as a wetland of international ecological importance and as home to many people who manage the land and water for their living. Large-scale flood control measures will not help communities to cope with the rigours of the floodplain and are only likely to alter the sensitive ecology of the area.

SUMMARY ASSIGNMENT

1. In Britain, the rain falls mainly in the north and west but the largest centres of population are in the south-east. Explain
   (a) why the pattern of rainfall occurs
   (b) the strategies employed to ensure that the population of the south-east receives adequate supplies of clean drinking water.

2. The figure below shows a breakdown of how water is used in England and Wales.

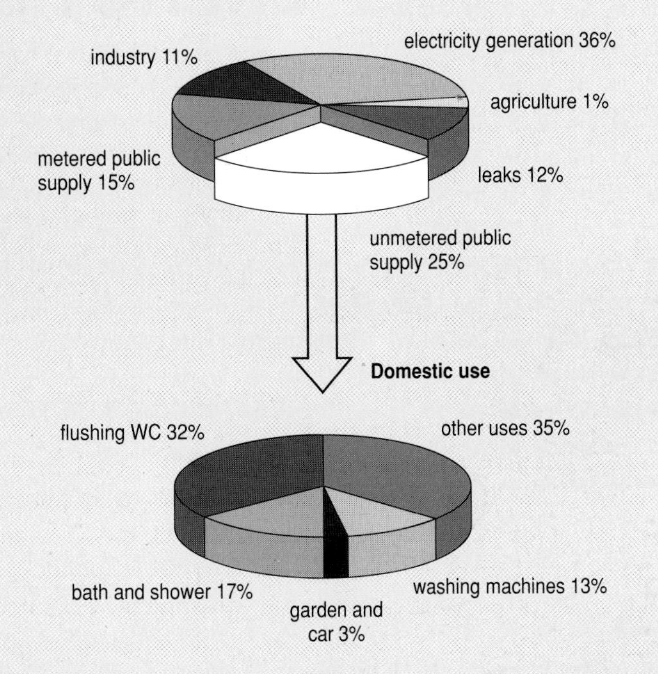

   (a) Describe **two** major ways in which water is used in a coal-fired power station.
   (b) Give **one** advantage and **one** disadvantage of installing water meters in houses.
   (c) The table below gives figures for water use and population in England and Wales for 1971 and predicted values for 2001.

	1971	2001	growth rate per annum
population (10^6)	48.6	57	0.53%
total water consumption (10^6 m^3 per day)	42.7	83.3	2.3%

   (i) Calculate the expected water use per day per person in 2001.
   (ii) Give **one** reason why water demand is growing faster than population.

   (d) Due to the building of new reservoirs and the decline in heavy industry, there is now a surplus of water for the north-west of England. The south-east of England is predicted to have a shortfall by the year 2020.
   (i) Give **two** reasons why a shortage is expected in the south-east of England.
   (ii) Describe **two** methods by which this shortage could be met.

NEAB June 1995

Chapter 12

ENERGY RESOURCES

Energy plays a central role in many aspects of modern society. The use of energy has enabled developed economies to attain a high standard of living for much of their population, and developing countries are keen to follow suit. However, every energy type has an impact on the environment; for example, acid rain (Chapter 10) presents serious international problems, and the enhanced greenhouse effect (Chapter 1) threatens to disrupt world climate. Much of the developed world relies on finite supplies of fossil fuel which, by definition, run out.

This chapter will consider the challenges which we face in sustainably meeting our energy needs.

LEARNING OBJECTIVES

After completing the work in this chapter you will be able to:

1. describe energy use in developed and developing countries
2. compare the advantages and disadvantages of renewable and non-renewable energy
3. describe the pollution acts and organisations involved
4. explain future energy demands and the need for energy conservation measures.

12.1 ENERGY SOURCES

All of the energy types available to humans can be associated with one of the following:

- **solar activity**, which powers the wind and the water cycle, is converted into biomass via photosynthesis and is exploited directly for heat and light (Chapter 1)
- **the gravitational pull of the moon and the sun**, which causes the tides
- **nuclear power**, which releases the energy within radioactive atoms
- the **earth's internal heat**, which is exploited as geothermal power.

Energy can be classified into renewable and non-renewable forms (Table 12.1). **Renewable energy** supplies are those which cannot be used up, whereas **non-renewable energy**, once used, is depleted. Most energy used in developed countries comes from non-renewable sources. However, there is a great need for the use of renewable forms to increase because non-renewable resources are effectively finite (they are being used at a much faster rate than they are produced). The environmental effects of generating non-renewable energy are considered to be more harmful than those of generating renewable energy. At some point in the future all of our energy demands will have to come from renewable forms of energy.

Table 12.1 Sources of non-renewable/renewable forms of energy

Renewable	Non-renewable
biofuels	coal
wind	oil
photo-voltaic cells	natural gas
passive heat and light	synfuels
ocean thermal	peat
wave	
hydro	

QUESTIONS

12.1 **(a)** Explain how the energy transferred from coal-powered electricity-generating plants originally came from the sun.

(b) Name the two processes involved in the origin of hydro-electric power.

12.2 Suggest why most energy used in developed countries comes from non-renewable sources.

Fossil fuels

Fossil fuels include coal, oil, natural gas, synthetic fuels (for example gas from coal) and peat. Oil, natural gas and other liquid fossil fuels are jointly grouped together as **petroleum**. The energy released from burning fossil fuels is obtained from solar radiation that has been converted into biomass via photosynthesis (Chapter 5) and then stored in fossil form.

Coal

Coal is a sedimentary rock (Chapter 3) which forms from the slow transformation of plant organic matter (Figure 12.1). Much of the UK's coal, such as that in the South Wales coalfield, was originally deposited over 280 million years ago in the Carboniferous geological period.

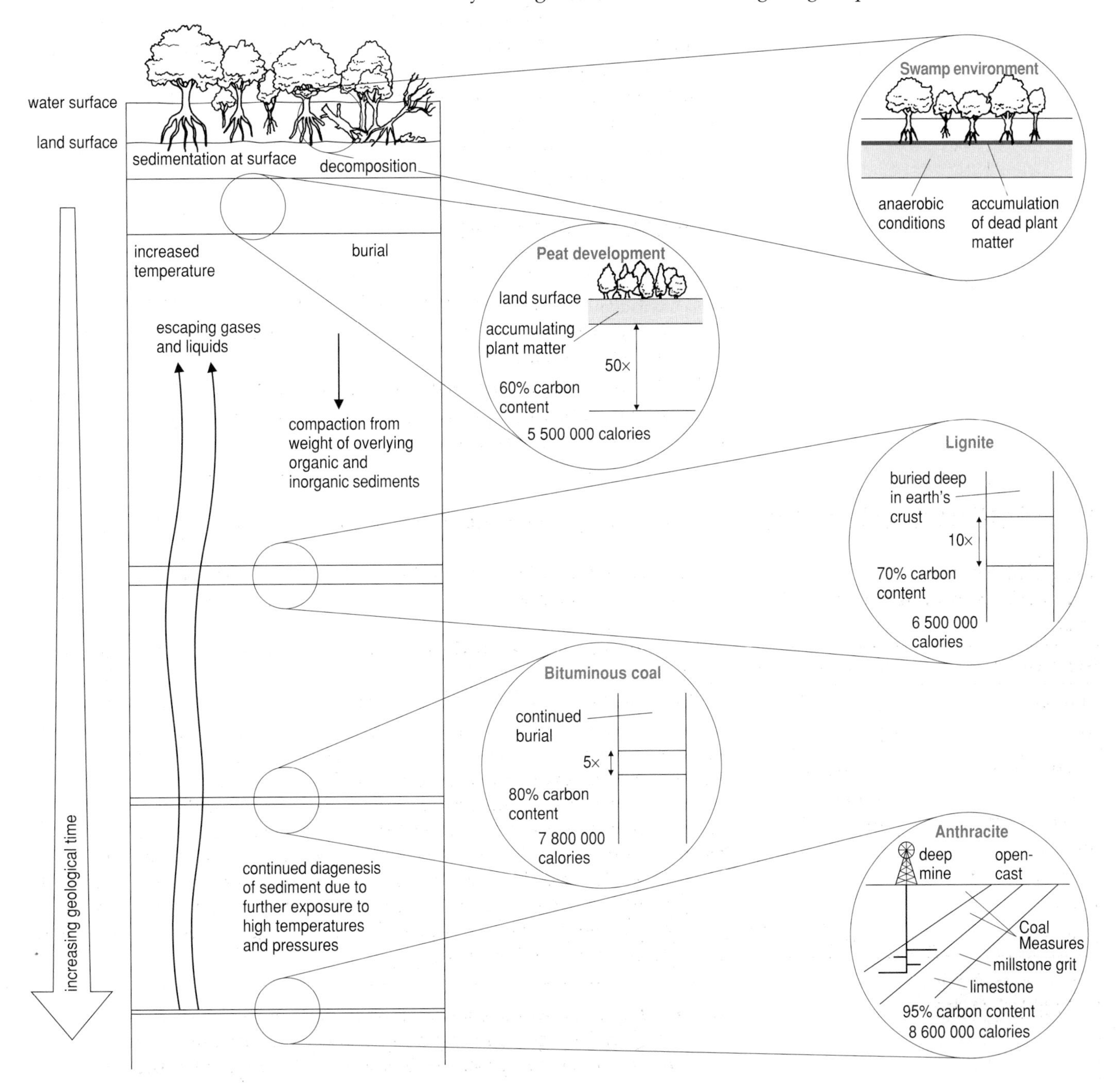

Fig 12.1 The formation of coal.
Coal forms through the process of coalification. There are three types of coal that are classified by carbon concentration (lignite, bituminous and anthracite). As carbon concentration increases, so too does the calorific value – the amount of energy that can be released by burning a kilogram of coal.

The starting point for coalification, the process by which coal is formed, is the accumulation of dead plant matter, which is then subjected to bacterial decay under anaerobic (oxygen-free) conditions. The anaerobic conditions of coastal swamps such as Okefeenokee in Georgia, USA, for example, ensure that bacterial decay is incomplete. The carbon (C) concentration of the material increases as the more volatile components such as oxygen (O_2), hydrogen (H_2) and methane (CH_4) are released.

The decaying matter is subsequently buried by more plant material and eventually by inorganic sediment. Burial results in metamorphosis because of the higher pressure and temperature. Throughout the burial, the carbon content is gradually increased by further removal of other components, so that the remaining matter is a mix of carbon, clay minerals, carbonates, salt and iron pyrites.

Oil

Oil is a hydrocarbon, that is, it is a compound of hydrogen and carbon. The oil extracted from the earth is called 'crude oil' and needs to be processed before it can be used because it contains impurities such as nitrogen, oxygen, sulphur and some trace elements. Oil is formed under ocean and sea beds from the decomposition of microscopic marine life called phytoplankton. This process is called **maturation** and is similar to coalification in that it results in a high carbon concentration and occurs with increased temperature and pressure over a geological time-scale. Most oil formations exploited today are 60 to 100 million years old. They were derived in the late Tertiary and Cretaceous periods.

Oil migrates towards the surface from a **source rock** to a **reservoir rock**, where it is contained by an impermeable layer called the **cap rock**.

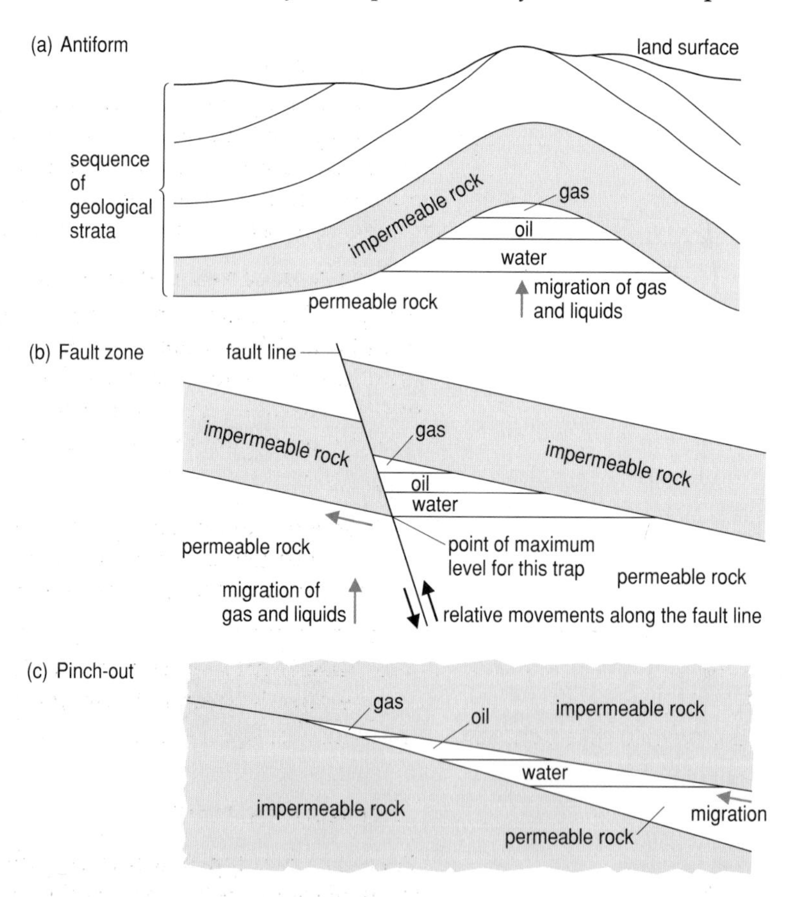

Fig 12.2 Oil and gas traps.
There are various types of cap rocks. Common examples are antiforms (an upward curving, arch-like geological structure), faults and pinch-outs (a gradually narrowing sedimentary structure).

The world's oil reserves are very unevenly distributed, resulting in a **production-consumption imbalance**. As with many other natural resources, oil is produced and exported from countries that use little of it and, in general, the countries that consume the oil do not have large reserves (Figure 12.3).

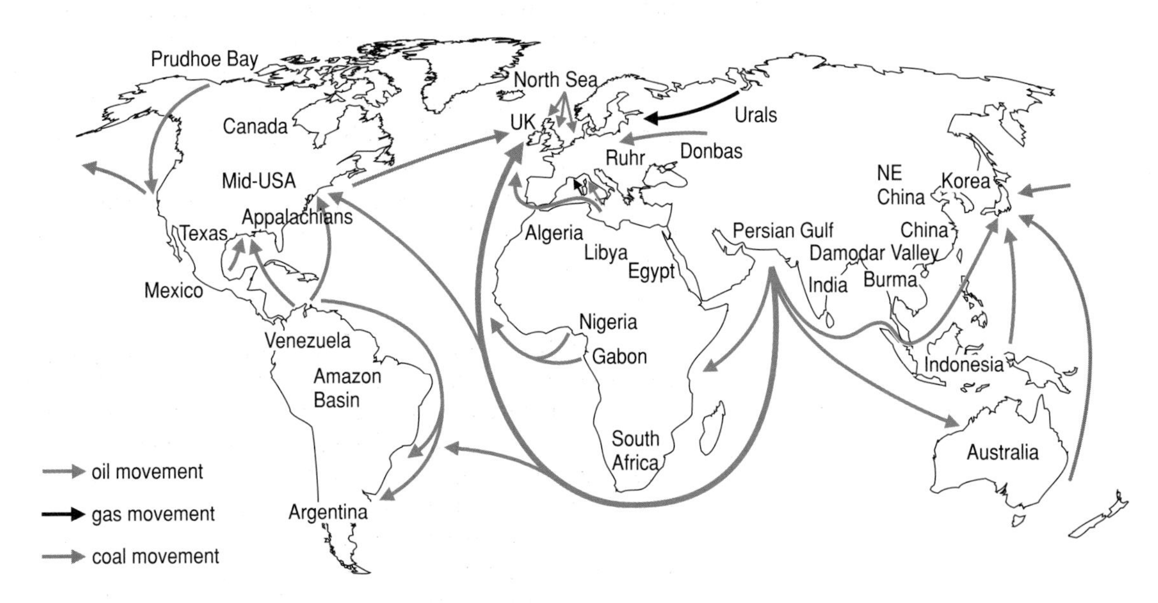

Fig 12.3 Location and movement of world fossil fuels.
Some major areas of oil exportation are the Middle East, West Africa, Alaska, Canada and South America. However, Japan, Western Europe and America are the main consumers.

Natural gas

Natural gas is produced either through coalification (called 'dry natural gas') or through maturation (called 'wet natural gas'). Dry natural gas sources are preferred because the more economically vital oil deposits are often affected during exploitation of wet natural gas. The same geological formations that trap oil are associated with gas reserves (Figure 12.3).

Other sources of petroleum

It is possible, though not often economically feasible, to extract petroleum from **tar** and **bituminous sands** which develop from the evaporation of oil near the surface (for example Athabasca in Alberta, Canada, where an average of 45 000 barrels a day is produced). Although inefficient, it is also possible to produce petroleum from coal (**synfuels**) to avoid dependence on imports.

Peat

Peat develops at an early stage of coalification (Figure 12.1). 'Brown coal' is used extensively as a fuel in Eastern Europe. Peat is still widely used as a domestic fuel and to generate electricity in Ireland.

QUESTION

12.3 Most fossil fuels contain sulphur. Explain why it is preferable to remove this before burning the fuels.

Nuclear power

There are two types of nuclear power: fusion and fission. **Nuclear fusion** takes place in the sun and is the process whereby two atoms literally fuse together under extremely high temperatures to form a single atom (Figure 12.4 a). Extreme temperatures (100 million °C) are needed to overcome the repelling force between the two atoms.

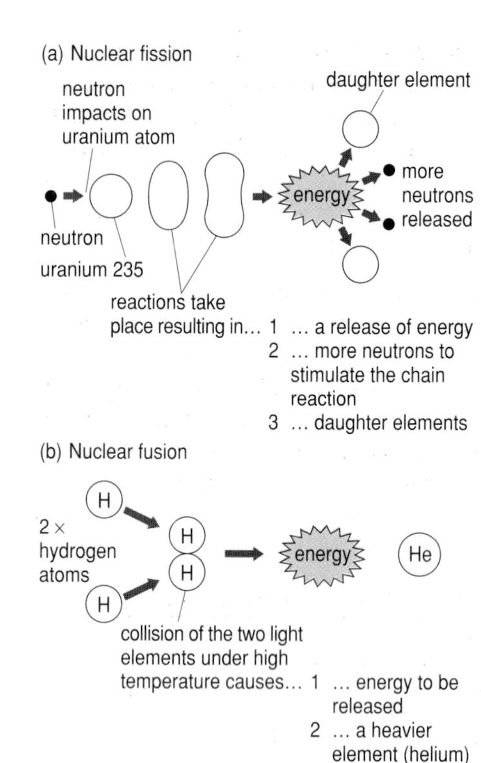

Fig 12.4 Nuclear power.

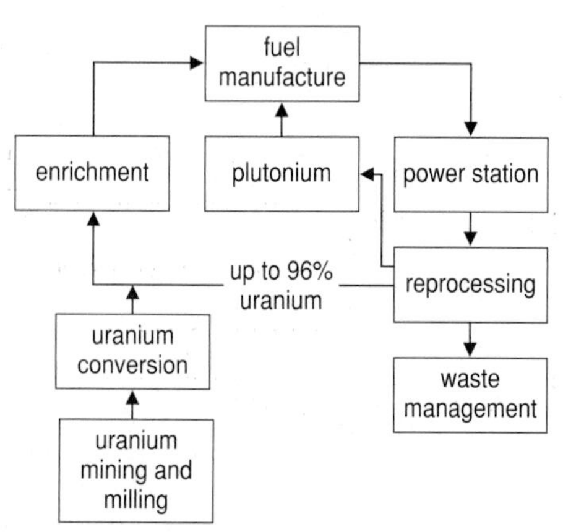

Fig 12.5 Nuclear fuel cycle.

Nuclear fusion is an extremely attractive potential energy source because it releases huge amounts of energy (Chapter 1). However, it is difficult and expensive to replicate the solar conditions on earth, as the high temperatures required are extremely difficult to contain. European research funds and expertise have been coordinated into the Joint European Torus (JET) fusion experiment, in Culham, Oxfordshire, but a breakthrough still appears remote.

We use **nuclear fission** to produce electricity. Uranium atoms are bombarded with neutrons. The nuclei of the uranium atoms absorb the neutron, and this causes the uranium atom to split, releasing energy and more neutrons, which then split more uranium atoms, and so on. This ongoing process is called the **nuclear fission chain reaction**.

Natural uranium is made up of two **isotopes** – different forms of an element which contain the same numbers of protons and electrons but different numbers of neutrons. The two isotopes are uranium 238 – which makes up 99.3% of the uranium – and uranium 235, which makes up the other 0.7%. The result of fission of these two isotopes is shown:

$$^{235}U + \text{neutron} \rightarrow {}^{236}U \rightarrow {}^{95}Y + {}^{139}I + \text{neutron} + 3.2 \times 10^{-11}\ J$$

(rapid breakdown) yttrium iodine

$$^{238}U + \text{neutron} \rightarrow {}^{239}U \rightarrow {}^{239}Pu \rightarrow {}^{100}Zr + {}^{137}Xe + \text{neutron} + 3.4 \times 10^{-11} J$$

(rapid breakdown) plutonium zirconium xenon

Yttrium, iodine, plutonium, zirconium and xenon are all examples of fission products which are highly radioactive and which decay, releasing energy and more neutrons as they do so. Fast breeder nuclear reactors are able to use the plutonium produced, effectively increasing the efficiency of the process, but it should also be noted that plutonium can also be used to produce nuclear weapons.

Spent nuclear fuel consists of 96% unused uranium 235, 1% plutonium 239 and 3% waste, so there is a lot of potentially useful uranium going to waste if it is disposed of. The unused uranium can be recovered and then reused by **reprocessing** the spent fuel. This reduces the amount of uranium that has to be mined, as well as decreasing the amount of spent fuel which requires expensive disposal. Reprocessing plants, such as the Thorp plant at Sellafield, convert low-grade, used fuel to higher-grade, reusable fuel.

There are considerable reserves of uranium that could supply world demand for the conceivable future (Chapter 15). The largest single reserve is in the United States, but other major reserves exist in Canada and Australia. There are no uranium reserves in the UK.

The **nuclear fuel cycle** involves the manufacture of the fuel (mining and processing), energy production, the reprocessing of the spent fuel and the recycling of the reprocessed fuel (Figure 12.5).

Electricity is generated by harnessing the heat produced from the fission reaction to create steam, which is used to turn a turbine attached to a generator. Within the plant, a graphite moderator encases the fuel rods which slows down the neutrons so that they can be absorbed, and the reactor is surrounded by masses of concrete which prevent harmful radiation from escaping. There are several designs of nuclear plants:

- **Magnox reactors** (so named because a magnesium alloy is used to encase the fuel) were the first to be used in this country (for example Hinkley Point, Somerset; Trawsfynydd, Merionethshire; Calder Hall, near Windscale).
- **Advanced Gas-cooled Reactors** (AGR) (for example Dungeness B) were built because the magnox reactors proved to be very expensive to build and not as efficient as first imagined.
- **Pressurised Water Reactors** (PWR) (for example Sizewell B) have proved to be the most economically successful.
- **Fast breeder reactors** are so called because the reaction is sustained by 'fast neutrons', that is, neutrons that have retained most of their energy

since being produced in the fission process. These reactors are able to use other radioisotopes such as plutonium and thorium as well as uranium, thus more effectively using reprocessed spent fuel. They manage to produce more fuel than they consume by surrounding the core with material which can absorb escaping neutrons and hence become suitable for use as fuel itself. As yet, no commercial fast breeders are in operation, although some prototype plants have been built.

There are over 434 nuclear power plants in operation today, although this figure will drop as old plants are decommissioned as they reach the end of their useful lives (around 40 years). Despite the fact that large uranium reserves remain unused, nuclear power is in decline; old, decommissioned plants are not being replaced. Sweden, for example, has agreed that it will not use nuclear power after 2010, and in the United States, no new stations have been ordered since 1979. The advantages and disadvantages of nuclear power are summarised in Table 12.2. Pollution from nuclear power stations is discussed in Section 12.4.

Table 12.2 The advantages and disadvantages of nuclear power

Advantages	Disadvantages
less CO_2 emissions than fossil-fuel power stations	environmental consequences of radiation
long lifetime of uranium as a resource	linked with nuclear weapons
	high start-up costs (in the last 30 years, over \$200 000 million in start-up costs worldwide)

QUESTIONS

12.4 Explain the benefits of reprocessing uranium in the nuclear fuel cycle.

12.5 Why could the fact that the fast breeder reactor produces more fuel than it consumes become a problem?

12.6 Nuclear fusion involves the combination of hydrogen atoms to form helium. Suggest two advantages this would have over nuclear fission as a source of energy.

12.7 The diagram shows the essential features of an advanced gas-cooled nuclear reactor.

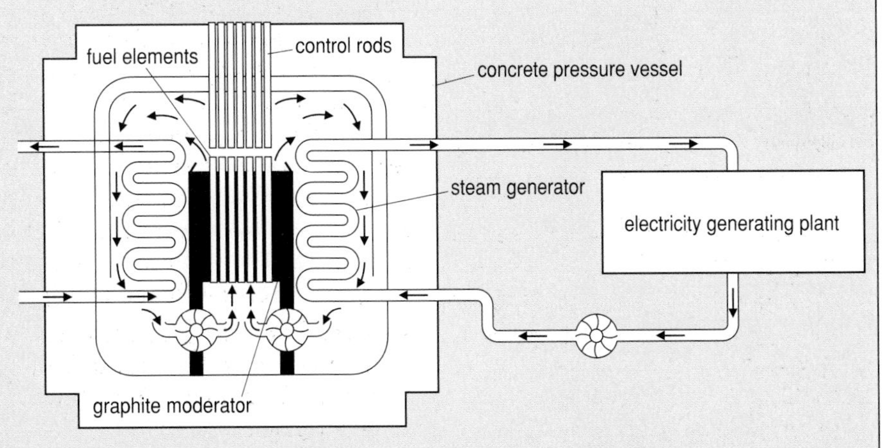

J. F. Allen, *Energy Resources for a Changing World* (CUP) 1992

(a) (i) Name a suitable fuel for this reactor.

(ii) What is the function of the graphite moderator?

(b) What is meant by *nuclear fuel reprocessing*?

AEB June 1995

12.8 The figure shows the range of products that can be produced from coal.

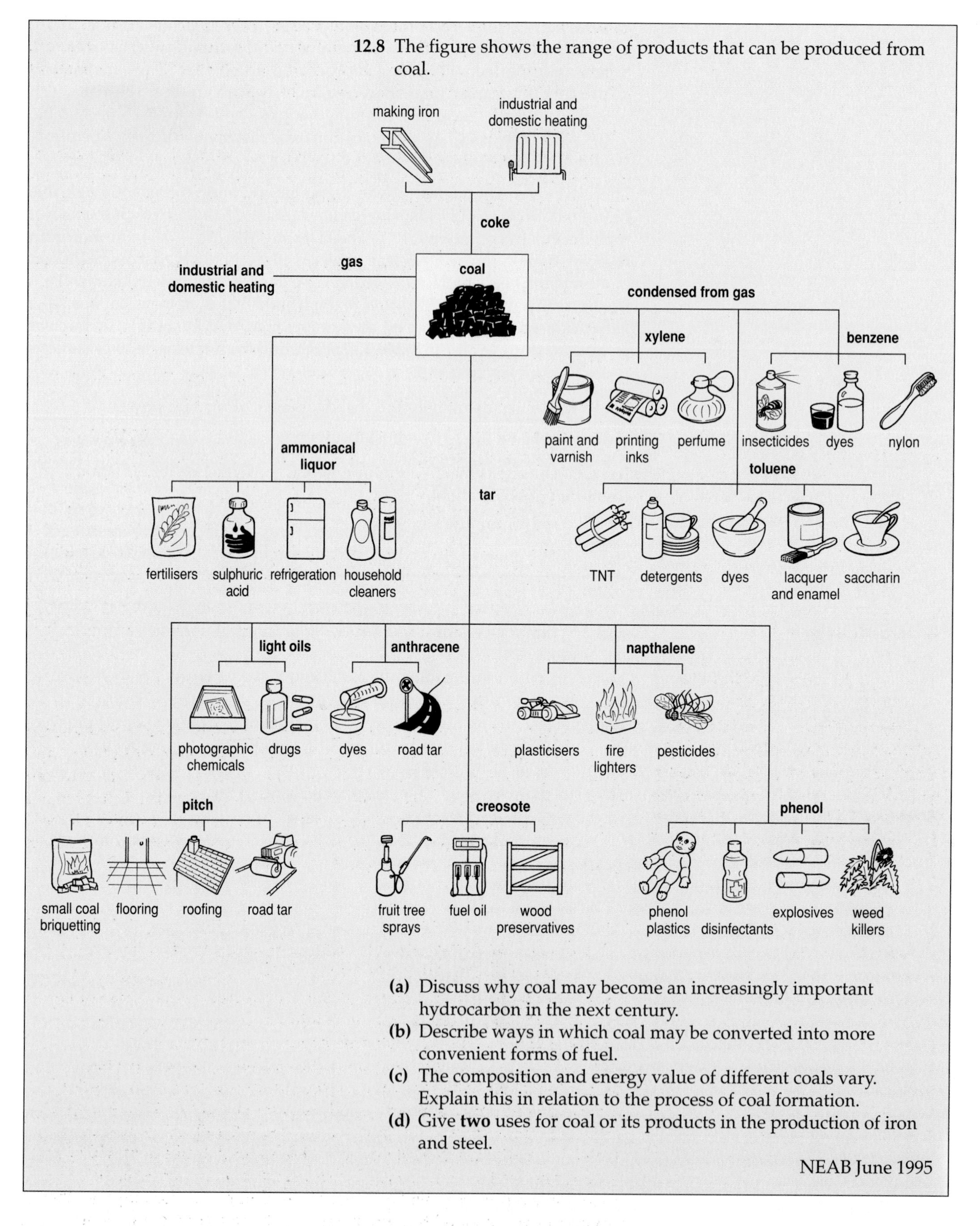

(a) Discuss why coal may become an increasingly important hydrocarbon in the next century.

(b) Describe ways in which coal may be converted into more convenient forms of fuel.

(c) The composition and energy value of different coals vary. Explain this in relation to the process of coal formation.

(d) Give **two** uses for coal or its products in the production of iron and steel.

NEAB June 1995

Renewable forms of energy

Recent technological improvements have enabled renewable technologies to become economically competitive with conventional fossil fuel energy generation. Despite this, renewable energy still plays a minor role in generating electricity in the energy-intensive developed world. There are numerous types of renewable energy, many of which will play important roles in an integrated energy strategy for the future.

Photo-thermal

Solar radiation can be used directly to heat water which is usually used for heating. Mirrors are used to concentrate the sun's radiation onto a water-filled, coiled black pipe. The water within the pipe can heat up to 60°C.

Passive solar energy

Passive solar techniques use buildings to capture, store and distribute solar energy. A building which is to be heated passively by solar energy may, for example, have large south-facing windows with small north-facing windows. The technology for this is well established but is not as yet incorporated into UK building regulations. Several examples of passive solar design already exist in the UK, such as South Staffordshire Water Company Headquarters and 'Pennylands' estate in Milton Keynes.

QUESTION

12.9 To answer this question you will also need to read Chapter 3.

(a) Explain why the buildings in Pennylands Estate at Milton Keynes have large south-facing windows but small north-facing windows.

(b) Suggest which household uses of hot water could be generated by photo-thermal techniques and state the advantages of doing this.

Fig 12.6 A photo-voltaic roof array on a solar energy house in Denmark.

Photo-voltaic conversion

Photo-voltaic cells convert solar radiation into electricity via chemical energy. They are based on silica, which is the most common element in the earth's crust, and are silent because they have no moving parts. The greatest potential for solar power is in the tropical developing world, but there is a large potential for use in the UK. In the UK, solar cells are currently only used in isolation and do not contribute to the **national grid**, the electricity distribution network which covers the country.

CASE STUDY

The David Bellamy Study Centre, at the Centre for Alternative Technology, Machynlleth

This ambitious £1 000 000 project is set to be the forerunner of its kind in the UK, involving photo-voltaic generation of electricity and many energy-saving techniques, including energy storage in batteries, although this is generally inefficient. The building has a passive solar design to maximise the heat, and the light from the sun will save 87% of the energy bill from a similarly-sized conventional building. The largest integrated photo-voltaic roof array in the UK (100 m^2) will produce 17 000 kWh (kilowatt hours) per year. Photo-thermal collectors will be used for hot water. Only materials that used minimum amounts of energy when produced will be used. The building will also employ other environmentally sound techniques such as the collection of rain water for use in toilets, natural sewage disposal in reed beds and the use of sustainably produced timber (Chapter 13).

Biofuels (biomass and biogas)

Biofuels can be defined as crops or organic wastes that are used for fuel. The chemical energy produced via photosynthesis and stored in green plants can be burnt in solid form, roasted into charcoal, and then burnt or converted into ethanol or biogas (through similar anaerobic conditions that are necessary for coalification). Wood is the world's oldest and most common source of energy. It is a special form of renewable energy because it is potentially exhaustible if not managed correctly.

Energy crops are those that are grown specifically for energy generation. Planting willow (*Salix* sp.), alder (*Alnus* sp.) and other fast growing, high carbon content plants is a useful way of using surplus farm land (Chapter 14) because it may improve the local habitat and it provides a stored, renewable energy source that can be used when necessary. About 40 ha of willow provides enough electricity for 5000 houses. Burning the crop releases carbon dioxide, but the growing plants absorb this gas during photosynthesis, hence energy generation by this method may have little net effect on atmospheric carbon dioxide levels (Chapter 1).

Waste can also be burned to generate energy. **Straw**, for example, supplies 1.5% of Denmark's energy needs. Another possible source is **wood waste** (sawdust and cut-offs), of which only one-third of the UK's total is currently used.

Biogas is created from the decomposition of organic matter, usually in landfill sites (Chapter 16). Methane produced in this way is often wastefully released to the atmosphere to prevent explosion, but adds to the greenhouse effect.

Wave power

Although still at the research stage, there are several techniques by which the kinetic energy of a wave can be utilised, both on-shore and off-shore.

There is a great potential for wave power in the UK, where it is estimated that up to 33% of our electricity could come from this source. Off-shore sites would create difficulties in transmission of the energy to the mainland but would produce regular electricity and have less environmental impact than, for example, tidal generation.

Waste to energy

It is possible to generate electricity by incinerating domestic waste, which has a high calorific value because of the plastics incorporated within it. Burning this waste reduces pressure on landfill sites but does release pollutants into the atmosphere (Chapter 16).

Tidal power

The tides are controlled by the gravitational forces from the sun and moon, the geomorphology of ocean basins, and the Coriolis force. The tidal cycle is very predictable, with a frequency of approximately 12.5 hours, and is therefore a very reliable energy source. However, the tides do not always coincide with the peak demand for electricity, and so electricity will sometimes be generated in the middle of the night when the demand is low. A tidal barrage is a barrier constructed across an estuary containing a series of sluice gates and turbines at the base of the barrier. The sluice gates allow the flow of water through the barrage to be controlled.

The tides produce a difference in water levels on each side of the barrage and, once released, this water drives the turbines which produce electricity. The difference in levels – the tidal range – is crucial. The energy that can be generated is proportional to the difference, but too large a difference creates operating difficulties. The optimum range for efficient power generation is between 5 and 15 metres. A very good tidal site is one where there is a high tidal range and one which is close to an urban centre. The Severn estuary is one of the best sites in the world.

There are three ways in which energy generation can occur:

- **Single-action outflow (ebb) generation** occurs when the tide is leaving the estuary.
- **Flood generation**, when the estuary is filling, is not as efficient as ebb generation but is useful when combined with a neighbouring ebb generation station to provide a better generation base.
- **Two-way generation** combines ebb and flood generation.

In addition, a storage impoundment located in the middle of the estuary may be used to generate electricity when the normal tidal cycle is 'out of peak'. Pumping to raise water levels can also occur, which, despite using power, can increase output by up to 10%.

At present, electricity is not generated from tidal power in this country, even though the resource is estimated at 105 Twh (terawatt hours) per year, which is 20% of the UK's current electricity demand and nearly 50% of the total European tidal resource. There are very few tidal power stations in the world. The 20-year-old La Rance tidal plant can use either a single action outflow or two-way generation, with or without pumping depending on the grid requirements. The most likely locations in this country are small-scale sites on the Conwy, Wyre and Duddon estuaries and larger-scale plants on the Severn and the Mersey estuaries.

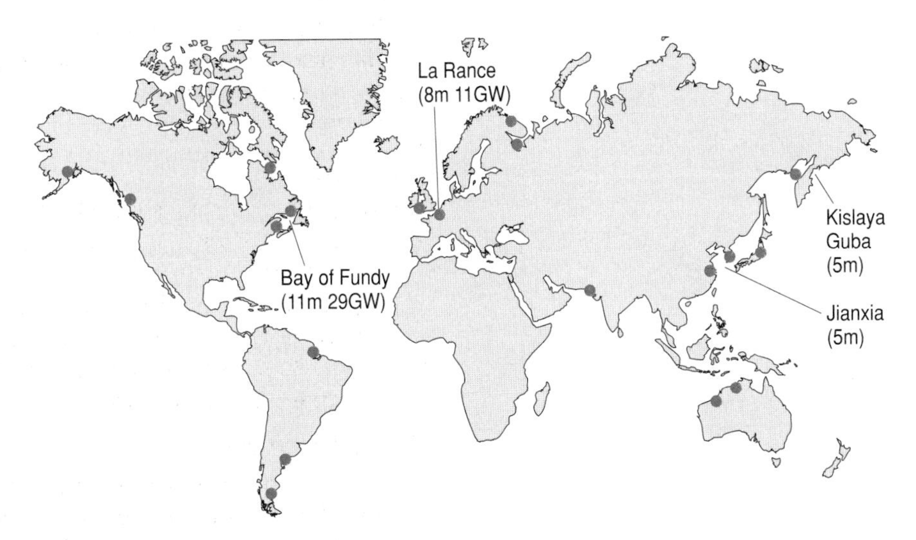

Fig 12.7 World's tidal barrages (the four largest are named).

Ocean Thermal Energy Conversion (OTEC)

Within 30° latitude of the equator, a permanent thermocline (zone of uniform temperature) exists in the ocean with a difference of 20°C between the surface (25°C) and lower waters (5°C). The heat of the surface water can be used to evaporate low boiling point liquids, such as chlorofluorocarbons (CFC) and ammonia (NH_3). The resulting vapour can then be used to drive turbines before being condensed with the cooler, lower waters. The enclosed system would be on a very large scale, covering around 600 m vertical water depth and a diameter of 20 m.

OTEC is inefficient because the speed of the vapour through the turbines is low and energy is used to pump cold water towards the surface. So far, only pilot plants have been built.

QUESTION

12.10 (a) Explain the purpose of (i) pumping and (ii) a secondary impoundment in tidal power generation.

(b) Describe the arguments against utilising ocean thermal energy.

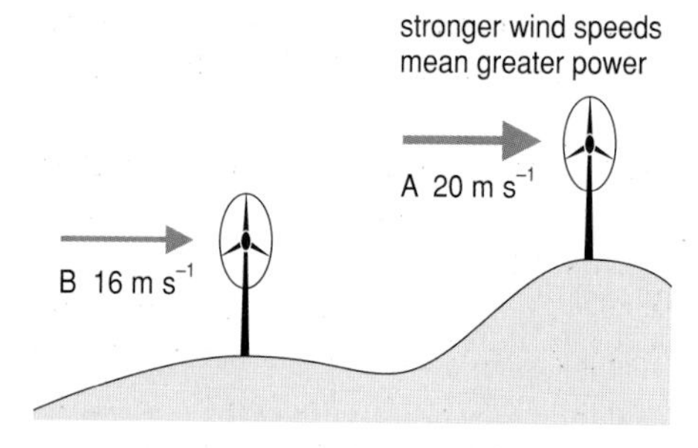

Fig 12.8 The effect of altitude on power output. The best wind sites are at an altitude where the power output is increased four-fold by an increase of twice the altitude.

Fig 12.9 Wind farms in England and Wales, 1993.

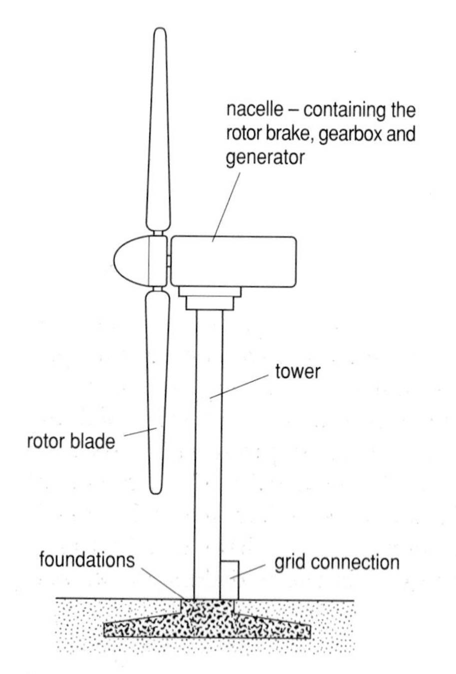

Fig 12.10 A typical wind turbine.

Hydro-power

Unlike the other forms of renewable energy, hydro-electric schemes already account for a significant percentage of the world's primary energy consumption (6.7%).

Hydro-electricity accounts for around 2% of the UK's electricity, which comes mostly from Scotland. The largest scheme is the Dinorwig hydro-electric pumped storage scheme, near Ffestiniog in Snowdonia, Wales. This scheme, opened in 1984, was built into constructed caves in Elidir mountain and involved excavating 3 million tonnes of rock from 16 km of tunnels. The system works by storing water in a lake at altitude which can be released to turn turbines when electricity is required. The water is pumped back to the top of the mountain from a second reservoir at the bottom when demand is low, using off-peak nuclear generated electricity.

The Energy Technology Support Unit (ETSU) believes that there is around 700 MW of small-scale, unexploited hydro-power schemes in the UK. Small-scale systems are useful for supplying local demand in areas away from the national grid.

Wind power

Windmills have been used for centuries in this country for grinding corn or pumping water in the Fenlands. Wind power can consist of a stand-alone turbine (Figure 12.10) or as a wind farm that generates electricity for the national grid (Figure 12.9).

Whereas many renewable energies are still developing, the technology for wind power is already proven. Electricity is generated by blades which turn a gear mechanism attached to a generator. Although large blades are more efficient, they have a greater visual impact. The power output is also directly proportional to the cube of the wind velocity. To be economical, wind farms are restricted to areas where the average annual wind speed is over 5.5m s^{-1}. Wind power is not as efficient as other types of energy generators. Because turbines have to be spaced apart to prevent reduction in wind speed towards the middle of wind farms, large areas of land are required.

The UK has the best wind resource in Europe. It is estimated that 10% to 20% of the UK's electricity demand could be met by on-shore wind power, even if many areas are not used. The development of off-shore floating platforms would exploit a much larger resource and have a minimal effect on the environment.

On average, wind farms only operate for one-third of the time because the wind is either too fast, where there is a danger that the blades will break, or too slow, and there is no generation on cold, still winter mornings, when the power is needed most. Wind power is, therefore, an unreliable energy source which must be used in conjunction with other energy forms.

Table 12.3 Advantages and disadvantages of wind power (see also page 171)

Advantages	Disadvantages
reduced dependence on finite fossil fuels	visual intrusion, noise, and possible effects on wildlife (birds)
pollution-free	television and radio interference
injection of money to rural farmers	inefficient and unreliable
land use does not conflict with present farming operations	land use conflicts owing to the large areas of land required
	safety

Geothermal power

Geothermal power uses energy generated within the earth's core. It is the radioactive decay of the minerals such as uranium within the rocks in the crust, for example granite under Dartmoor, which produce this effect. Alternatively, in Iceland water is pumped through basaltic lava that is still cooling to achieve the same effect.

The heat which is trapped by geothermal power stations is therefore either from radioactive decay of elements *in situ* in the crust or rocks which have been erupted from a depth of 120 km at the most.

This heat can be utilised by two different methods: the hot '**aquifer**' method simply involves extracting hot water confined in an aquifer layer (Chapter 11). This can be used to heat buildings or can be converted into steam to generate electricity. In the second method, water is heated after injection into a highly fractured, hot rock (for example granite intrusions in Cornwall). The highly fractured nature of the rock enables a lot of water to be stored within it. Removal of the steam is then controlled via another borehole.

There are no plants in the UK that generate electricity directly from this method, although geothermal power has a long history in other countries, most notably in Iceland. However, there are several locations in the UK that could be developed for heat extraction from a hot aquifer source. One such scheme has been developed in Southampton City Centre, which heats several buildings and thus reduces the demand on electricity supply in the local area.

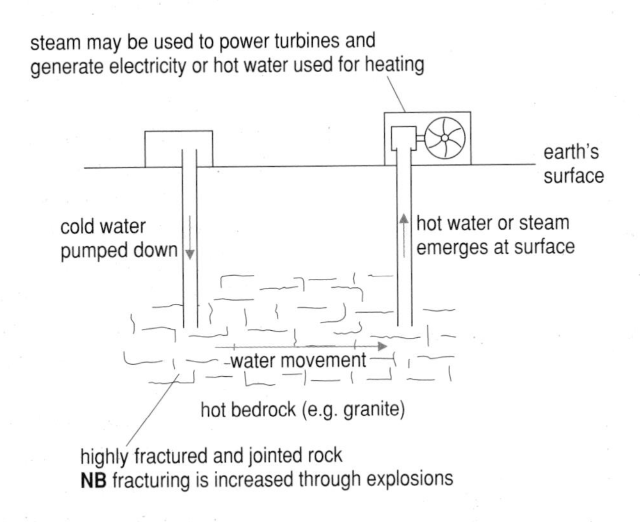

Fig 12.11 Cross-section of a 'hot rocks' geothermal plant.

12.2 ENERGY CONSUMPTION

Energy consumption per capita (per person) is much greater in developed countries than in developing countries, and such countries are reliant upon concentrated sources of fossil fuels or uranium. Yet the energy requirements of over half the world's population are still met from the combustion of dung, agricultural residues, charcoal and wood. Charcoal, produced by roasting wood in simple earth kilns, is used because it has twice the energy content of wood. The process of making charcoal reduces the carbon in wood so that it is less oxidised, that is, the carbon is in a purer form, so that when it burns, it is reoxidised and gives out more energy. It burns with less smoke and provides a more constant heat.

The use of wood is most common in low-income, rural areas of less developed countries. Wood collected from nearby trees is the only means to cook food, provide light and heat, and fuel local industries (for example brick making and some food preparation). This is one of the major causes of deforestation (Chapter 13).

The remote nature of many rural villages throughout the less developed world is most suited to small-scale, independent electricity generation (for example solar power) because of the distance that a grid network would have to cover. The people of cities in less developed countries either buy wood or bottled gas or buy electricity if the household is comparatively wealthy.

Table 12.4 Comparison of wood used for energy in selected countries

Country	Wood as a percentage of total energy consumption
Mali	97
Rwanda	96
Tanzania	94
Ethiopia	93
Kenya	74
Tunisia	42
United Kingdom	<1

12.3 FUTURE ENERGY DEMAND AND ENERGY CONSERVATION

Projected demands

Predicting future energy demand upon which policies can be based has proved to be very difficult. For example, after the Second World War, nuclear power was heralded as the ideal endless energy source. However, because of environmental problems and economic arguments, it is likely that nuclear generation in the year 2000 will only be 10% of the target figure quoted by the International Atomic Agency in 1974. World energy demand is set to increase dramatically as poorer countries develop. However, in the UK, energy demand will only increase slightly because economic growth is currently low, the industrial base is changing away from energy-intensive 'smoke stack industries', there will be further efficiency gains (for example house improvements), and metering and other controls will cause lower usage of energy.

In the early 1980s, the most important energy use was industry, followed by household use, transport, public and private sector services and, finally, agriculture. However, within the last decade, energy use has shifted so that in the 1990s transport, which is entirely dependent upon fossil fuels, is the dominant energy user (Chapter 10).

Governments can influence energy use in several ways. For example, the electricity bill of 1989 introduced the 'non-fossil fuel obligation' (NFFO). The NFFO is subsidised on fuel bills and is designed to ensure that electricity companies generate a certain percentage of electricity through non-fossil fuel sources, part of which must be from renewable sources, although the majority comes from nuclear power.

There is no single organisation responsible for energy policy. Instead, the decisions are made by several bodies within government. The Environment White Paper, *This Common Inheritance – Britain's Environmental Strategy* (1990), stated that there was a need to ensure that 'we are not undoing in one area what we are trying to do in another'. This means formulating policies which will be in harmony with one another.

QUESTION

12.11 Explain why a transport policy based on more roads contradicts current thinking to reduce dependence on fossil fuels.

Table 12.5 The cost of generating electricity

Energy type	Cost/pence per kilowatt hour at 1991 prices
coal	low 3.5, high 4.0
gas	low 2.3, high 2.8
nuclear	low 5.0, high 7.5
wind on-shore	low 2.9, high 5.2
wind off-shore	approx 8.0
hydro-power	6.0
waste to energy	6.55
landfill gas	5.7

Note: Both nuclear power and renewable energy receive a discount on price which is paid for by the NFFO.

Economics plays a very important role in determining a future energy strategy. Table 12.5 compares generating costs of various energy types. There is now no doubt that many renewable energy forms are economically competitive with the more conventional energy forms.

Conserving energy

An increase in economic growth need not result in increased energy use. Many people argue that policies should focus on energy conservation and not on new ways of producing energy. Most domestic energy expenditure is used in heating space, therefore passive solar designs and insulation (for example double glazing and cavity wall insulation) have a dramatic effect on domestic use. Tighter legislation can dramatically improve the situation. It is estimated that between 1970 and 1989 there was a 32% efficiency gain in domestic energy use based on the amount of energy delivered per household, and two-thirds of this came from insulation (89% of homes now have loft insulation). Appliances can also be made more efficient. For example, modern washing machines save on energy and water, and it is estimated that if every household in the UK converted to long-lasting, energy-efficient light bulbs, then it would save the electricity of approximately two large power stations, save the user money on the electricity bill, and prevent large volumes of waste light bulbs (that is, fewer bulbs and reusable adaptors).

Payback times

The **payback period** is the amount of time that an energy conservation measure takes to pay for itself by reducing energy bills. Cavity wall insulation may, for example, have a payback time of four to ten years, depending on the lifestyle of the user and the efficiency of the house. However, simple payback periods do not take into account rises in fuel prices, the expected lifetime of the item, interest rates, or the fact that people may not mind the increased cost owing to the environmental benefits.

QUESTION

12.12 Study Table 12.6 on payback periods.

(a) What effect would a rise in fuel prices have on the payback period for double-glazing?

Table 12.6 Payback periods

Measure	Payback period / years
draught proofing	1–6
condensing boiler	1–5
loft insulation	1–5
cavity wall-fill	4–10
double-glazing (heated living room)	6–25
double-glazing (whole house)	9–85

(b) Suggest why the payback period for double glazing is so variable.

(c) If you were asked for advice on the best two energy conservation measures for a UK householder to invest in, which two would you choose and why?

12.4 ENERGY AND POLLUTION

Nuclear power

During the 1950s, nuclear power was heralded as the ideal energy source. However, this is now questioned for a number of reasons. **Radiation** may be deliberately discharged or leaked during accidents. There are several types of radiation, each of which has a different effect. The three types of **nuclear waste** also present their own problems. The practice of dumping waste at sea was outlawed in 1983 by the London Dumping Convention, but was strongly opposed by the British government which has since authorised the dumping of over 950 000 curies of radioactive waste into the world's oceans. **High-level** nuclear waste, such as spent fuel rods, gives off large amounts of ionizing radiation. No methods for disposal of high-level waste are guaranteed safe and permanent. **Intermediate-level** waste includes irradiated fuel cladding and reactor components, and **low-level** waste includes items such as discarded protective clothing.

CASE STUDY

Chernobyl

The accident at Chernobyl, in the Ukraine, a former republic of the former USSR, was caused by human error. The scientists in charge switched off the safety measures in order to carry out an unauthorised experiment. The pressure that built up as a result of this experiment simply became too great, and at 1.23 a.m. on Saturday 26 April 1986 two near-simultaneous explosions from Unit 4 of the plant split the building apart. The Russian authorities only admitted the accident when Swedish scientists recorded high radiation levels over two days later. The radioactive cloud spread out from Chernobyl across Europe.

More radioactivity was released than from both Nagasaki and Hiroshima combined (the two Second World War bombings). Over 200 000 people had to be resettled, 500 000 people were irradiated and 4.8 million people were harmed in some way. The cancer deaths in the Ukraine have increased by a third since the accident. More than 50 500 square miles were contaminated with radioactive caesium^{137}. The total cost of the clean-up operation was estimated at 9 billion roubles, although probably 250 billion roubles have been spent in total, including all knock-on effects (for example the cost of generating electricity from a new source). In 1995 the Russian authorities admitted that 125 000 people had died to date as a result of the Chernobyl accident. Places as far away as Britain were affected when the radioactivity moved with the prevailing atmospheric conditions and was deposited in rain. The main impact on Britain was that farmers could not sell their goods if they were found to be highly radioactive.

Since Chernobyl, nuclear power in the former USSR has declined. The targets for 1990 were that 77 million kilowatts of energy be produced each year by the nuclear programme. In 1989, only 37.4 million kilowatts were produced, less than half the projected figure.

Fossil fuels

Combustion of fossil fuels may release carbon monoxide, carbon dioxide, sulphur dioxide, nitrogen oxides, particulates, hydrocarbons and water vapour. However, gas is far cleaner than coal, releasing only half as much carbon and over 1000 times less sulphur dioxide (SO_2) per unit of energy. The environmental consequences of such emissions are examined in detail in Chapter 10, but a summary of their effects are given in Table 12.7.

Fig 12.12 A coal-fired power station on Humberside.

Table 12.7 Air pollution from fossil fuels

Emission	Effect
carbon dioxide (CO_2)	contributes to acid rain and greenhouse effect
sulphur dioxide (SO_2)	contributes to acid rain
nitrous oxide (N_2O)	greenhouse gas
nitrogen monoxide (NO)	oxidises to NO_2
nitrogen dioxide (NO_2)	contributes to acid rain and photochemical smog
hydrocarbons	contributes to photochemical smog
particulates (e.g. fly ash, arsenic, lead, cadmium)	may enter lungs with various harmful effects

Since all fossil fuels contain carbon, albeit in different proportions (Figure 12.13), combustion always releases carbon monoxide or carbon dioxide.

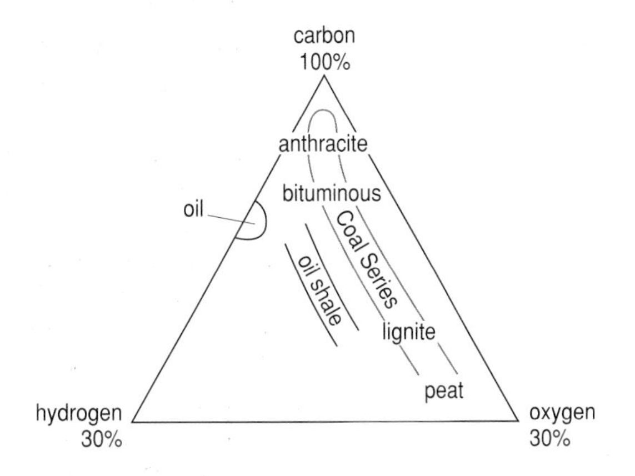

Fig 12.13 Compositional relationships between coals, oil shale and crude oil.

The impact of using fossil fuels such as coal is, in fact, very wide-ranging (Figure. 12.14).

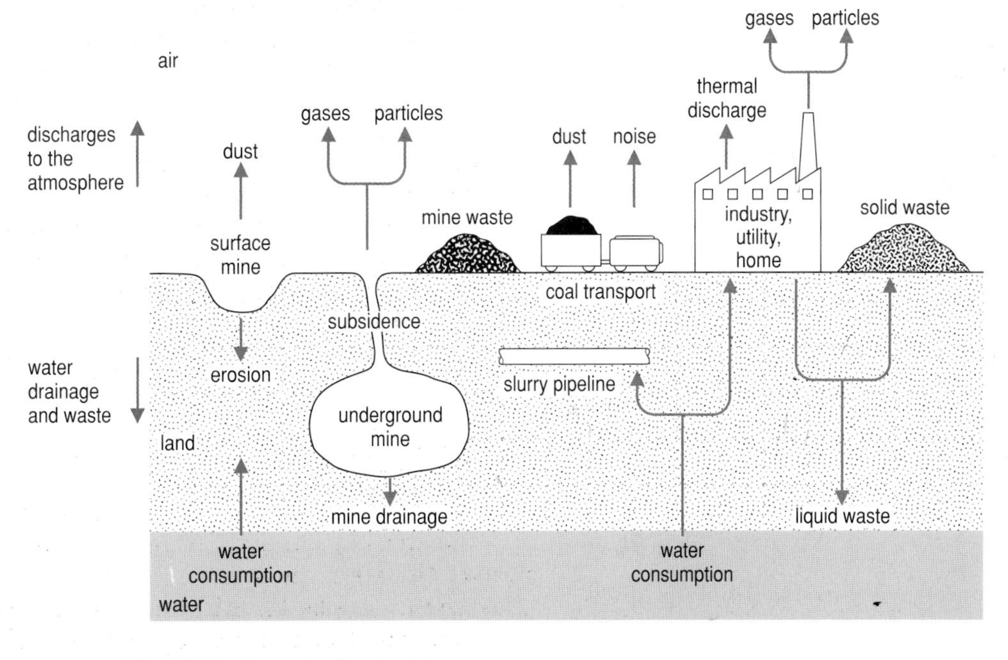

Fig 12.14 The cycle of coal use.

Fossil fuels discharge pollution into rivers, the sea and onto the land, as well as into the atmosphere. Processing coal produces huge quantities of waste spoil, which is simply dumped as **spoil heaps**.

Fly ash is very fine-sized particles produced during coal burning. Fly ash is often dumped at sea where it forms a hard crust on the sea floor, which is difficult for organisms to penetrate. If fly ash is dumped in a landfill site (Chapter 16), then it is liable to leak into groundwater because of its texture and size.

Fossil fuel and nuclear power stations are built next to large water bodies because water is required for cooling. The water is released back into the water body at a higher temperature than the surroundings. Thus **thermal pollution** induces changes in the local ecosystem.

Oil is released into the environment via extraction, processing and transportation. Highly publicised accidents do have a major impact on a localised area, but it is through regular and legal discharges that most oil enters and pollutes the environment. These discharges are necessary in order to clean out the huge ships, some of which have 500 000 tonne capacities (Chapter 10).

It is clear that conventional energy types all have major impacts upon the environment; however, the use of renewable energy also has serious consequences.

Solar power

Apart from manufacturing and disposal, solar power is totally pollution-free. There is little chance of exhausting silica resources, from which photo-voltaic cells are made, because it is the most abundant element within the earth's crust. Over its lifetime, a single photo-voltaic cell will:

- generate more electricity than an equivalent-sized piece of uranium fuel
- prevent 30 kg CO_2 being emitted into the atmosphere
- prevent 88 g SO_2 being released into the atmosphere
- prevent the use of 14 kg oil.

Tidal power

In contrast to solar power, tidal power from a tidal barrage across an estuary has an environmental impact over the entire estuary. The coastal processes, tidal levels, water quality and hydrological regime will all be altered so that the physical character and ecology of the estuary will change.

Tide levels

Behind a barrage, low-tide levels would be at the former mid-tide levels and there would be a decrease in the tidal amplitude. Extreme high or low tides will not occur. This will permanently flood mud flats where wading birds feed and will also prevent the annual flooding of upper marsh areas where birds nest and breed. The salinity of these upper marsh areas will therefore reduce and a natural succession to a more terrestrial environment will occur.

Waves

Reduced wave velocities behind the barrage may reduce erosion, but a lower tidal amplitude means that wave action is concentrated on a smaller area of cliff. The rate of erosion may increase. Lower velocities will cause increased deposition, which may make it necessary for navigation channels to be dredged.

Water quality

The higher water levels will improve all recreational activities and therefore higher standards of water quality may be required. There will be a dramatic increase in the 'flushing time', which is the time a body of water takes to move out of the estuary. This will cause an increased impact of pollution because pollutants released into the estuary will remain for longer. It is estimated that the flushing time for the Mersey estuary would increase from 30 to 40 or even 50 days if a barrage was built. Behind a barrage, salt concentrations generally decrease, which is good for the surrounding agriculture, but will change the natural ecosystems (Chapter 7).

Light penetration

The lower velocity will cause less sediment within the water column and therefore more light penetration. This will cause pollutants to be broken down more quickly and also benefit plant life through the increased rate of photosynthesis. However, it will also cause increased micro-organism growth (for example algal blooms), which can clog fish gills and, eventually, reduce the dissolved oxygen content of water.

Fish

There is concern that a barrage would act as a barrier to migratory fish. Fish could be physically harmed if they pass through the turbines. To combat this, a fish pass, used in combination with deterrents (lights and sound), could be constructed. This could help species such as salmon which swim in the top two metres of the water column.

Impacts on bird populations

- There may be a permanent loss of some inter-tidal feeding areas and nesting areas.
- Siltation will lower the numbers of invertebrate populations on which to feed.
- Increased competitiveness between bird species may cause the increase of predator-avoidance by prey.
- Eventually more mud flats will be created to replace those lost. These new mud flats will house increased amounts of food.
- Improved oxygen concentrations will allow invertebrates to colonise new areas.
- A general increase in biological productivity will result, increasing biomass and providing more food.

QUESTION

12.13 Why do some environmentalists regard tidal barrages as a mixed blessing?

Wind power

Wind power uses no fuel and produces no harmful emissions or wastes apart from those produced during manufacture and disposal. It is estimated that 15 × 400 kW turbines save 6700 tonnes of coal, 18 200 tonnes of carbon dioxide and 103 tonnes of sulphur dioxide. However, wind power, which is already part of the British countryside, has become a very controversial energy type because of numerous problems.

Noise

Wind farms generate noise from the 'swishing' of the blades and the 'hum' of the generator within the nacelle (Figure 12.16). Noise is most noticeable in light winds and at night when there is low background noise. It is possible to compare the noise with common sounds (Figure 12.15). This should be viewed with caution, however, because the problem is not simply the level of noise but the fact that the noise occurs in usually quiet areas.

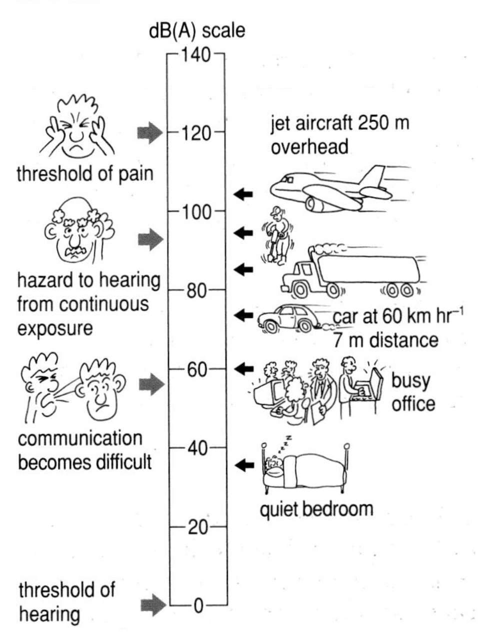

Fig 12.15 Noise values on the dB(A) scale.
A wind farm produces a constant noise, but this noise is only as loud as many familiar sounds. The problem is that this noise intrudes into usually quiet areas.

Visual appearance

Although some people think that wind farms add interest to the landscape, many others consider wind farms unattractive. The old-style windmills blended into the landscape, but the modern white towers do not. Wind farms are temporary structures which last for 30 years and can be removed when they have served their purpose. The visual intrusion of wind farms is caused by:

- the size; tall turbines generate more electricity, and therefore fewer turbines will be required to meet the demand, but tall turbines have a greater visual impact; there is a dilemma whether or not to build fewer, efficient, tall turbines or more medium-sized machines; for example, 12 × 100 m tall machines produce the same amount of electricity as 54 × 50 m medium machines
- the arrangement of turbines (rows or scattered across a site)
- the number and the area that is covered; the larger the area, the bigger the impact
- the colour; a white tower and green grid connection box are considered to be the best
- the number of blades; three-bladed machines are preferred to two-bladed
- the cumulative effect of neighbouring wind farms
- the nature of the surrounding area; summit sites can be seen from all around but are more efficient.

Fig 12.16 A wind farm.

If the target of generating 20% of the UK's electricity through on-shore wind power is achieved, it will cover an area of land 6000 km^2, which represents a major impact on the landscape. Much of the area that is suitable for wind farms is also of high landscape value (Areas of Outstanding Natural Beauty or National Parks – Chapter 18) and therefore there is a conflict of land use between energy generation and landscape protection (Chapter 17). At present, national park authorities are opposed to the siting of wind farms within or close to their boundaries.

Safety hazard

The blades can fall off, and the ice which forms on motionless blades can also fly off. If a wind farm is close to a major road, the turning blades can be a distraction to drivers.

Wildlife

The major effect on wildlife occurs during construction. The only possible long-term effect is on birds, which may be disturbed or fly into the blades. In 1988, a Royal Society for the Protection of Birds (RSPB) study concluded that a turbine on the Isle of Orkney had 'no measurable impact on birds, which seem oblivious to the large turbine blades'.

Land use

The actual land used for turbines and access roads is approxmately 1% of the total land area covered by the wind farm, and thus the land use can continue as before (for example sheep farming).

QUESTION

12.14 Produce a table which summarises the data provided, comparing the advantages and disadvantages of wind power.

Hydro-power

Dams have a major impact on the environment. Some effects are beneficial, such as providing recreational facility, but there are also many disadvantages.

Hydrological effects

Dams control river flow and therefore can prevent serious flooding and yield year-round water for irrigation schemes. They are susceptible to siltation, rendering the dam useless and causing the water table to rise (Chapter 11). The water is stationary and so reaches a higher temperature, which promotes micro-organism growth. This growth, which is also promoted by the presence of nutrients trapped behind the dam, causes an increased biological oxygen demand, which eventually means that fish and other organisms cannot compete for the available oxygen, and they perish (Chapter 10).

Pedological effects

Soils downstream are starved of the fertile silt deposits that become trapped behind the dam.

Geomorphic effects

The lower sediment content downstream of the dam means that the river is using less energy to carry material, which results in an increased rate of erosion. Earthquakes can be caused by the weight of the dam on the earth's crust (Chapter 3).

Climatological effects

Large dams may increase local humidity (Chapter 2).

Wildlife

There is a considerable amount of habitat destruction in dam construction. For example, the Balbina Dam in Brazil destroyed 2400 km^2 of virgin tropical rainforest. The dam is unpassable to migratory fish, and wildlife can be damaged by micro-organism growth.

A great effort was put into reducing the environmental impact of Dinorwig hydro-electric power station, discussed in this chapter, section 12.1. It is partly on the site of an old mine, so many people think that the view has improved. The substation was built inside the mountain to avoid visual intrusion, and 11 km of cables are underground. The reservoir at Llyn Peris, which had to be deepened, was excavated from the bottom to minimise the appearance of an obtrusive dam. Much of the waste was disposed of in old slate workings. However, some environmental impacts, such as the draining of lakes and the diversion streams, was unavoidable.

SUMMARY ASSIGNMENT

1. Construct a summary table stating two advantages and two disadvantages of each type of renewable energy.
2. List six ways in which an average household could reduce its annual energy consumption.

Chapter 13

FORESTRY

Forests cover 30% of the planet and play an essential role in maintaining atmospheric composition (Chapter 1) and climatic and nutrient cycles (Chapter 8), and provide a habitat for thousands of species of plants and animals.

> **LEARNING OBJECTIVES**
>
> After completing the work in this chapter you will be able to:
>
> 1. describe the major roles of the world's forests
> 2. describe the role of forests in economic development
> 3. explain why some management practices may harm the environment and how conflicts can be resolved
> 4. describe the special value of and threats to tropical forests
> 5. explain the significance of good forest management for soil and water quality
> 6. describe the new initiatives in community forestry in the UK.

13.1 THE PURPOSE OF FORESTS

The major roles of the world's forests are as follows:

Atmospheric regulation

During photosynthesis, trees absorb carbon dioxide and release oxygen (Chapter 5), thereby helping to regulate the atmospheric composition of these two gases. Carbon dioxide is a greenhouse gas and afforestation – the planting of trees – can help to reduce the greenhouse effect (Chapter 1).

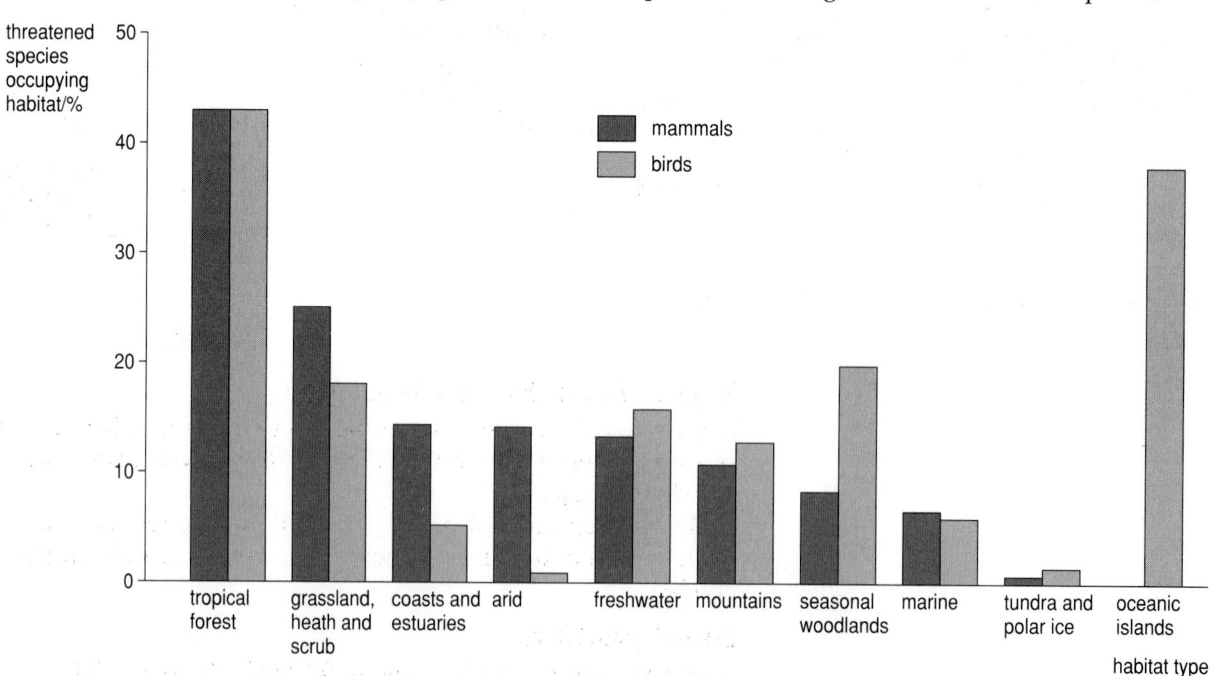

Fig 13.1 Habitat distribution of threatened mammals and birds.

Habitat

Forests provide an irreplaceable habitat for many species. Species diversity is greatest in the moist tropical forests which, although covering only 6% of the earth's surface, contain between 50% and 90% of all species of plant and animal. Destruction of forest and woodlands is the reason that so many species of birds, mammals and insects are now threatened with extinction (Figure 13.1).

Soil conservation

Trees may help to reduce soil erosion in a number of ways (Chapters 4 and 14). Through interception, trees reduce the velocity and erosive power of rainfall. Because some of the intercepted water evaporates from the tree surfaces before it can reach the soil, the total amount of water reaching the soil is reduced. Tree roots help to physically bind the soil together and, by providing organic matter in the form of decaying leaves, help to generate humus which has a binding effect (Chapter 4).

As Figure 13.2 shows, these effects are particularly important in catchments, where soil erosion may lead to silting of reservoirs, and on steep slopes, where the erodability of soils may be high.

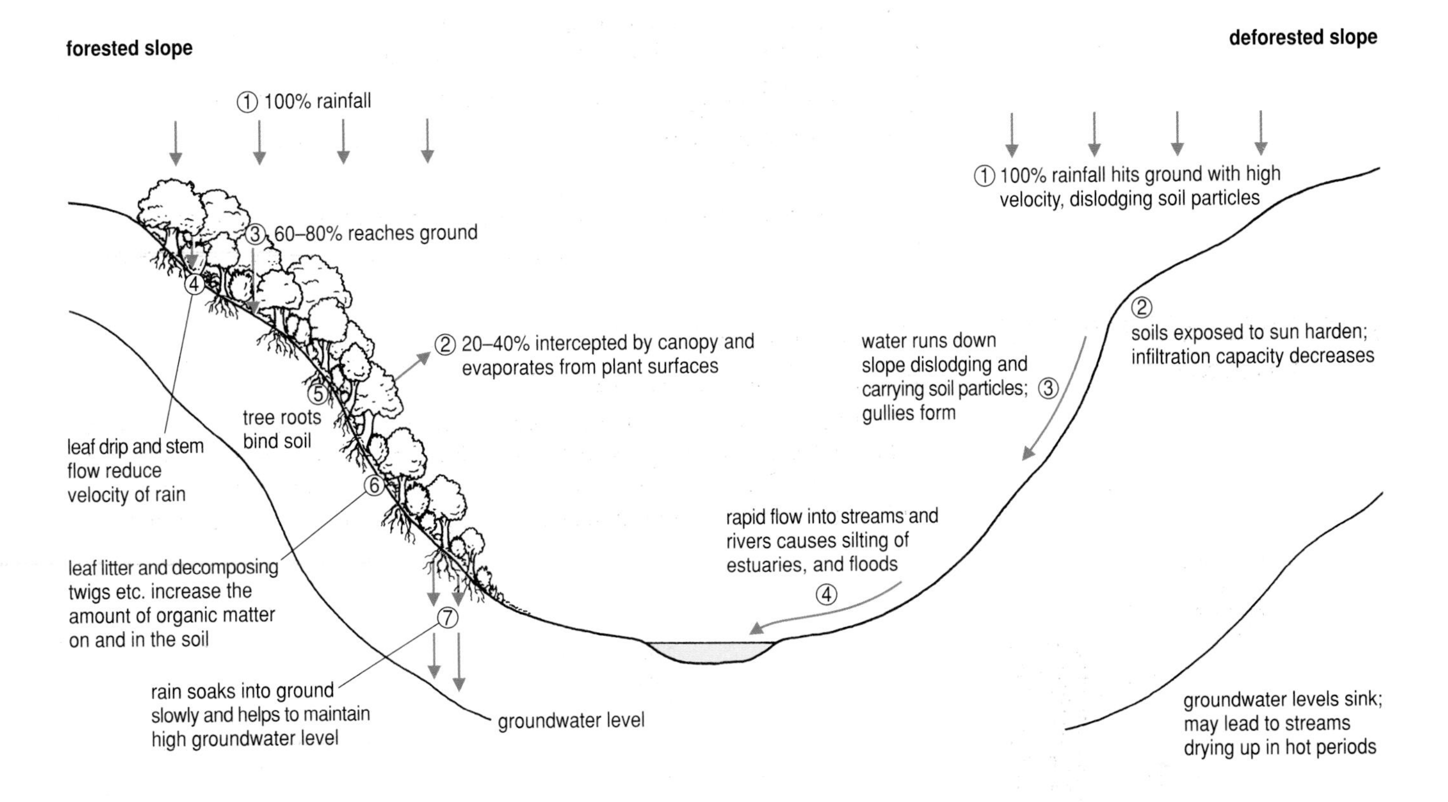

Fig 13.2 The effects of forests on soil erosion and hydrology.

Regulation of hydrological cycle

Large areas of forest release moisture into the atmosphere by transpiration and may increase humidity sufficiently to generate rain. These are the so-called rainforests.

By intercepting and holding rainfall, trees effectively slow down the release of water back into the soil or waterways, so reducing the chances of flooding.

Forest products

Trees and forests provide many useful products (Table 13.1).

Table 13.1 Forest products

Product	Example of use	Type of forest/tree
timber	construction of buildings	conifer plantation
poles	fencing	e.g. Sitka spruce
veneer	furniture	mahogany/hardwoods, e.g. oak, cherry
pulp	paper-making	conifers, e.g. Sitka spruce
chips	fibreboard, chipboard	conifers
firewood/charcoal	cooking, heating	any that is available
leaves	fodder for cattle	any palatable
medicines	anti-cancer drugs	1400 species of tropical forest plant have anti-cancer properties
food	nuts, fruit, honey	fruit trees
gums, resins	latex	rubber tree
dyes	tannin for leather	oak

Generation of microclimate

Forests moderate local climates, generally producing moister and less variable conditions than areas in the same region that do not have forests. Trees in hedgerows and woodlands can be used to provide shelter for cattle and to reduce windspeed around crops.

Recreation and amenity

Woods and forests have inspired many of our artists, musicians and poets and, as people continue to take more and longer holidays and retirement ages fall, it is likely that the importance of forests for recreation will increase. Many forests are specifically managed to provide opportunities for outdoor recreation (Table 13.2).

Table 13.2 Forests and recreation

Activity	Advantage of trees or forest environment
camping and caravanning	sense of privacy
archery	tranquillity of range/traditional associations
shooting	trees provide habitat for birds or deer, hide traps and absorbs noise
orienteering	restricted visibility puts emphasis on map-reading skills
mountain-biking	circular routes provided by roads and rides with little danger from traffic
cyclocross/rallycross	roads may provide challenging circuit; trees absorb noise and large numbers of spectators
horse-riding	tranquil, safe routes
rambling	tranquillity
picnicking	trees provide shelter and can absorb large numbers while retaining sense of seclusion
wildlife study	habitat of many species

Fig 13.3 Forest visitors.

Forests may also provide amenities such as nature trails and interpretation facilities such as notice boards giving details of local wildlife. Urban woodlands may act as psychological oases, areas of relative calm within noisy, crowded cities.

Many tropical and developing countries are using their forests and associated wildlife as a basis for developing ecotourism – low-impact tourism bringing in valuable foreign exchange which can then be used to conserve the natural environment of these countries (Chapter 18).

Forests always provide jobs and, although mechanisation has reduced the significance of this in developed economies, it remains a vital contributor to the development of a country.

QUESTION

13.1 Outline the arguments for and against the afforestation of a catchment.

13.2 FORESTRY AND ECONOMIC DEVELOPMENT

Forests provide many different products, on a renewable basis, which can lead to the development of a number of separate industries (Figure 13.4). This creates jobs and wealth for developing countries. Figure 13.5 shows the

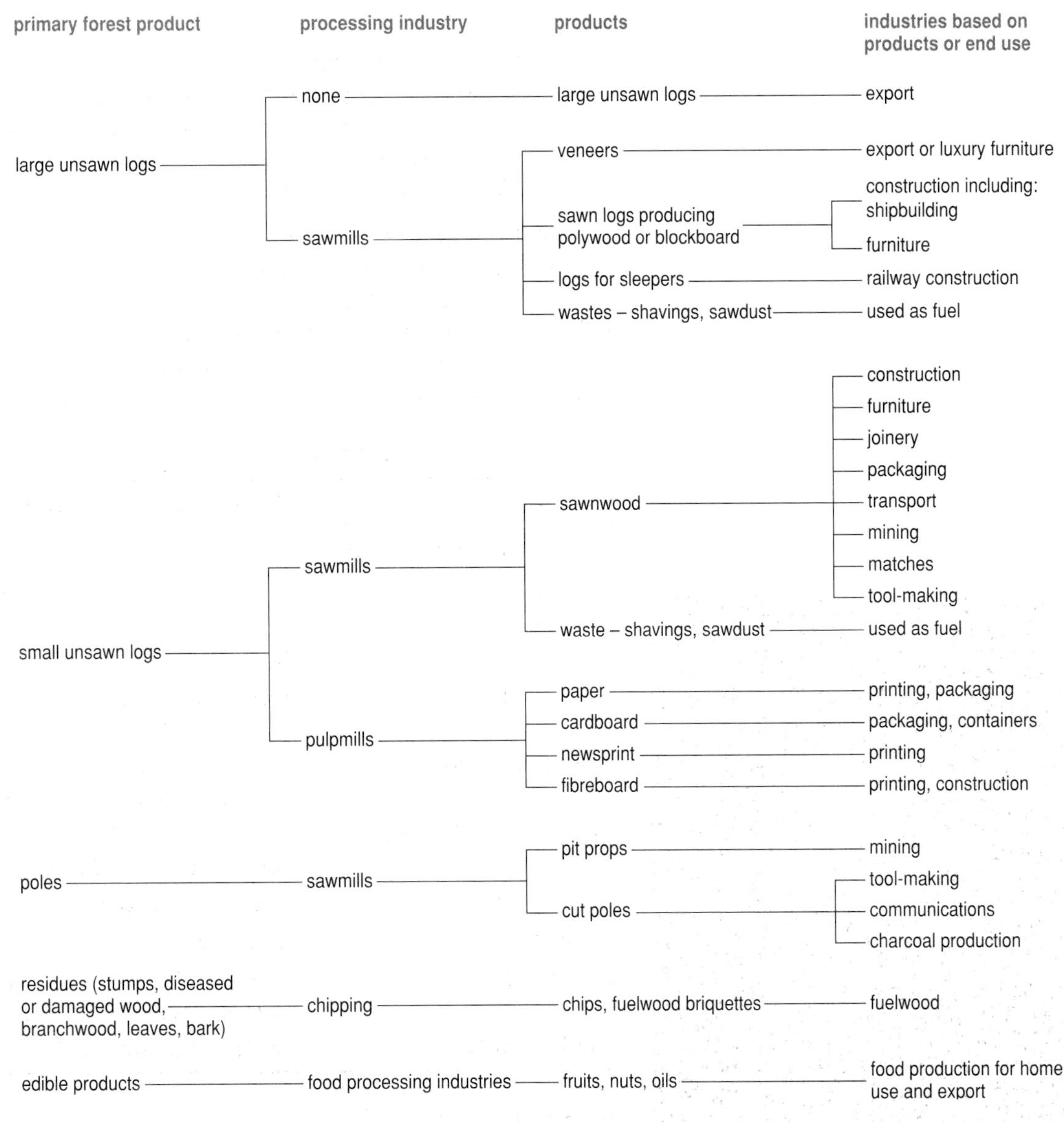

Fig 13.4 A simplified guide to forest industries.

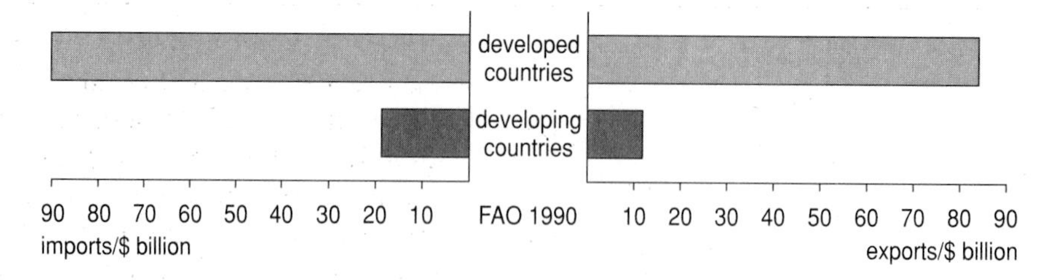

Fig 13.5 World trade in forest products.

pattern of trade in forest products in developed and developing countries. Developed countries produce and consume most forest products. In the past, developing countries such as Indonesia and Malaysia have tended to export precious raw materials to countries in western Europe, such as the UK. Ironically, they have then had to spend precious money importing the processed and more valuable products which the western countries had made from the raw materials. Such countries are now developing their own processing industries and therefore raising the value of the products they export.

As with most other products, per capita consumption of forest products, particularly items such as paper and cardboard, increases as national income increases. As a country develops, the production of goods increases, and many of these goods use wood, either directly in their manufacture or in their packaging. Thus as a low-income country develops, demand for wood and wood products both for domestic use and for export increases rapidly. Without proper management, this could result in extensive deforestation.

Most of the tropical rainforest is in developing countries, so it is natural for these countries to exploit this resource to raise money for development. Many have now felled a large proportion of their natural forests. Recently, developed countries have been critical of this attitude towards the tropical forests, arguing that because of their global functions and importance, these forests should be conserved or at least managed in such a way that they are not destroyed. In response, developing countries have argued that:

- the forests are theirs to exploit as they wish
- developed countries are being hypocritical, firstly because it is they who are creating the demand and therefore raising the price of the tropical timber and products, and secondly because developed countries have often destroyed much of their own forests
- revenue from exploiting the forests will trickle down and raise the standard of living of everyone in the country, including the people who actually live in the forest.

ANALYSIS

Boycott mahogany...save the rainforests

Tropical rainforests exist in South and Central America, Africa, South-East Asia and on some of the Pacific Islands. There are several different types of moist tropical forest, but all grow in areas of high temperature and rainfall. Recycling of nutrients is extremely rapid. Consequently, above-ground biomass tends to be very high and, in contrast to temperate forests, most of the mineral elements are in the vegetation (Figure 13.6). Soils are thin, acidic and low in nutrients, and soil quality quickly deteriorates following deforestation.

Because of the immense variety of habitats and niches, and the age of the tropical rainforest, species diversity is incredibly high. It has been estimated that between a half and two-thirds of all terrestrial species of plants and insects live in the tropical rainforests.

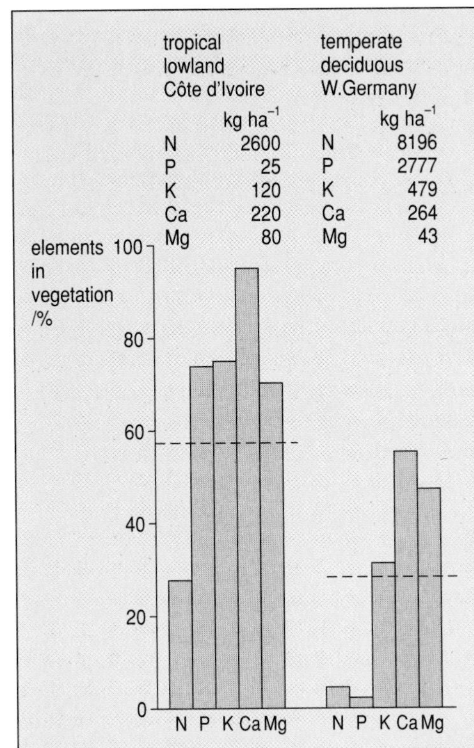

Fig 13.6 Percentage of nutrients in the vegetation of tropical and temperate woodlands.

Fig 13.7 Deforestation patterns in Rondonja State, Brazil, seen from space. The region now shows extensive deforestation alongside the main and side roads and river courses. (Shuttle Mission, 1992)

However, tropical rainforests are under threat. Less than a half of the original forest area now remains. The main causes of deforestation are the building of roads to open up virgin forests to loggers and settlers (Figure 13.7), the extraction of logs of a relatively few valuable species and the conversion of forest to farmland. Traditional shifting agriculture is not harmful to the forest, provided population densities remain low and fallow periods remain long. However, landless peasants, forcibly moved to make way for plantations, dams and other development projects, have effectively reduced the fallow period (Figure 13.8).

The Brazilian rainforest is often the centre of media attention. In the southern Amazon the main cause of forest destruction is the extraction of mahogany (*Swietenia macrophylla*), which is exported mostly to the USA and the UK for the manufacture of luxury furniture goods. This tree is so valuable that logging companies are prepared to build roads and burn large areas of what they regard as worthless forest to enable extraction to occur. The richest legally available sources of mahogany have now been exploited, so logging companies have moved into areas which have been traditionally set aside for the protection of native Indians and wildlife. There is little or no evidence of trickle-down of wealth or benefits to the Indians who live in these forests.

The individual mahogany trees are highly valuable but widely dispersed, so the amount of damage which logging causes is out of all proportion to the volume of wood actually removed. One study in the state of Para in the southern Amazon region found that for each mahogany tree removed, 28 other trees were destroyed and 1450 m^2 of forest were significantly damaged. Genetic diversity is also harmed, conservationists argue, because only the best trees, the straightest and tallest, are taken.

In November 1994, at the Convention on International Trade in Endangered Species (CITES) in Florida, the Dutch government proposed that mahogany should be added to Appendix 2 of the convention, which would force timber companies to obtain export licences from the Brazilian government, which would then be required to ensure that this trade did not harm the species. The international tropical timber industry fiercely opposed the proposal arguing that conservationists' estimate of the number of mahogany trees left in Brazil was unreliable. The true number is very difficult to calculate, since the species, growing only in small clumps of similar age, is unevenly distributed over vast areas.

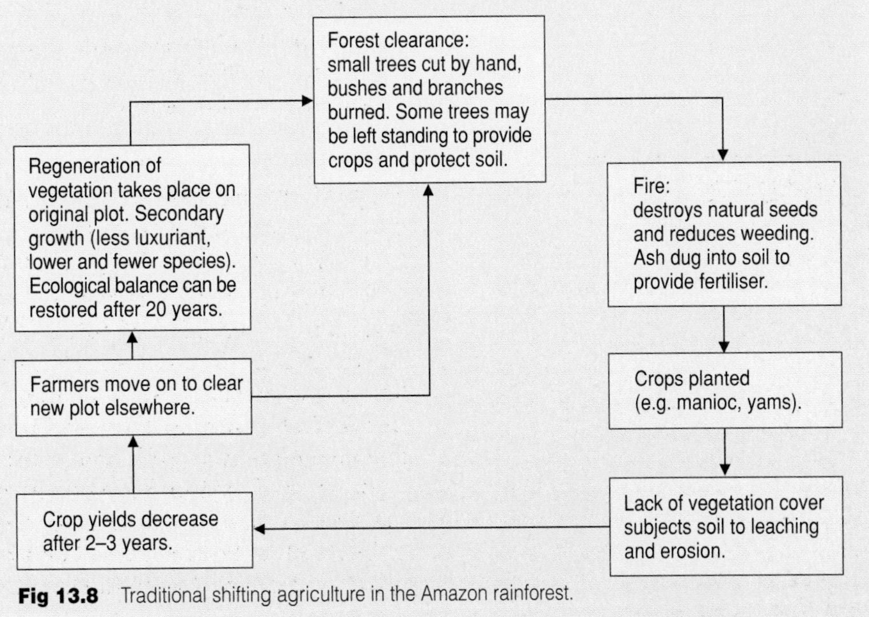

Fig 13.8 Traditional shifting agriculture in the Amazon rainforest.

Logging companies have claimed that they are operating in an environmentally responsible way by establishing mahogany plantations which will make up for the trees extracted from the forests. Conservationists argue that this does nothing to restore the diversity of the exploited forests, and that the extent of plantations is insignificant given the scale of extraction. In any case, plantations could not provide an effective substitute for at least 30 years.

Fig 13.9 Use of mahogany.

Britain is, in terms of its value, the largest importer of tropical timbers in Europe. Major uses include doors, window frames, furniture, and fitted kitchens and bathrooms. Britain's consumers can therefore have a significant impact on what happens in the rainforests of Brazil. Friends of the Earth, a leading environmental pressure group, launched a Boycott Tropical Timbers campaign, arguing that the rate of destruction would slow down if the public refused to buy products made from mahogany.

This campaign was supported by José Lutzenberger, the former Brazilian Secretary of State for the Environment who, in April 1987, wrote this open letter to British consumers.

The trade in Brazilian mahogany is out of control. In 1992 most of the timber leaving this country for Britain will come, illegally, from Indian and biological reserves. By buying Brazilian timber, you in Britain are threatening many of the Amazon's indigenous peoples with extinction.

The cutters are not only ransacking the forests in these protected areas to supply you with kitchens and lavatory seats: in many places they are also killing the Indians. The diseases lumbermen introduce have turned into epidemics.

Timber traders in many parts of the Amazon wield more money and power than most government departments ...there is little we can do to stop the supply, so it is up to the people of Britain and other First World countries to stop the demand. Britain uses 52 per cent of the mahogany Brazil produces. Please stop this trade; you are dealing with human lives...

Signed
José A Lutzenberger
Porto Alagre, April 30, 1992

In 1994, five major DIY retailers in Britain, led by B&Q, agreed to stop selling products made from mahogany. It is claimed that all of the wood sold by these retailers now comes from 'sustainable' sources.

1. Suggest why recycling of nutrients is so rapid in the tropical rainforests.
2. Explain why deforestation in the tropics has more immediate consequences for soil quality than deforestation in temperate areas.
3. Why has shifting cultivation become a problem rather than a way of managing the forest?
4. What evidence is there that road development leads to greater forest destruction?
5. Outline one technique which might be used to estimate the number of mahogany trees in an area of 1000 km.
6. Explain how a boycott of products made from tropical timbers might slow down forest destruction.
7. What is meant by the term 'sustainable management'?

13.3 FORESTRY IN THE UK – A SHORT HISTORY

Over 80% of Britain was once covered in forests, but today that figure stands at 10%, one of the lowest in Europe. Up until the end of the nineteenth century, a process of almost continuous deforestation took place, and today only a small fraction of the original forests remain. In the lowlands, trees were felled to provide timber for the construction of buildings and ships, for firewood, for charcoal production and for making tools. The uplands were deforested to provide pasture for sheep and cattle. At the beginning of the twentieth century, it is estimated, only 5% of Britain remained as forest.

The First World War was a sharp reminder of the crucial role of forests as a strategic reserve. Wood was essential to open and maintain the mines which were needed to supply the iron and steel industry essential for the war effort. The Forestry Commission was set up by the government in 1919. Their task was to afforest thousands of hectares to provide a strategic reserve in case there was another war. However, the best land was always used for agriculture, and this forced the Forestry Commission to carry out much of their planting in the uplands of Scotland and Wales. The uplands are cold, wet and exposed, and soils are usually acidic and low in nutrients (Chapter 4). Given these conditions, foresters turned to exotic tree species such as Sitka spruce, Norway spruce and Lodgepole pine, all of which were capable of good growth in the harsh conditions.

For the next 40 years afforestation of the uplands was actively encouraged both by the Forestry Commission and by private forest management companies. The 1947 Town and Country Planning Act introduced legislation which meant that existing land uses could not be changed without a lengthy process of checks and controls. If the change of land use was not considered to be in the national or local interest, the proposed change of land use was usually turned down. However, forestry (and agriculture) were considered important to the national interest, and they were not included in the legislation of the 1947 Act. A change of land use to forestry or agriculture did not need planning permission, and that is still largely the case today. Between 1919 and the early 1980s, forestry was almost solely concerned with growing timber as quickly and efficiently as possible; other possible objectives such as recreation, conservation or amenity were given little or no thought.

As stockpiles of nuclear weapons grew, the strategic justification for afforestation weakened. Attention therefore turned towards forestry's employment potential. It was argued that forestry was potentially a very valuable source of jobs in rural areas where there was little alternative employment. By providing jobs, particularly for young people, it was suggested that forestry could slow down rural depopulation and help maintain communities. Figure 13.10 shows some of the features of employment in the Forestry Commission over the last 50 years.

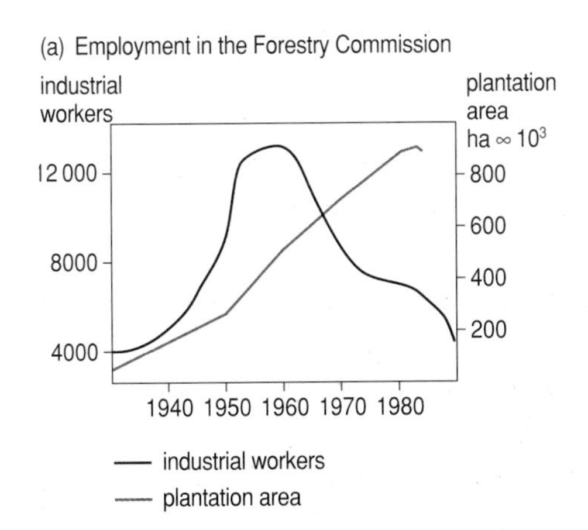

Fig 13.10 Forestry in the UK.

QUESTION

13.2 (a) Describe the trends shown in Figure 13.10 a.

(b) Suggest an explanation for the trends shown.

(c) What do the data shown in Figure 13.10 b suggest about the stability of employment offered?

The environmental implications of afforesting huge areas of land with coniferous monoculture first began to be appreciated in the 1960s. Over the next 20 years it became clear that plantations could cause serious damage to landscape (Chapter 17), nature conservation (Chapter 18), and water and soil quality. The main characteristics and the environmental implications of such plantations are shown in Table 13.3.

Table 13.3 Environmental conflicts of conifer monocultures

Characteristic	Reason	Environmental conflict
exotic conifer	faster growth rate than native broadleaves	few associated insects or birds, therefore poor for wildlife
monoculture – only one species or genotype grown over a large area	easier management – planting, weeding, fertilising, felling, marketing	since all the trees are the same species, they all demand the same nutrients, which may cause the soil to become deficient in these nutrients; artificial fertilisers are therefore needed; may lead to pest epidemics, therefore pesticides needed
all trees are the same age	easier management	gives little variation in height of trees, reducing number of possible habitats
grown in straight lines	easier management – planting, restocking, weeding, fertilising, felling	looks odd in landscape of gentle contours
dense planting pattern	ensures rapid canopy-closure – reducing light to forest floor; reduces branch growth, therefore reducing incidence of knots which weaken and reduce the value of timber	makes plantation impenetrable and dark
clear-felled (all felled at same time)	easier felling and marketing	represents a drastic change to landscape and to any fauna/flora living in or near the plantation; will drastically change microclimate (Chapter 3); may result in huge increase in soil erosion

Many of the environmental problems stem from using even-aged monocultures of exotic conifers, which grow quickly to a harvestable crop within 40 years. This kind of approach can only be justified if other objectives such as nature conservation or recreation, or the environmental damage caused by such plantations, are not an issue. Until recently, foresters have tended to manage individual woodlands for just one objective, either for timber production or for nature conservation or for recreation, but not for more than one of these reasons at the same time. Figure 13.11 shows that this can lead to forests of very different natures and appearances.

Fig 13.11 Types of forests.

ANALYSIS

Conflicts in multi-purpose forests

Using Figure 13.11, Table 13.3 and your own knowledge, complete Table 13.4, which can then be used to identify some of the possible conflicts (and solutions) in multi-purpose woodland management.

Table 13.4 Conflicts in woodland management

	Primary reason for woodland management		
	Timber production	**Nature conservation**	**Recreation**
Tree species	exotic conifer for rapid growth rate		
Tree age		mixed to allow for variation of habitats	
Planting density	dense to allow for rapid canopy enclosure – reducing light to forest floor, thus reducing branch growth and incidence of knots which weaken/ reduce the value of timber		
Planting pattern			varied to create a more scenic and natural atmosphere
Open areas?	no, will reduce production		
Fencing around or within woodland	yes, to stop deer		
Weeding carried out? If so, how?	yes, to reduce competition for light, nutrients, etc.; herbicides		
Dead trees left?	no		
Fertilisers used?	yes		
Will the woodland be thinned?	if thinnings can be sold		
Timing of operations, e.g. thinning	economic considerations – demand, availability of labour		
Interpretation facilities?		possibly	

Although it is possible to manage a woodland for several objectives simultaneously, such multi-purpose woodlands do not allow any one objective to be maximised. Instead, foresters aim to optimise several objectives.

QUESTION

13.3 Why would timber production be reduced in a woodland that was also managed for
(a) nature conservation
(b) recreation?

The Forestry Commission (FC)

The Forestry Commission was established in 1919 to plant enough trees to ensure adequate supplies of timber in the event of another war. Despite having to sell off many of its woodlands in recent years, the FC remains the largest landowner in the UK. The FC is split into two parts: the Forestry Enterprise is responsible for the growth and sale of timber to forest industries, and the Forest Authority is responsible for advising the government on forest policy, undertaking research and producing technical reports. The Forest Authority is also responsible for paying grants to encourage private individuals and organisations to plant trees.

Forestry in practice

Different approaches to forestry practice can have very different impacts upon the environment. The two examples which follow illustrate this.

CASE STUDY

Afforestation of a peat bog in Caithness

The blanket peat bogs covering Caithness and Sutherland in the far north of Scotland are the largest in Europe and, on this scale, are a unique habitat. The acidity and low nutrient status of the bogs mean that plant communities which grow there are highly specialised and many species are not found elsewhere in the world. The bogs have now been designated a World Heritage Site (WHS) by the International Union for the Conservation of Nature (IUCN) (Chapter 18). These sites have been under considerable pressure by afforestation.

Turning a peat bog into a forest

1. The area is waterlogged, so it is ploughed and deep drains are cut into the peat.
2. Sitka spruce, grown from seed in a nursery for two to three years, are planted at one metre intervals ($1\ m^2 = 2500$ trees ha^{-1}).
3. Fertilisers such as NPK may be applied aerially.
4. The area is beaten up five years later; any dead or sickly trees are replaced.
5. After 15 to 20 years, the plantation may be thinned. Weak, dead or diseased trees are removed to give more space, light, water and nutrients to those that remain. However, 85% of FC plantations cannot be thinned, as thinning creates holes in the canopy, which allows the very strong winds to devastate the plantation.
6. The forest may be aerially sprayed again with fertilisers or pesticide.
7. The forest is clear-felled after 40 to 60 years.

Consequences

Peat is irreversibly dried. The bogs are a habitat of specialised upland bird community, including waders such as greenshank. Afforestation destroys this habitat. Deep ploughing and drainage causes increased sediment to run into streams. Soil erosion follows deforestation.

CASE STUDY

Coppice with standards woodland management in Kent

Coppice is an ancient form of woodland management, involving the periodic cutting of small trees which then produce many shoots rather than one main trunk. The standards are coppice trees that have been allowed to grow to maturity. A woodland is split into compartments according to the number of years in the coppice cycle. For example, if coppice is going to be cut every ten years, perhaps to provide fencing poles and firewood, then there will be ten compartments (Figure 13.12). This ensures that some timber is available every year.

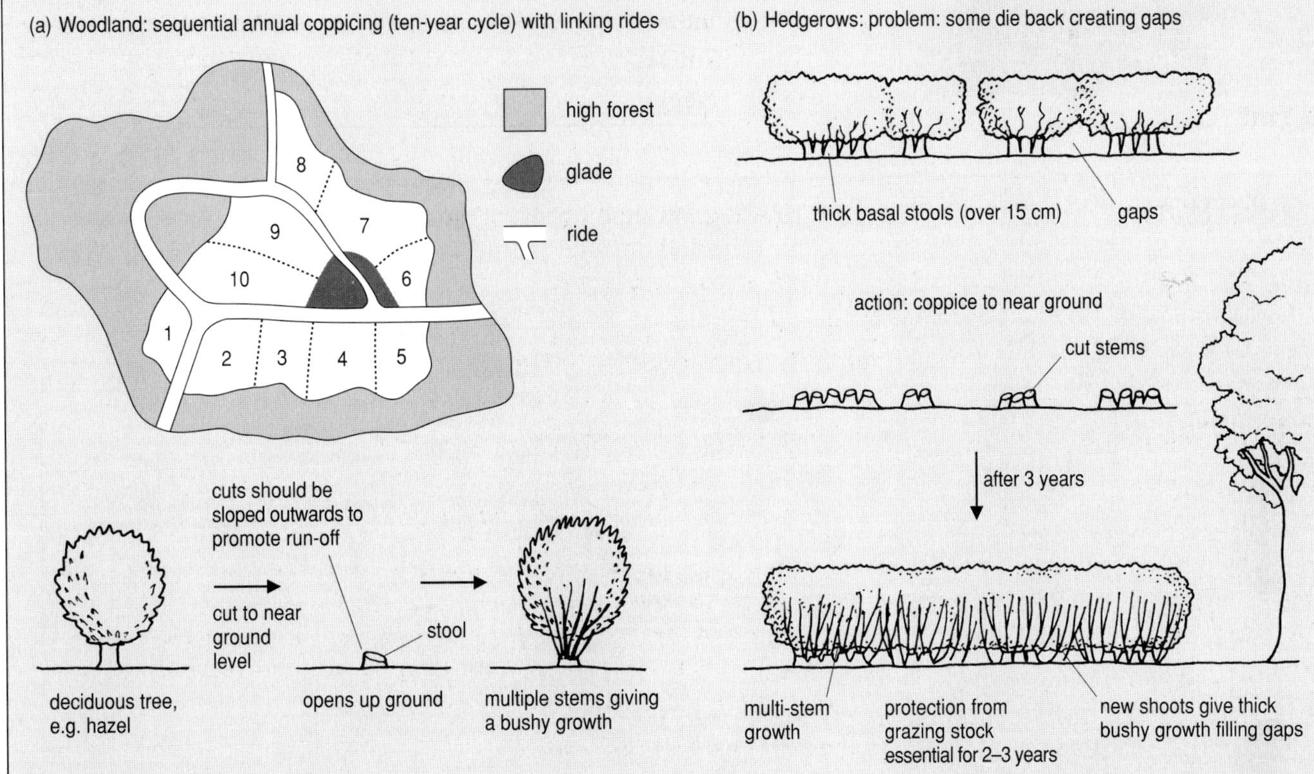

Fig 13.12 Coppice with standards.

A coppice woodland may contain trees of several species – oak, ash, hazel and chestnut are commonly coppiced – but will certainly contain trees of very different ages and sizes. This, in turn, ensures that different parts of the woodland floor receive very different light regimes and that ground flora is usually diverse. By varying the length of the rotation, different-sized poles can be harvested and standards can be felled if large timber is required.

Consequences

The wildlife and conservation value of such woodlands is high because deciduous, often native, species are used. There is a variety of tree ages and heights. Light regimes to the woodland floor are varied; in some areas there may be extensive ground cover. The woodland is never clear-felled, so woodland cover is continuous.

Two types of woodland yield different products and, in any case, deciduous species such as hazel would never survive in the bogs. However, paper and pulp production does not have to mean coniferous monocultures – mixed plantations of both broadleaved and coniferous species supply pulp mills in England. Over the last ten years, afforestation has become more concentrated in the lowlands, with a much greater use of broadleaves.

Fig 13.13 Collecting fuelwood.

Fuelwood use

For millions of people in developing countries, wood is almost the only available source of fuel for heating and cooking. Deforestation (the chopping down of forests), often over huge areas, is one consequence.

Table 13.5 Fuelwood use

Country	Percentage of energy derived from wood
Brazil	33
India	36
Tunisia	42
Nepal	98

QUESTION

13.4 The diagram shows some of the relationships between population pressure and demand for fuelwood and agricultural land in Nepal.

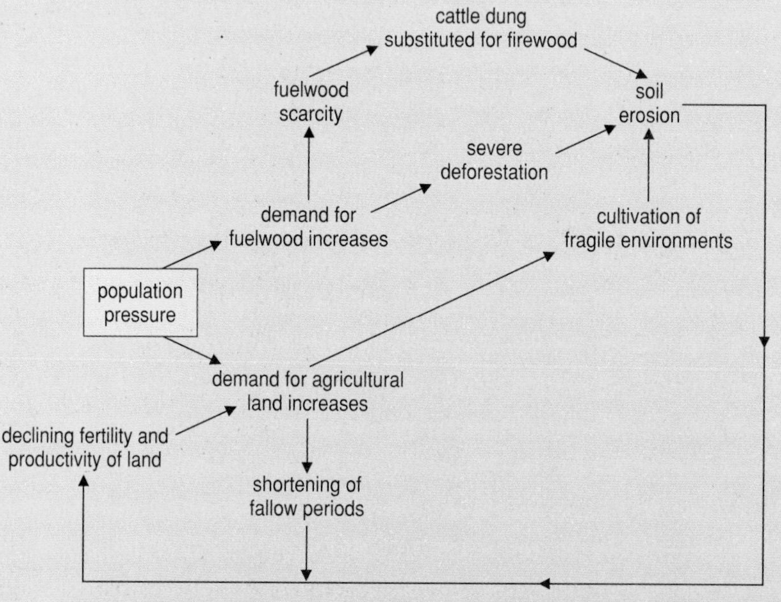

Source: World Wildlife Fund for Nature *Data Support for Education*, 1990

(a) Explain why the use of cattle dung as a fuel might lead to increased soil erosion.

(b) Using information in the diagram, explain the term *positive feedback*.

AEB June 1995

13.4 NEW DEVELOPMENTS

In 1987 the Countryside Commission published a policy document which recommended a move away from the sole timber production objective of forestry. Two new initiatives were discussed: the creation of **community forests** around some of the major cities in England, and the establishment of a major new **National Forest** in the Midlands.

Community forests

This is a joint Forestry Commission/Countryside Commission initiative. Planting has begun in twelve community forests, all of which are around major urban conurbations (Figure 13.14). Although they are called forests, there will be much more to them than just trees; the aim is to create a patchwork of wooded areas, open farmland, lakes and other recreation facilities. Only 30% to 60% of the area will be planted with trees, so that the following objectives may be achieved:

- to improve the landscape around the cities
- to provide increased opportunities for access, sport, outdoor recreation, cultural and educational events
- to protect areas of high landscape, historical, archaeological or nature conservation value
- to provide increased opportunities for farm diversification (Chapter 14)
- to provide a source of local timber and to stimulate wood-based industries.

Forest	Community Forest creation dates	Area /hectares	Existing tree cover/%	Population within 20 km/millions
Great North Forest (north-east Durham)	1990	16 000	8.0	1.5
Forest of Mercia (south Staffs)	1990	23 000	6.4	4.0
Thames Chase (east of London)	1990	9 800	8.0	3.0
Watling Chase (south Herts)	1991	16 250	7.8	3.0
Marston Vale (Bedford)	1991	16 000	5.0	0.4
Great Western Forest (Swindon)	1991	40 000	2.8	0.75
Cleveland	1991	24 800	6.8	1.0
South Yorkshire	1992	39 400	7.4	1.4
The Greenwood (Nottingham)	1991	44 000	10.0	1.6
Red Rose Forest (Greater Manchester)	1991	75 700	3.9	4.7
Mersey	1991	92 500	3.9	2.0
Bristol/Avon Forest	1992	41 500	5.0	0.9

Fig 13.14 Community forests.

Most of the land that is targeted to become community forest is and will remain privately owned, mostly by farmers. Reaching voluntary agreements with landowners has not been easy; there is little financial aid in the form of grants to tempt farmers to participate, and owners have been worried about the long-term nature of such a change in land use.

The National Forest

In 1990 the Countryside Commission announced that a new National Forest, covering 194 square miles, was to be created across the derelict mining areas of Leicestershire, Derbyshire and Staffordshire. This is a similar concept to community forests – the aim is to achieve 30% to 50% tree-cover without any change of land ownership (30 million trees will be planted) and to convert the unsightly, mining-scarred landscape into a mosaic of wood and heathland. Thirty million people will be in easy reach of the forest, which is intended to become the major focus for outdoor recreation in the Midlands. Besides providing valuable habitats for plants, birds and insects, many jobs will be created in the leisure and forest industries, and there are also plans to build a wood-fuelled power station, using willow and poplar grown in the forest.

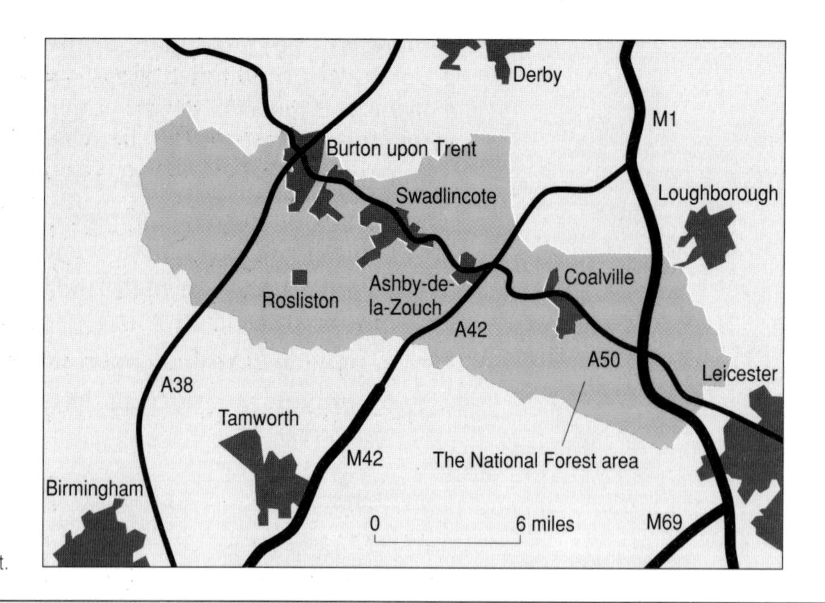

Fig 13.15 The National Forest.

SUMMARY ASSIGNMENT

1. 'The purpose of forests in 2000 is very different from that in 1919.' Discuss this statement.
2. How can afforestation aid the development of a country?
3. **(a)** The diagram shows some of the changes in water chemistry which follow deforestation in a temperate habitat.

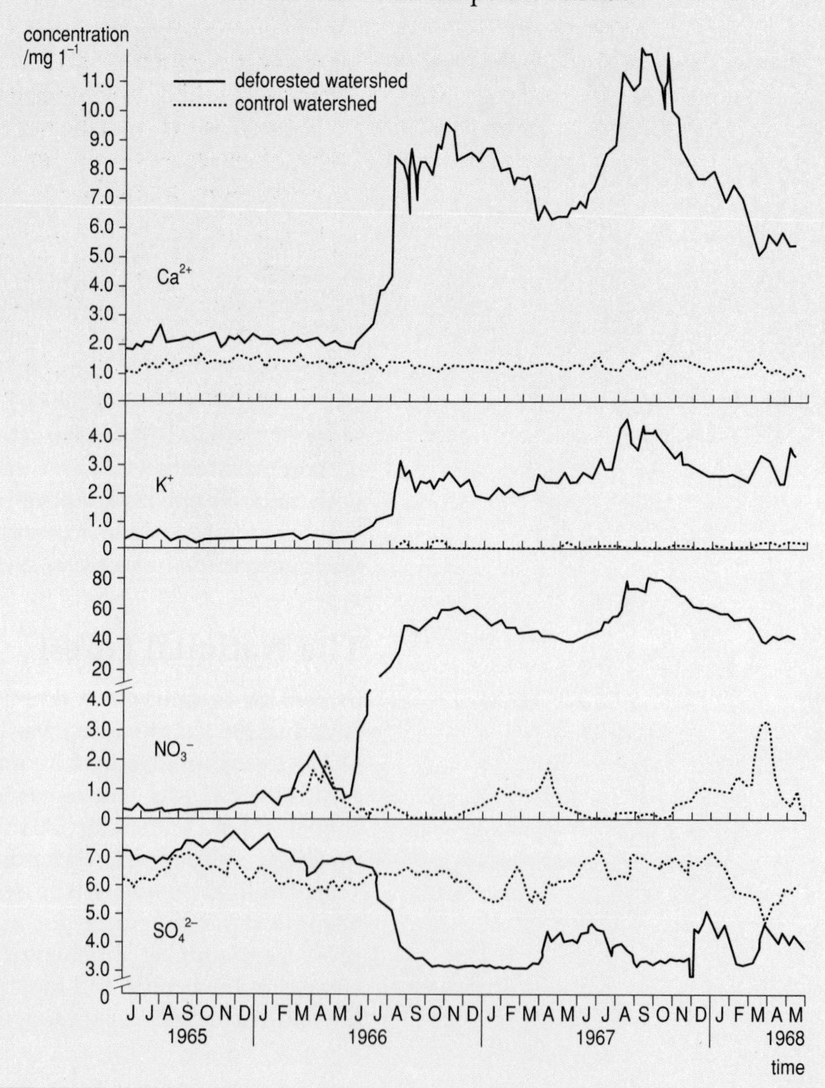

(i) From the graphs, state when deforestation is most likely to have occurred.

(ii) Describe the changes that have taken place since deforestation and suggest how these may have taken place.

(iii) How would you expect the nitrate (NO_3^-) trace in the diagram to change if the vegetation removal were from a taiga (sub-Arctic woodland) habitat rather than a temperate deciduous forest?

(b) The diagram shows the distribution of nitrogen in organic matter present in different ecosystems.

Sub-Arctic woodland (taiga)

above ground
root
soil

Temperate deciduous forest

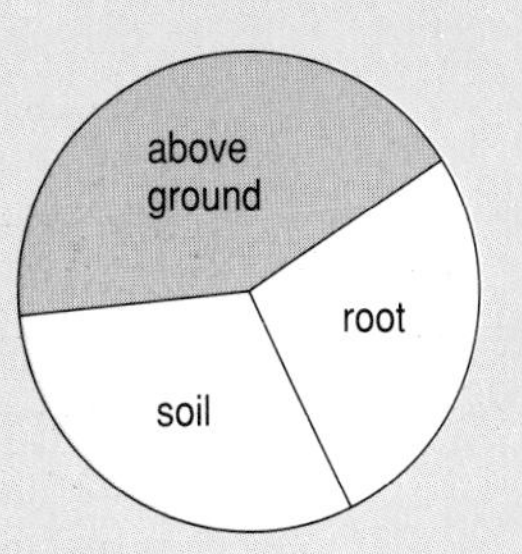

Equatorial forest

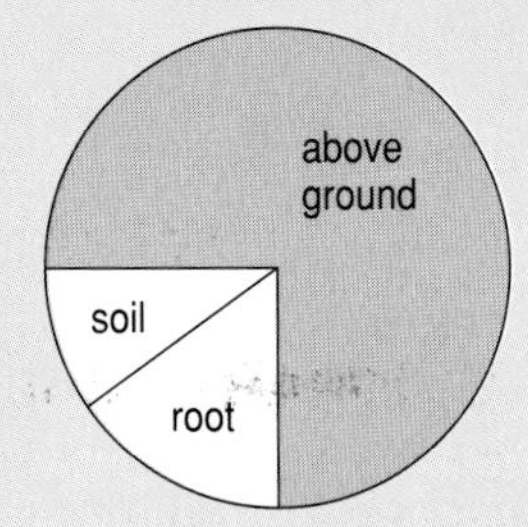

Account for the differences in nitrogen in the above-ground matter in the different ecosystems.

ULSEB June 1990

4. A forest owner has decided to adopt management practices which will allow him to achieve three aims:

- to produce saleable timber
- to provide habitats for wildlife
- to allow access for quiet recreation.

Using the information above and your own knowledge, explain the reasoning behind each of the following practices:

(a) initially planting trees very close together

(b) removing 15% of the trees after ten years

(c) leaving strips of land unplanted near streams

(d) planting several different species in one area.

Chapter 14

AGRICULTURE

Food production is the oldest industry of all. Without a source of food and energy, there can be no life, and it is perhaps not surprising that humans have drastically altered the natural environment in order to grow more food. Food production has increased enormously over the last 50 years, but such gains have been at a high cost to the environment. European countries now have huge surpluses of staple foods and have begun to look at ways of decreasing production and making agriculture more environmentally friendly. Meanwhile, many developing countries are still far from sufficient in basic foodstuffs. This may result from uncontrollable climatic problems or because many developing countries are forced to earn foreign currency by growing cash crops such as tea and coffee instead of the cereals that could feed their own people. Whatever the reason, such countries look set to implement the same environmentally damaging and perhaps unsustainable techniques that the West has used.

LEARNING OBJECTIVES

After completing the work in this chapter you will be able to:

1. explain how huge increases in agricultural productivity have been achieved
2. explain how this has caused environmental damage
3. describe recent initiatives to decrease production and reduce environmental damage
4. outline the principles of organic agriculture
5. describe the features of a more sustainable agriculture.

14.1 MORE FOOD AT ANY COST

The Second World War reminded many countries of the importance of maintaining self-sufficiency in basic foods. Between 1945 and the mid-1980s, the overriding aim of agricultural policy was to increase production. The nature of agriculture in the UK changed more during these 50 years than it had done in the previous two centuries. Such changes had profound environmental effects (Figure 14.1).

Mechanisation

The use of machinery to replace manual labour had begun before the Second World War but rapidly accelerated thereafter (Figure 14.2). Widespread use of machines offered immediate advantages, and productivity was increased by the use of larger and more effective implements, such as deep-cutting ploughs. Land which had previously been used to provide fodder for draft animals could now be sown with crops for direct human consumption.

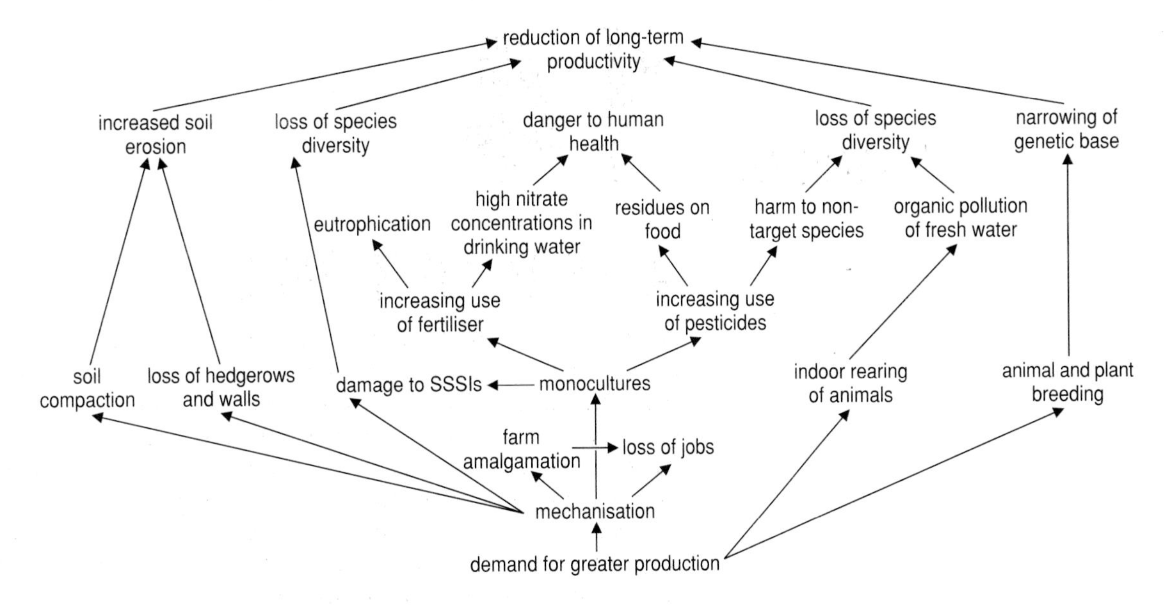

Fig 14.1 Environmental problems caused by agricultural intensification.

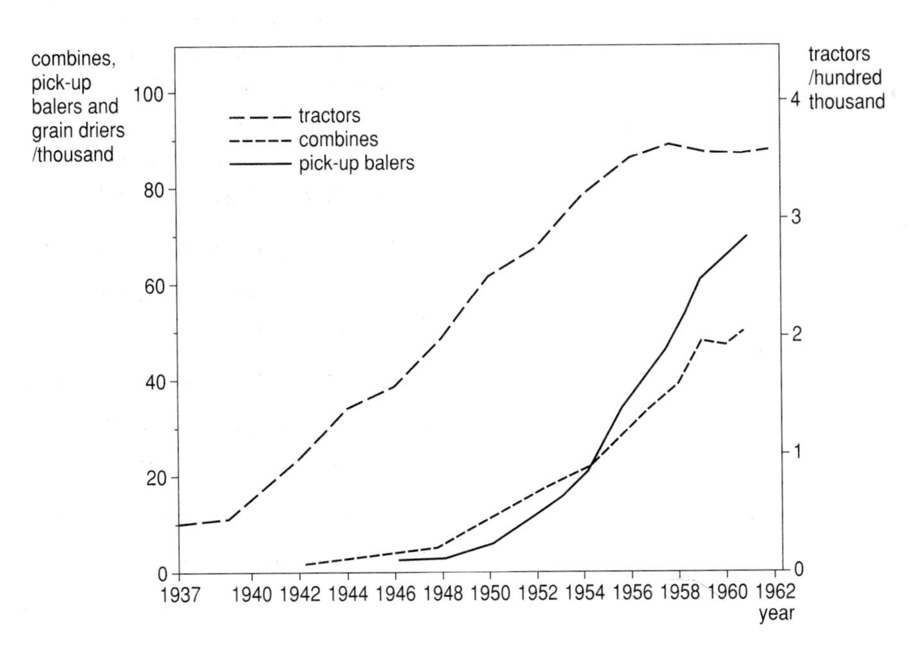

Fig 14.2 Farming and mechanisation.

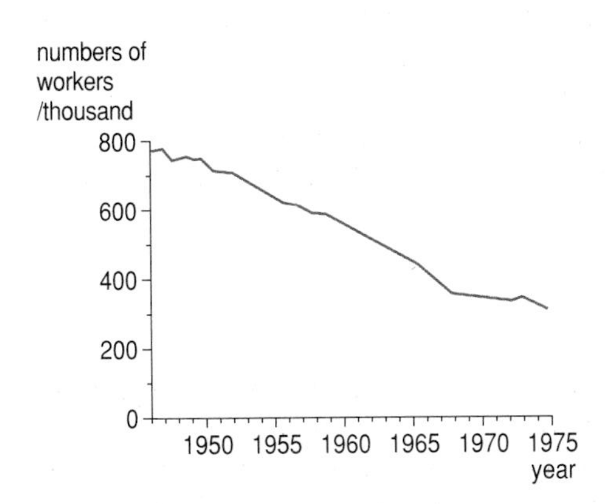

Fig 14.3 Changes in the agricultural labour force in the UK, 1945–75.

Jobs, such as harvesting, could now be done in a fraction of the time that manual labour would take. By making such operations less dependent on the weather, wastage of crops was reduced, further increasing productivity. However, all of this meant there was a rapidly declining demand for manual labour on farms, and agricultural unemployment rose sharply (Figure 14.3).

Mechanisation also led to the increasing use of monoculture – large areas of land devoted to a single crop, species, strain or even genotype. Once a farmer had purchased an expensive, specialised piece of machinery, he would usually try to recoup the cost as soon as possible by planting large areas of the crop concerned. As discussed in Chapter 13, **monocultures** may increase the risk of pest epidemics, leading to increased use of pesticide or of deficiencies of particular soil nutrients, further increasing the use of fertilisers. Furthermore, large machines (and the trend has been for size to increase) are only efficient when making long straight runs – combine harvesters and huge ploughs cannot work around corners – so huge fields were created by destroying field boundaries, such as hedgerows, walls and ditches.

Since specialised machinery is usually designed to work one particular crop, it is understandable that a farmer who has invested in such a machine would want to devote as much of his farm as possible to that crop. **Farm specialisation** was therefore encouraged and less profitable crops were eliminated. This led to farm **amalgamation** (the merging of small farms to farm larger farm units) (Table 14.1).

Table 14.1 Farm size changes in Great Britain, 1950–1987

	Farm size groups /ha (% change)			No. of farms /× 10^3		% change
	<20	20–100	>100	1950	1987	
England	–66	–41	+113	317.6	155.8	–51
Wales	–67	–20	+763	55.1	28.7	–48
Scotland	–81	–45	+317	74.8	30.9	–59
Great Britain	–69	–39	+154	447.5	215.4	–52

Artificial chemicals

Mechanisation also accelerated the use of pesticides and artificial fertilisers (Figure 14.4). Farmyard manure was increasingly replaced by compound fertilisers, which simultaneously supplied nitrogen, phosphorus and potassium (NPK). Consumption of nitrogen, in particular, rose rapidly after the war and dramatically contributed to increasing yield. At the same time, a wide array of pesticides, including insecticides, herbicides and fungicides, were developed which, like the new fertilisers, could be easily and effectively sprayed by the new machinery. Use of pesticides and fertilisers caused serious (but different) environmental problems, but these did not become apparent until the mid-1960s and early 1970s.

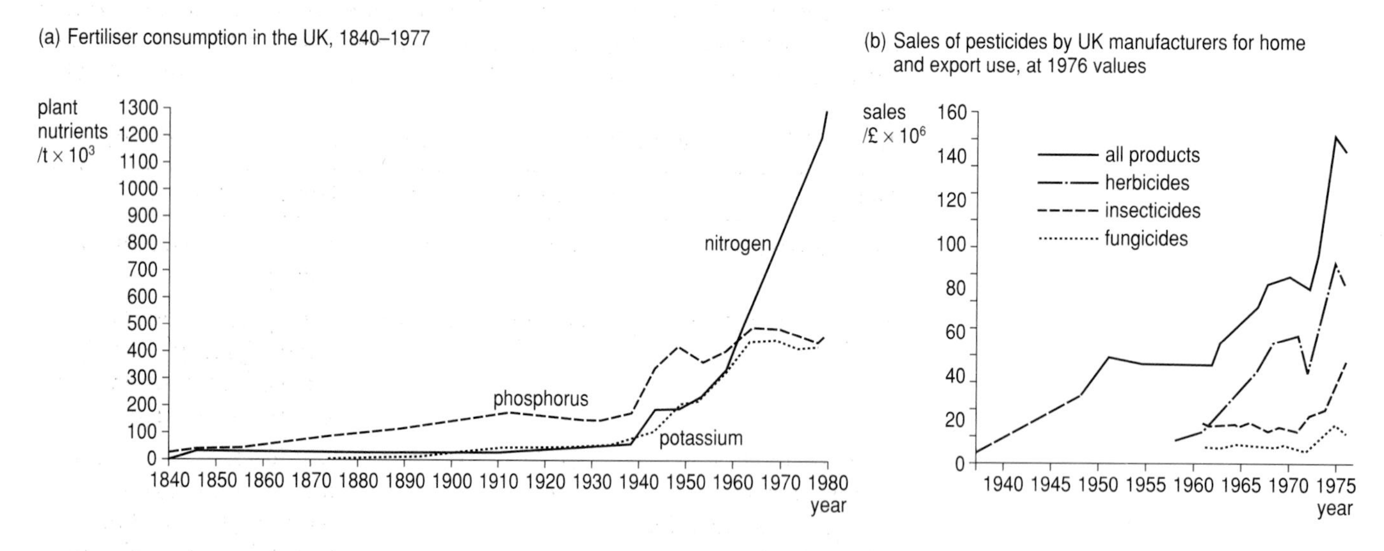

Fig 14.4 Chemicals in agriculture.

Animal and plant breeding

Scientists have made rapid advances in developing crops and animals with improved characteristics. These include increased yields, better disease resistance, improved digestibility of cereals, frost-resistant crops and improved resistance to trampling and wind-blow.

Intensive rearing of animals

Demand for higher yield also led to the development of intensive systems of meat and milk production. Intensive beef production may involve keeping the cattle housed in sheds throughout their lives. Movement is restricted and diet (mainly cereals and silage) is carefully controlled. Average weight gains of over one kilogram a day mean that the cattle may be slaughtered after only 11 months. Besides animal welfare concerns, the main environmental problem has concerned the disposal of slurry and silage effluent which would have devastating effects were they to enter rivers (Chapter 10).

Intensive livestock production often involves the use of hormones designed to increase growth or, as in the case of BST (bovine somatotrophin), to increase milk production. Such use may have potentially harmful effects on cow health, and in any case, critics argue, milk surpluses already exist, so why bother? The use of antibiotics as growth promoters and to treat diseases such as *Salmonella* is another cause for concern. *Salmonella* strains are becoming increasingly resistant to antibiotics, and this poses problems for both livestock and human health.

QUESTION

14.1 Why might antibiotic-resistant *Salmonella* be a danger to human health?

Genetic engineering

Genetic engineering has many varied actual and potential uses in agriculture. Using genetic engineering, useful genes can be transferred from one organism to another. A gene is simply a section of DNA which carries the code for a particular characteristic. Identifying particular genes is extremely difficult, but once a useful gene has been identified, for example a gene for resistance to a fungus, it can be transferred to other plants or animals, making them resistant to the fungus too. Thus organisms can be created which contain combinations of genetic material that do not occur naturally. Some of the main agricultural applications of the technology are shown in Table 14.2.

Table 14.2 Possible agricultural applications of genetic engineering

Gene	Recipient of gene	Gene function
frost resistance	strawberry plants	reduces chance of frost damage
herbicide resistance	wide range of crops	allows less carefully targeted spraying; only the weeds (which lack the gene) will be affected
nitrogen fixation (ability to convert nitrogen gas into a form which the plant can use to make proteins)	non-legumes	would allow crops to fix their own nitrogen, reducing the need for expensive and polluting nitrate fertilisers
insect toxin production	crop plants	gene codes for toxin, which is lethal to a wide range of insects
control of lipid production	chickens	reduces cholesterol content of eggs

The release of genetically altered organisms into the environment raises some concerns. Genetically engineered organisms might mutate (spontaneously change during copying), producing changed organisms which might cause totally unforeseen effects. The introduction of genetically altered organisms into an ecosystem may disturb natural population balances. Genetically altered fish which show faster growth have now been produced, but their release into wild, unaltered populations might upset natural predator–prey balances, leading to the displacement of some species. A similar effect might occur if, as suggested, Atlantic salmon were given 'anti-freeze genes' which allowed them to survive colder waters.

While this technology seems to have benefits for humanity, some of the suggested uses may make existing environmental problems worse. The insertion of genes for herbicide resistance into crops may encourage careless spraying which could lead to the destruction of valuable, non-target species. Worryingly, some of the research is concerned with developing crop tolerance for herbicides, such as paraquat, which is extremely toxic to humans and 2,4–D, a known carcinogen. Much research has gone towards the development of crops which are dependent on high inputs of fertilisers and pesticides. The companies which undertake the genetic research often produce these chemicals as well. Unfortunately relatively little research has been done on developing crops which require low levels of inputs.

Out of an estimated 10 000 edible plant species on the planet, about 150 have been widely cultivated, but only 29 now account for 90% of the world's food products. Eight species of cereal provide half of the world's energy supply. Such reliance on a small number of species may be short-sighted. Single-species monocultures of these cereals represent a potential feast for the pests of that species.

During the Irish potato famine in 1846, thousands of people died when the potato harvest was destroyed by blight due to a particular fungal species. If a wider range of varieties had been grown, such a disaster might have been avoided because some varieties might have been resistant to the infection. By maintaining genetic diversity – by growing many different species and varieties of crops – we are, in effect, taking out a biological insurance policy against disaster.

To summarise, the main trends in agriculture in developed countries from 1945 to the mid-1980s were:

- **Intensification**
  Increasing inputs such as fertilisers and pesticides, and increasing mechanisation to increase yields per hectare. This made agriculture much more dependent on other industries which supplied inputs, such as chemicals and fuel.
- **Concentration**
  Fewer, larger farms.
- **Specialisation**
  Reduced range of activities (crops, grain or animals raised), with a focus on those that were most profitable.

The benefits of these developments were obvious; food production dramatically increased, fewer people suffered malnutrition, and diet-related deficiency diseases, such as rickets, declined. Environmentally, however, the harmful effects of these trends were wide and varied (Figure 14.1). The biological effects of pesticides, fertilisers and organic pollution have been discussed in Chapter 10, but the main sources and effects of agricultural pollution are summarised in Table 14.3.

Table 14.3 Summary of agricultural pollution problems

Pollutant	Environmental effect
pesticides	may kill non-target, beneficial organisms such as pollinators; may leave toxic residues on food; may contaminate rainfall; may contaminate groundwater, and levels may exceed drinking water standards
nitrates and phosphates	may cause eutrophication of lakes and rivers
nitrates	may contaminate drinking water and cause methaemoglobinaemia or convert to carcinogenic nitrates in stomach
silage wastes	will cause huge Biological Oxygen Demand (BOD) if it enters freshwater, killing most aerobes
organic livestock wastes (faeces and urine)	as for silage, but may also contain hormones, antibodies and heavy metals, which may contaminate fresh water
methane from livestock	principal greenhouse gas
nitrous oxide from breakdown of nitrogen fertilisers	principal greenhouse gas and contributor to stratospheric ozone depletion

In 1993 there were over 2500 reported farm pollution incidents in England. Most were associated with the disposal of silage wastes and animal wastes and the use of fertilisers and pesticides.

Table 14.4 UK products which exceeded EU maximum residue limits (MRL)

Product	Pesticide residue	Year of survey	Total samples	Proportion exceeding MRL (EU or CAC)	
Fruit and vegetables	All	1981–84	1649	1.8%	EC/CAC
Soft fruits	All	1982	137	55.2%	EC/CAC
Lettuce	Thiram	1979–80	103	42.0%	EC
	Iprodione	1979–80	217	20.0%	CAC
	Dimethoate	1979–80	199	1.0%	EC
	Quintozene	1979–80	98	16.3%	CAC
	HCB*	1979–80	64	25.0%	CAC
UK produced	All	1981–82	–	2.0%	EC/CAC
imported	All	1981–82	–	0.7%	EC/CAC
Nuts	Bromine	1984	45	40.0%†	CAC†
Brassicas/carrots	DDT	1981	13	30.7%	EC
	DDT	1982–83	121	4.9%	EC
Strawberries	DDT	1985	236	0.4%	EC/CAC
Brassicas	DDT	1985	197	4.0%	EC
	DDT	1987	225	1.0%	EC
Apples/pears/plums	Propargite	1982	24	8.3%	EC
Potatoes	Tecnazene	1985–86	67	55.2%	CAC
Maize (imported)	Dieldrin	1981	80	2.5%	CAC
Lamb	Lindane	1984	500	0.8%	CAC
Chicken	Dieldrin	1984–85	122	2.5%	EC/CAC
Pork/poultry imported from China	DDT	1985–87	26	77.0%	EC
	α-HCH	1985–87	26	85.0%	EC

* HCB is a contaminant of Quintozene

† There is no MRL for bromine, but the guideline level of 0.01 mg kg^{-1} was exceeded in 40% of the samples

The National Rivers Authority (NRA), which was established by the 1989 Water Act, is responsible for the quality of the nation's rivers. The NRA monitors river water quality, offers advice to farmers on how to reduce the risk of pollution, and has the power to clean up pollution and prosecute polluters. The Ministry of Agriculture, Fisheries and Food (MAFF) publishes good practice guidelines which advise farmers on how to store and dispose of pollutants.

Pesticide contamination of drinking water is monitored by the National Rivers Authority. Similarly, the Ministry of Agriculture, Fisheries and Food undertakes routine tests to determine whether or not pesticide residues exceed maximum recommended levels. Although most residues that are detected are at a low level, the maximum residue levels are sometimes exceeded, particularly as a consequence of post-harvesting treatment, especially fumigation, to increase storage life (Table 14.4). What is certain is that the vast majority of people in the UK already have residues of pesticides, such as organochlorides, stored in their fatty tissues. We do not know what effect, if any, they have on our health, but it is perhaps reassuring that as the use of such pesticides has been prohibited or slowed, residues in the soil have fallen.

ANALYSIS

Agricultural changes

Table 14.5 Changes in field boundary lengths, 1947–85/ km $\times 10^3$

	1947	1969	1980	1985
hedgerows	662	578	534	507
fences	162	170	175	183
walls	101	98	96	94
banks	142	132	125	121
open ditches	116	111	107	107

Table 14.6 Changes in land use, 1947–80/ km $\times 10^3$

	1947	1969	1980
broadleaf woodland	7.2	6.1	5.5
coniferous woodland	0.8	2.6	3.1
mixed woodland	1.0	1.4	1.3
total	**9.0**	**10.0**	**9.9**
semi-natural vegetation	13.3	10.5	9.7
farmed land	97.6	96.1	95.7
water/wetland	1.7	1.4	1.4
built-upon land	6.2	9.0	10.1
other land	8.7	12.2	13.6
total	**130.3**	**130.3**	**130.3**

Tables 14.5 and 14.6 show some of the changes to important landscape features in England since 1947.

1. Calculate the percentage decrease in length of hedgerows from 1947 to 1985.
2. Suggest how loss of hedgerows might lead to:
   (a) increased agricultural production
   (b) decreased agricultural production.
3. In England, the total area of farmed land decreased over the period, yet agricultural production increased dramatically. Summarise the factors which made this possible.

Soil erosion

A fertile soil is an important physical input into agriculture (Chapter 4). Many modern agricultural practices have led to increased **soil erosion** (the loss of vital topsoil by agents of erosion such as wind and water). Organic matter is vital to soil structure (Chapter 4). Unlike organic fertilisers, inorganic fertilisers, which are now more commonly used, add no organic matter to the soil. Reliance on such fertilisers may mean that the soil is not bound together by humus and this can lead to soil erosion. Any vegetation that covers the soil will tend to reduce soil erosion, so agricultural soils are most vulnerable when the crop has been harvested and the soil is left bare. Hedgerows decrease wind speeds, so their removal may accelerate erosion, as will overgrazing. Finally, the use of heavy machinery will lead to soil compaction and loss of structure, which may speed up soil loss.

14.2 MANAGING THE SOIL

The objectives of soil management are to maintain nutrient levels and optimum soil structure, to control pests and diseases, and to prevent erosion of the soil.

Maintaining nutrient levels

In order to maximise crop growth, farmers must try to maintain optimum levels of the macro- and micronutrients that plants require. An excess of any nutrient may be as harmful as a deficiency. Artificial and organic fertilisers or legumes (Chapter 6) are used for this purpose, but it is not enough just to ensure that the soil contains nutrients, they must be available for absorption, and this is largely determined by soil pH. Peak solubility of most elements occurs in slightly acidic soils. Soils that are too acidic can be treated with lime, which may also help to improve structure.

Maintaining optimum soil structure

Soil structure refers to the type and arrangement of peds. Clay soils, composed of large, plate-like peds, may prove impenetrable to plant roots, and can be broken up by ploughing. Conversely, very coarse-grained, sandy soils will have very high porosity (many air spaces) and may require the incorporation of organic matter such as treated sewage waste or irrigation in order to ensure that the crop receives sufficient water.

Controlling pests and diseases

Pest control can be achieved through cultural techniques, biological control or pesticides. Organic methods of pest control are considered later in this chapter (Table 14.10).

Preventing erosion of the soil

As previously discussed, many modern farming practices have dramatically accelerated the process of soil erosion. **Rainsplash** – the dislodging of soil particles and compaction of the soil caused by raindrop impact – can be reduced by ensuring that some kind of crop is maintained on the land at all times. The water on a slope can either infiltrate the soil and enter into the **groundwater flow**, move through the pore spaces of the soil itself as **sub-surface flow (interflow)**, or it can run downslope as **surface runoff**. Groundwater and sub-surface flow can erode fine material, but this is not as important as erosion by surface runoff.

There are four forms of surface runoff. **Overland flow** or **sheet erosion** occurs during intense rainstorms or after a prolonged period of rainfall when the soil is so saturated with water that precipitation is forced to flow on the surface. In this form of surface flow, the water is not channelled. **Rill erosion** occurs in small, temporary channels. **Gully erosion** occurs through larger, more permanent channels. They form as a result of an increase in water or a decrease in the ability of the channel to hold that water, and **stream bank erosion** occurs when the permanent channel cuts into the banks.

Table 14.7 Management techniques used to minimise soil loss

Method	Description	Effects
terraces	a series of flat 'steps' up a slope; allows steep slopes to be cultivated	lowers the gradient and shortens the slope, thus reducing the amount and speed of runoff
contour bunding	simple mounds of earth across a slope	prevents surface runoff; prevents wind erosion if planted with trees and grasses
contour ploughing	ploughing along the slope, not up the slope, thus producing a series of ridges and furrows	slows down runoff; only effective against low-intensity rainfall on low gradients
tillage techniques	preparation of the soil by overturning	need to use the correct method at the correct time (e.g. chisel plough, duck-foot cultivator)
shifting cultivation	traditional method of preventing soil erosion in tropical areas; land is prepared, used for only a few years, and then left to fallow (recover)	erosion is low due to the remaining organic content when the land is cleared; the system fails if too many crops are grown for too long or if the fallow period is not long enough for the land to recover
multiple cropping (or sequential cropping)	growing two crops at a time, or one crop after another, in areas that have enough water, temperature and nutrients	produces a permanent cover and two crops a year (e.g. maize and cassava in Nigeria)
high-density cropping	cropping close together	reduces crop yield but increases the cover
strip cropping	the alternation of crops with grass or cereals in bands 15–45 m across	lowers wind velocity and traps moving soil particles; only protects if the wind is blowing in the correct direction
cover crops	a fast-growing plant which covers the soil after clearance or cropping; planting can occur just before cropping so that there is a permanent cover	produces a permanent cover, when the soil would otherwise be left open to the elements; retains nutrients which would otherwise be lost through leaching
windbreaks	a barrier to the wind	alters the microclimate; only protects if the wind is in the right direction
mulches	the addition of substances such as straw, manures and sewage sludge	improves the growth of cover crops; adds to the organic content; prevents rainsplash, surface flow, and lowers the wind
synthetic stabilisers	the addition of substances such as polyvinyl alcohol	temporarily binds the soil surface together
irrigation	the careful use of water	reduces soil desiccation; misuse can induce salinisation (the concentration of salt in the topsoil), cause a lowering of the water-table which can lead to desertification, or induce anaerobic (gley) conditions

Fig 14.5 Surface runoff can be reduced by terraces, such as these in Nepal.

Surface runoff can be reduced by: changing the slope, for example through **terracing** (creating a series of 20 to 30 m wide 'steps', which lowers the gradient); increasing the natural resistance of the soil, for example by ploughing; and altering the hydrological regime, for example by establishing drains or by ensuring that the surface is always covered by vegetation.

Wind erosion is particularly prominent in areas that have low rainfall (where soils are dry), strong dry winds and a low humidity. Soils with a weak binding structure (for example sandy soils) with small, light particles (for example fenland soils) are particularly susceptible to wind-blow, but can be protected by hedgerows (Chapter 2) and by the addition of extra organic matter.

The removal of soil may also occur by **mass movements** (Chapter 3) – mudslides, slumps and soil creep – although this form of erosion generally occurs on slopes that are too steep for agriculture. Some of the most important techniques for soil conservation are shown in Table 14.7.

CASE STUDY

The dust-bowl

During the 1930s in the central plains of North America, the natural prairie grasslands were converted into wheatfields by settlers from the eastern coast. The crop failed within a few years because the climate in the region was too dry. The soil was left exposed to the action of the wind. Huge volumes of topsoil were lost in massive dust-storms (Figure 14.6).

Fig 14.6 The American dust-bowl.

The loss of topsoil had a huge socio-economic impact. Thousands of people were made unemployed and forced to migrate to areas such as California where they were exploited as cheap labour. This environmental disaster was caused by the mismanagement of agriculture.

ANALYSIS

Soil erosion

Figure 14.7 shows some typical farming practices which may accelerate soil erosion. The Universal Soil Loss Equation, which may be used to predict average annual field soil losses, is also shown.

Fig 14.7 Farming practices which may accelerate soil erosion.

$$A = 0.224 \times R \times K \times L \times S \times C \times P$$

where:

A = the soil loss kg $m^{-2} s^{-1}$
R = the rainfall erosivity factor
K = the soil erodibility factor
L = the slope length factor
S = the slope gradient factor
C = the cropping management factor
P = the erosion-control practice factor

1. List the factors which are likely to affect the erosion as a result of rainfall.
2. Using information from this chapter and from Chapter 4, explain why modern farming practices may increase soil erosion.

14.3 GOVERNMENT POLICY AND AGRICULTURAL CHANGE

The process of intensification described in this chapter, in section 14.1, could not have occurred without the active support of government and European policies. One of the key objectives of the 1947 Agricultural Act was to increase agricultural production and to achieve self-sufficiency in basic foodstuffs. To ensure this, the government implemented two kinds of measures: guaranteed prices and grants and subsidies.

Guaranteed prices

Farmers were offered a guaranteed price for a range of products including milk, cereal and beef. The food was sold to the consumers as usual, and the government made up any difference between the selling price and the price that they had guaranteed. This encouraged intensification and specialisation but not surpluses, since farmers only received guaranteed prices for products that they had actually sold.

Grants and subsidies

The government made available many grants and subsidies as a way of increasing efficiency and of raising farm incomes. Grants to purchase new

Fig 14.8 Harvesting grain.

machinery or subsidies on fertilisers encouraged mechanisation and intensification, respectively, while grants for draining wet areas for conversion to pasture led to habitat destruction.

British farmers continued to benefit from guaranteed prices and grants and subsidies from the **Common Agricultural Policy (CAP)** when Britain joined the European Economic Community (EEC) in 1973. The CAP tried to protect European farmers from cheap imports through imposing levies on imported food (for example New Zealand butter). Farmers' incomes were protected by **intervention buying**, that is, surpluses of products were bought by the EU to prevent prices from falling. These surpluses were then either stored, exported or destroyed, at great cost to the European consumer. European farmers could produce as much food as possible in the knowledge that the EU would buy up everything they produced at a pre-arranged price. This led to the notorious grain mountains and milk lakes of the 1970s.

By the 1980s it was clear that 40 years of agricultural intensification had caused serious harm to the rural landscape, to rural employment, to a wide range of plant and animal species, and to important scientific sites. Environmental pressure groups, such as the Campaign for the Preservation of Rural England (CPRE), along with statutory organisations, such as English Nature (formerly the Nature Conservancy Council), lobbied to publicise both the scale of the damage and the senselessness of European citizens having to pay large sums of money to store or destroy surplus food (Table 14.8).

Table 14.8 European food mountains, 1992

Foodstuff	Store / tonnes	Days' supply
butter	224 000	52
beef	629 000	38
cereals	19 612 000	–

In 1992 the government cut the guaranteed prices for cereals and livestock, and this, along with many other environmentally friendly initiatives, has reduced both agricultural production and environmental damage.

14.4 NEW INITIATIVES

The most important initiatives designed to reduce production and environmental damage are outlined below.

Set-aside

Reducing production has become an important objective for the European Union. Under set-aside regulations, farmers are now required to cease agricultural production on part of their land (15% in 1992–93). Farmers are paid to keep set-aside land fallow.

From 1993–94, two types of set-aside were created: rotational and flexible. Under the rotational scheme, farmers have to set aside a different area of land each year and to bring the previous year's set-aside areas back into production (Figure 14.9). Any piece of land can be set aside once every six years.

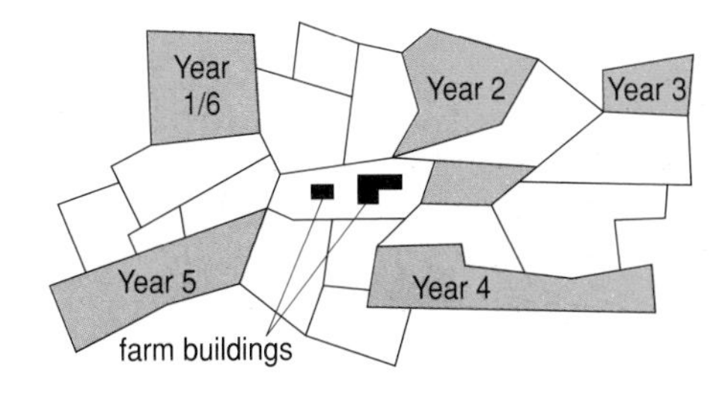

Fig 14.9 Rotational set-aside.

In flexible set-aside, farmers can leave the same piece of land fallow year after year. This is much more useful in terms of encouraging wildlife and nature conservation, but the major objective of rotational set-aside is to reduce production. However, because of fears that farmers would simply set aside their least productive land, the percentage of set-aside required will be 3% higher than for non-rotational set-aside.

Nitrate-sensitive areas (NSAs)

Nitrate-sensitive areas are areas where the nitrate concentration in drinking-water supplies exceeds, or is in danger of exceeding, the EU limit of 50 mg l^{-1} (Figure 14.10). Farmers in these 32 areas receive compensation for adopting practices which may help to reduce future nitrate levels. For example, farmers must agree to restrict their use of nitrate fertilisers, slurry and manure. In 1991 the EU Environment Council agreed on a Nitrates Directive, which required all member countries to designate all areas of land that drain into threatened supplies as Nitrate Vulnerable Zones. This changed a voluntary initiative into a compulsory one; farmers in these areas are forced to implement an action plan to limit future loss of nitrate from their land.

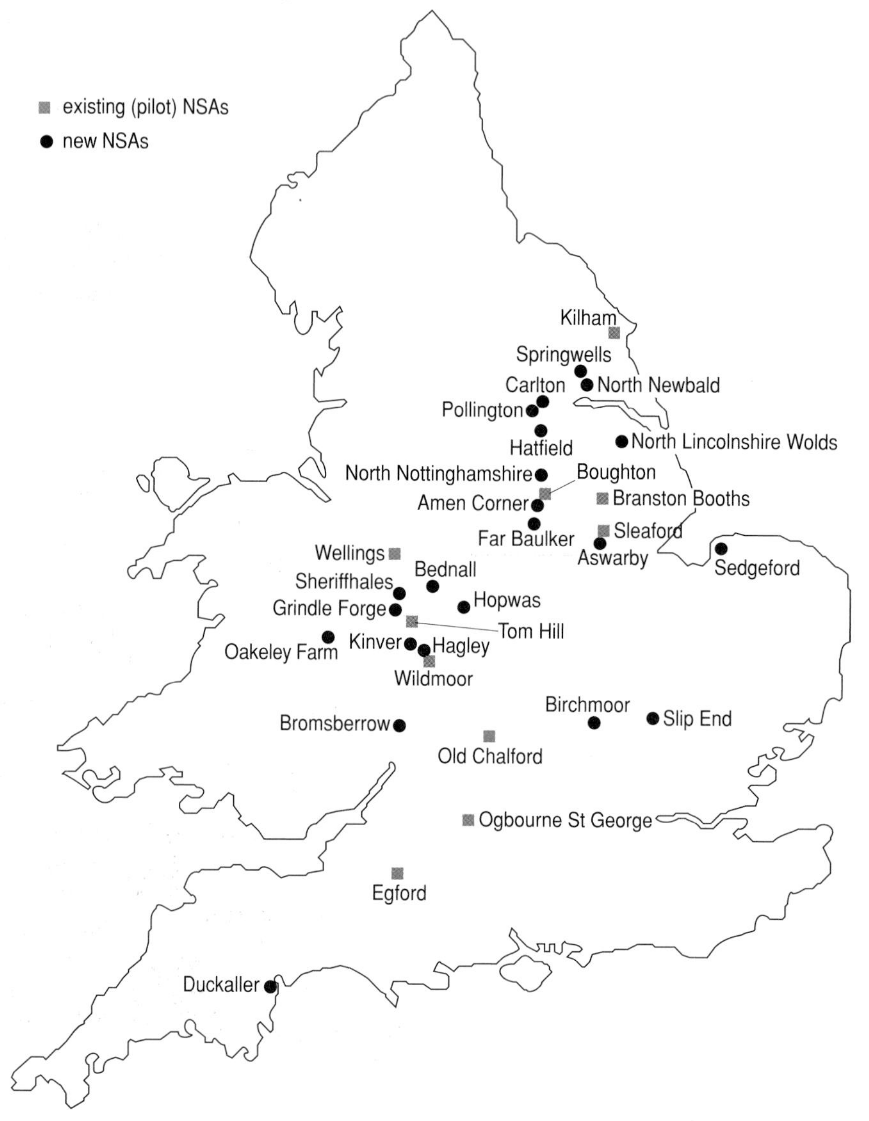

Fig 14.10 Nitrate-sensitive areas in England.

QUESTION

14.2 Explain how the following restrictions which apply to nitrate-sensitive areas may reduce loss of nitrogen from the land:
(a) avoid the use of legumes (Chapter 8)
(b) avoid the cultivation of land
(c) avoid any application of nitrate between August and February.

Environmentally Sensitive Areas (ESAs)

The Environmentally Sensitive Areas scheme was initiated in 1987 by the Ministry of Agriculture (Figure 14.11). Initially, 16 sites were chosen because of their beauty, rarity of habitat and historic interest. In 1993, six new areas were proposed. In all of these sites, farmers are compensated for using traditional farming practices which will help maintain the character of the region. Farmers will have to agree to limit organic and inorganic fertilisers, insecticides or herbicides, and to strictly control stocking densities of animals. Examples of management practices and suggestions for some of the newer ESAs are given in Table 14.9.

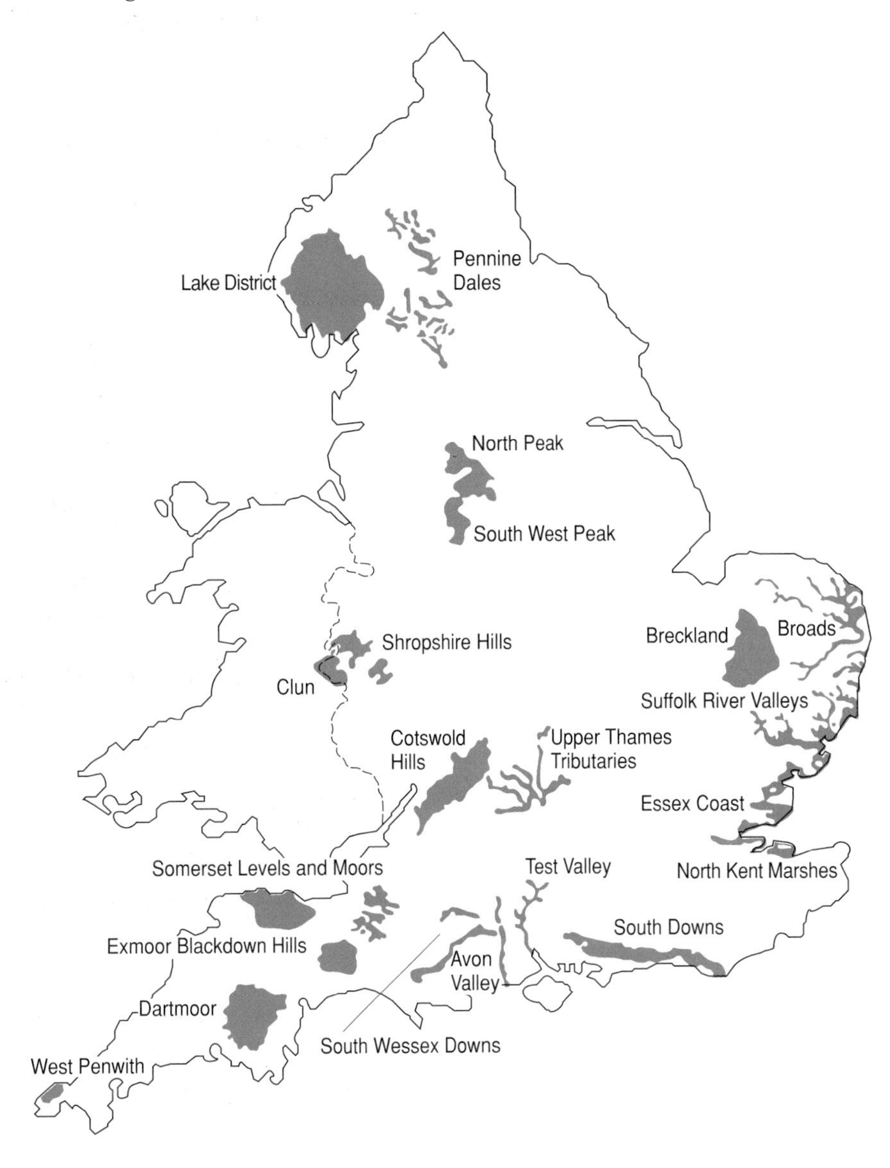

Fig 14.11 Environmentally Sensitive Areas in England and Wales.

Countryside stewardship scheme

This scheme, run by the Countryside Commission, is not restricted to particular areas but does target particular types of habitat, such as limestone, grasslands and waterside landscapes. Payments are made for a range of sensitive management activities and also for measures designed to improve public access. Some 115 square miles scattered around England are currently included in this scheme.

Table 14.9 **Management of Environmentally Sensitive Areas**

ESA	Why special?	Example of landscape feature	Means of protection
Dartmoor	very high landscape, geological and archaeological value	heather moorland, hedges, stone walls	limit stocking density; agree cutting and burning regimes for heather
Shropshire hills	high landscape value with important upland buildings and many archaeological and historic sites	heather moorland and patchwork of hedges between hay meadows and rough grazing	agree cutting and burning regimes for heather; maintain hedges, walls and banks; maintain permanent grassland with no excavation, cultivation, ploughing or reseeding
Essex coast	coastal wetlands offer important habitat for plants, invertebrates and birds; sites of Iron Age/ Roman salt production	marshlands, water-filled dykes and ditches	maintain water levels in dykes and ditches, and agree on a management plan

Hedgerow incentive scheme (HIS)

This scheme was launched in 1992 in response to growing concern about the decline of hedgerows in Britain. Payments are offered to encourage restoration and proper management of neglected hedgerows. Applicants are required to enter into ten-year agreements to manage all the hedgerows on their land in a manner generally sympathetic to wildlife and the landscape. This recognises the great importance of hedgerows as both wildlife habitats and attractive features in the rural landscape.

New habitat scheme

It is proposed to introduce a new scheme to encourage farmers to create wildlife habitats by taking land out of agricultural production, usually for 20 years, and managing it in an environmentally beneficial way. It is likely that the main focus of this scheme will be on waterside habitats in designated target areas and on intertidal habitats, particularly salt marsh.

Organic aid scheme

It is proposed to offer grant aid to farmers who wish to convert their land to organic production. Such conversion has major environmental benefits, as it reduces the amount of all types of chemicals being used.

Organic agriculture

When a crop is harvested, all the nutrients within the harvested parts are removed from that piece of land. If this was repeated year after year, soil fertility would rapidly decrease. Intensive agriculture attempts to solve this problem by adding artificial fertilisers but, as we have seen, this may lead to pollution and they are, in any case, produced using non-renewable resources such as oil. Organic agriculture attempts to use our ecological knowledge to maintain soil fertility; only natural substances such as animal manures or the use of legumes (Chapter 8) are employed, along with techniques such as crop rotation. The main features of organic agriculture are identified below.

Crop rotation

The purpose for which a particular piece of land is used is varied each year. For example, it may be used to grow cereals one year, legumes the next, followed by carrots, and then cereals again. This helps to prevent any particular species-specific pest from building up to economically harmful levels. Because different species have different rooting habits and nutrient demands, it also helps to prevent any particular nutrient from becoming deficient in the soil. Non-organic farming often encourages successive monocultures; resulting diseases are dealt with by pesticides, soil deficiencies by artificial fertilisers.

Organic fertilisers

Organic fertilisers such as manure or green manure (growing clover, a legume, for a year and ploughing it in) is an important principle of organic agriculture. Organic fertilisers tend to release their nutrients slowly, helping to prevent leaching losses and, unlike artificial fertilisers, improve both the structure and water-holding capacity of the soil. They also provide food for and encourage the growth of saprophytic bacteria and fungi, which aid nutrient recycling (Chapter 8).

Avoidance of artificial pesticides

Organic farmers prefer natural pesticides, biological control or cultural techniques to artificial pesticides. Pesticides, especially insecticides, have caused major ecological problems:

- They may often directly kill useful insects and sometimes birds and mammals, as well as the target pest. If the natural predators of the pest are killed, huge pest outbreaks may result. Many pesticides are also toxic to humans.
- Secondary pest outbreaks can arise. Usually a plant is attacked by more than one type of pest. If the insecticide is effective in eliminating all or almost all of one type of pest, then the population of a second pest species may explode.
- Insects become resistant. Within a large insect population, there may be a small number which possess genes to resist the pesticide. Such insects will then survive the application of pesticide. These individuals now face much reduced competition for food and, given the rapid reproduction rate of many insects, their population may increase very rapidly. Genetically determined resistance will be passed on to every offspring. To overcome these problems, organic farmers use natural techniques (Table 14.10).

Table 14.10 Examples of organic pest control

Technique	How it works
release of predators	may use natural predator–prey relationships, for example ladybirds on sap-sucking aphids micro-organisms may also be used, e.g. the use of *Bacillus thuringiensis* on caterpillars
trap crops	these are plants which are known to be attractive to the pest and which are used to lure the pest away from the actual crop
intercropping	two or more crops are grown together, either intimately within the same row or in adjacent rows; when onions and carrots are grown together, the strong smell of the onions masks that of the carrots which are thus given some protection from the carrot root fly; some plants are known to be repellent to some insect pests
tillage	physical uprooting and burial of weeds, or the unearthing of animal pests to make them more visible to predators such as birds

Farm diversification

Grants are available to enable farmers to diversify into non-agricultural activities, such as golf courses or horse-riding centres. Farmers are now being asked to produce much more than food (Table 14.11).

Table 14.11 Types of farm diversification

Direct marketing	Recreation	Accommodation	Woodland management	Unconventional products
farm shop	farmhouse teas	bed and breakfast	fuelwood	ostriches
pick your own	farm zoos	camping and caravans	small woodland products	fish
milk processing (e.g. ice cream sales, cheese production)	farm tours	holiday cottages		deer
	nature trails			goats
	riding			rarebreeds
	shooting			linseed
	golf			herbs
	wargames			

14.5 FEEDING THE WORLD

The agricultural revolution, which led to tremendous increases in food production, had five main elements:

- mechanisation was introduced
- the amount of land under cultivation increased
- the use of chemicals such as fertilisers and pesticides increased
- the use of irrigation increased
- sophisticated plant- and animal-breeding techniques were developed.

Such a revolution has occurred in almost all developed countries, and for many years this has allowed food production to keep pace or even ahead of population growth. Figure 14.12 shows that every region of the developing world has increased its food production since 1970, mainly as a consequence of increasing yield per hectare rather than because of an increase in land cropped. However, in most developing countries, per capita food production has fallen because population growth has exceeded growth in food

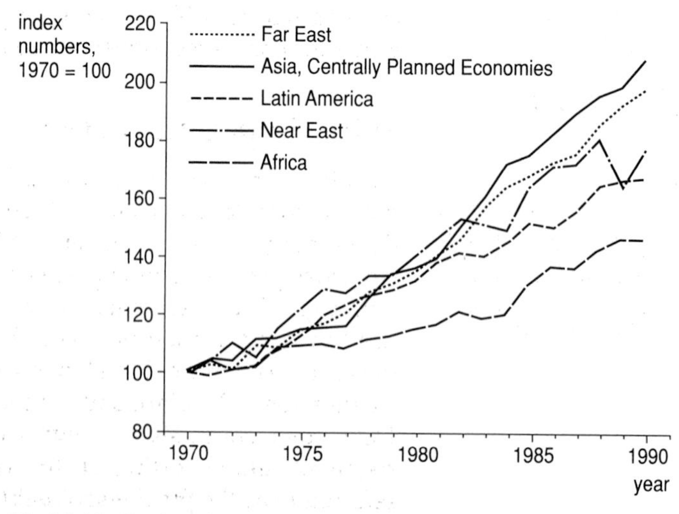

Fig 14.12 Food production in developing regions, 1970–90.

production; in Africa, per capita food production has fallen dramatically (Figure 14.13). In any case, a huge amount of land in developing countries is used to grow non-essential cash crops for export or are used to grow fodder crops, such as soya beans. Again, most of the fodder crops are exported to feed other nations' livestock (Figure 14.14).

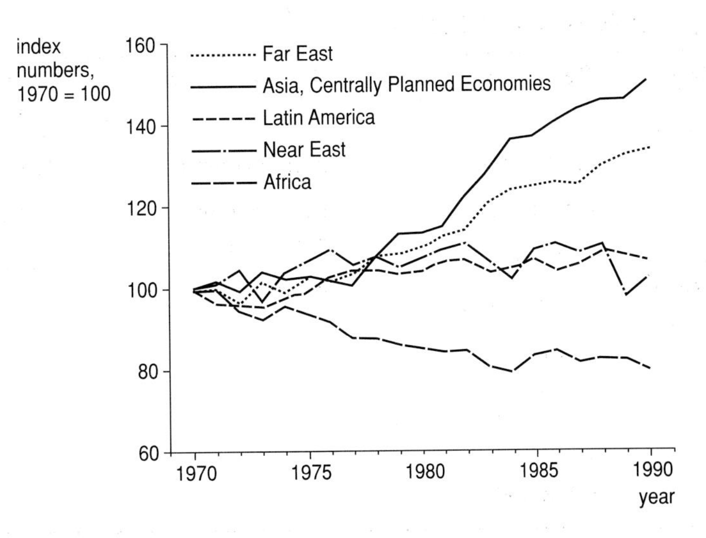

Fig 14.13 Per capita food production in developing regions, 1970–90.

(a) World cereal use, 1988–90

millions of tonnes

900

600

300

0

822

642

human use

animal use

(b) As a % of total cereal available, selected regions, 1990

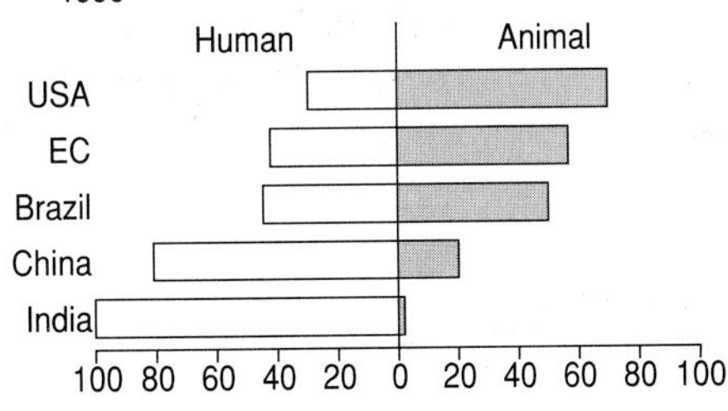

Fig 14.14 Grain for meat.

Farm animals consume nearly half the world's cereal produce. Growing grain to feed animals to turn them into meat is an efficient business – a hectare of cereals can produce five times more protein than a hectare devoted to meat production.

The world's population is growing by nearly 100 million people per year. Where will all their food come from? Future increases in yield will be difficult because much potential land has been severely degraded by soil erosion caused by overgrazing or salinisation (build-up of salt) due to inappropriate irrigation. Many experts suggest that the only way of providing new land is to convert some of the land that is currently being used to grow crops for livestock into land to grow crops for direct human consumption.

Earth's potential to grow more and more food is undeniably limited. Land is a finite resource, so too are the fossil fuels which power machinery and which are used to make chemical fertilisers. Over-reliance on monocultures will mean that pesticide-use remains high but that pests are becoming increasingly resistant to them. Many arid areas have increased food production through the use of irrigation, but much of this is not sustainable because groundwater reserves are being depleted. Genetic engineering is unlikely to increase productivity much further, and even exciting potential developments, such as the introduction of nitrogen-fixing genes into crops such as wheat, will not increase production, although modern, intensive agriculture is largely dependent on resources which are destined to run out.

Why do people starve?

Despite our success at producing food, people die every day from **malnutrition** (a lack of nutrients essential for health) and starvation. Very few developing countries are self-sufficient in basic foodstuffs. Instead, most countries produce and export specialised crops such as coffee, spices, fruit and sugar. The revenue earned from these exports may, theoretically, then be used to pay for imports of cereals, for example. One problem here is that developing countries have borrowed heavily from the richer West in order to finance their development – road building, investment in industry, etc. Developing countries now owe over one trillion dollars to the West, interest payments alone adding up to over fifty billion dollars annually. This is referred to as the **third world debt crisis**. Such debt has not only prevented many poor countries from developing self-sufficiency in basic foods – instead they continue to grow export crops to help pay off their debt – but it has also accelerated the rate at which precious natural resources, such as forests (Chapter 13) and ores (Chapter 15), have been exploited.

Recently there have been suggestions for a radical reorganisation of trade arrangements between the West and the developing world. Some believe that the only way the poorest developing countries are going to achieve sustainable self-sufficiency is for the West to cancel the present debts or at least to enter into debt-for-nature swaps (Chapter 18).

One in five people on this planet suffer from the effects of hunger or malnutrition. Hunger refers to a lack of food required for energy and basic nutritional needs, so that the person cannot lead a normal healthy life. **Malnutrition** is a lack of essential nutrients, such as vitamins, amino acids and minerals. The root cause of hunger and malnutrition is poverty. There is enough food to go around, but the world's poor do not have the money to pay for it nor the land on which to grow it. Further increases in agricultural productivity will do little to change this.

While hunger and malnutrition are so well established as to be taken for granted by the West, famine – a severe shortage of food accompanied by a significant increase in the death rate – has captured the attention of the media. Drought and war have been blamed for many of the famines that have occurred in Africa over the last thirty years. The Sahel region, which extends right across north Africa, is home to fifty million people who have traditionally practised **subsistence agriculture** (growing just enough food for the family or community) or pastoralism, the tending of cattle, sheep and goats. Even during the best of times, rainfall is seasonal and unpredictable and if rainfall simply fails to come at all, both crops and grasslands die.

To make matters worse, increasing human population in the region has led to unsustainable animal populations, which cause overgrazing and accelerated soil degradation. Farmers have abandoned their land and migrated into urban centres where they may be redirected into refugee camps. While the rains have returned to the Sahel, countries such as Ethiopia, Somalia and Mozambique face an almost constant threat of famine, largely as a result of civil wars.

This chapter began with an outline of how, over the last 50 years, agricultural productivity has been increased and of the problems which this has caused. What is needed now, one could argue, is not further increases in productivity but the development of sustainable forms of agriculture. Given that soils, climate and the form of societies in the West are very different from those in developing countries, it is to be expected that there will be differences in their forms of agriculture. However, certain key elements of sustainability should be found in both:

- Soil fertility should be maintained by utilising wastes and by harnessing natural cycles. Organic matter should be replenished by ploughing in crop residues and by using animal manures and legumes.
- Over-reliance on mechanisation should be avoided. Fossil fuels are, in the long term, unsustainable (Chapter 12). In contrast, human labour is a sustainable resource and, by reversing the trend of the last 50 years, labour-intensive agriculture will reduce unemployment.
- Populations of herbivores should be maintained within the carrying capacity of the ecosystem (Chapter 7). This will help to avoid the problems of overgrazing and the loss of valuable trees and woodland to provide fodder, for example in Nepal.
- Biodiversity must be maintained. Crop rotation should be practised more and the use of agroforestry – growing crops and trees together – could be developed (Chapter 13). Above all, we should avoid any further narrowing of the range of staple foods that feed the world.

SUMMARY ASSIGNMENT

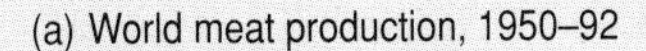

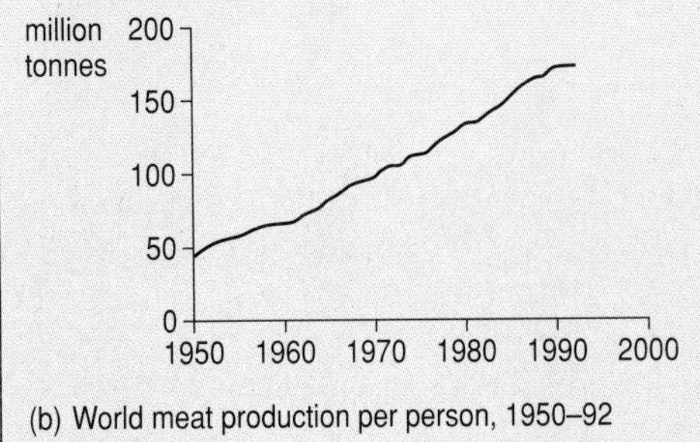

(b) World meat production per person, 1950–92

kilograms 40, 30, 20, 10, 0

1950 1960 1970 1980 1990 2000

(c) World meat production by type, 1950–92

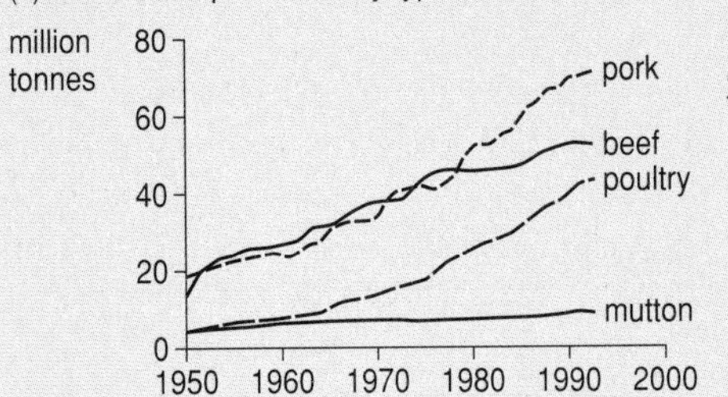

Fig 14.15 World meat production.

Table 14.12 Total world grain production 1950–92

Year	Total / million tonnes	Per capita / kg
1950	631	247
1955	759	273
1960	847	279
1965	917	274
1970	1096	296
1975	1250	306
1980	1447	325
1985	1664	343
1990	1780	336
1992	1745	318

1. Table 14.12 shows the total world grain production since 1950.
   (a) Plot a graph of these figures.
   (b) Describe the trends shown.
   (c) Calculate the percentage increase in per capita production between 1950 and 1992.
   (d) Despite these figures, many people in Africa face starvation or malnutrition-related diseases. Explain why this is still the case.
2. Figure 14.15 gives data on world meat production since 1950.
   (a) Summarise the factors which have enabled such an increase in production.
   (b) Production of pork and poultry appears to be increasing faster than beef and mutton. Suggest:
   (i) an economic reason which might explain this trend
   (ii) a health-related reason which might explain this trend.

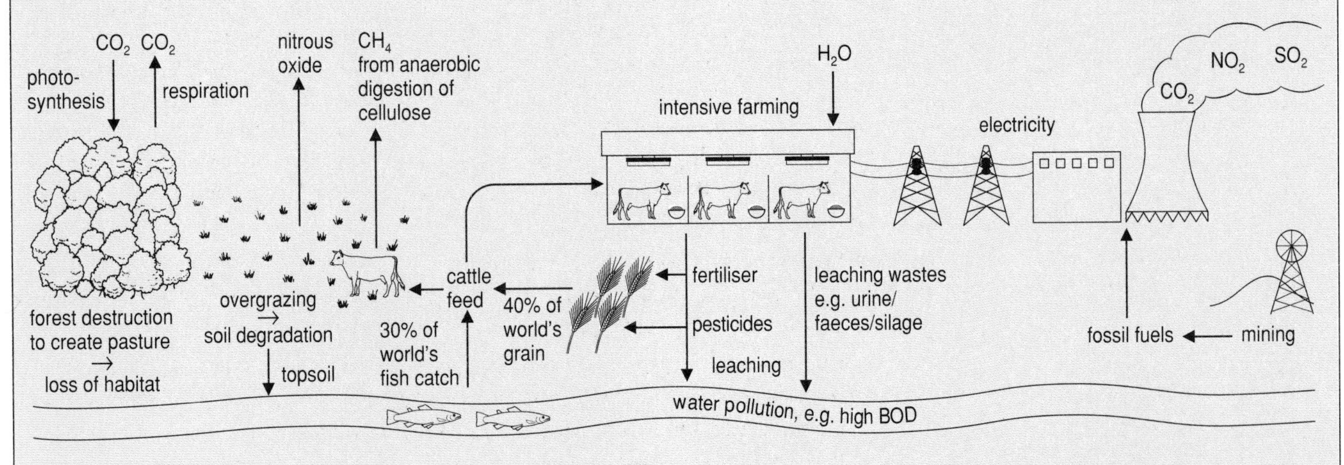

Fig 14.16 Meat production and the environment.

3. Figure 14.16 shows some of the possible harmful environmental implications of eating meat. Study the chart and suggest why:
   (a) increasing meat consumption may lead to an accelerated greenhouse effect
   (b) increasing meat consumption may lead to increased soil erosion.
4. Summarise the conservation value of government initiatives designed to reduce food production.
5. Explain the ecological principles which underlie organic agriculture.

6. In agricultural ecosystems careful management of the soil is essential for the success of the crop. The figure shows three typical soil types.

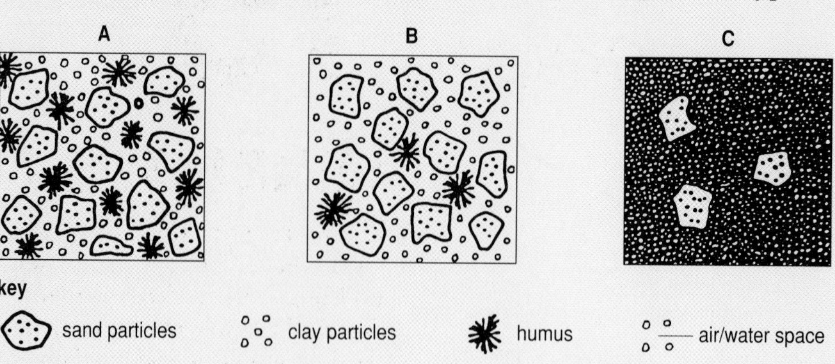

(a) Suggest which soil will be most fertile and briefly explain why.
(b) The use of artificial fertilisers is one method of improving soil nutrient status. Suggest and explain the value of **two** other methods that can be used to improve the soil nutrient status.
(c) Soil erosion is a serious problem in some parts of the world.
  (i) Name **two** agricultural practices which encourage soil erosion.
  (ii) Outline **two** problems caused by soil erosion.
(d) Describe how **each** of the following management practices can be used to prevent soil erosion.
  (i) Zero tillaging methods.
  (ii) Contour ploughing.

NEAB June 1995

Chapter 15

MINERALS AND METALS

It is almost impossible to imagine modern life without minerals and metals, since they provide the material for, among other things, machinery, buildings, roads, fertilisers and thousands of other chemical compounds. Demand for minerals and metals is increasing, as a result of both the growing world population and the rising per capita demand. The mining of the earth's natural resources is therefore accelerating, but it has accompanying environmental consequences. This chapter will consider the use and abuse of our mineral wealth.

LEARNING OBJECTIVES

After completing the work in this chapter you will be able to:

1. describe the formation, extraction and processing of ores
2. explain how geographical, technological, economic, political and environmental factors affect the mineral reserve
3. discuss the main environmental impacts of mining operations.

15.1 AVAILABILITY AND EXTRACTION

The modern energy-intensive society uses almost every type of rock for one purpose or another (Chapter 3). A **mineral** is a naturally-formed element or compound, often in the form of crystals. These may occur in concentrated **veins** or be **disseminated** (spread out) throughout the rock. A rock containing at least one mineral is called a **mineral deposit**. Table 15.1 shows some common minerals and their sources.

Table 15.1 Minerals and their sources

Rock / source of mineral	Main mineral
limestone	calcium carbonate $CaCO_3$
salt	halite NaCl
gypsum	gypsum $CaSO_4\ 2H_2O$
china clay	kaolinite $Al_4Si_4O_{10}(OH)_8$

Almost every rock can theoretically be used but, in practice, only some mineral deposits, called **ores**, can be exploited commercially. Which mineral deposits are classified as ores will depend on a number of factors, including the amount of the required material present in the ore, the geographical location (which will affect transport costs) and the ease of extraction. Even if an ore is identified, it may not be exploited because of environmental factors, political decisions and economic factors.

The use of minerals varies greatly between countries. The greatest use of minerals occurs in more developed countries, many of which have a large supply of minerals – known as a good **mineral base**. However, some countries, such as Japan, have a poor mineral base and therefore import minerals from elsewhere. The UK has had a long history of mining, but today the vast majority of minerals used in the UK are imported simply because we have exhausted most of the economic mineral deposits. The use of a mineral is determined by its properties. Table 15.2 shows some common metals alongside their properties.

Table 15.2 Metals and their properties

Metal	Principal ores	Properties	Use
aluminium (the most commonly traded metal)	bauxite	abrasive, light, strong, durable	aircraft, shipping and car industries; engine castings, windows
copper	copper pyrites, bornite	good conductor of heat and electricity	electric wiring, hot water pipes, coinage, pesticides alloy with zinc produces brass, alloy with tin produces bronze
mercury	cinnabar (solid)	the only liquid metal	medicine, ammunition and paint
iron	haematite, magnetite, limonite	heavy and magnetic	pigment in paper, wood preservative, catalyst, medicines and in steel
steel	an iron and carbon alloy	stronger alloys contain more carbon	construction, cars (0.2% carbon) and tools (0.8% carbon)

Mineral formation and distribution

Ore bodies may form in several ways; the most common processes are summarised in Table 15.3.

Table 15.3 Processes of mineral formation and distribution

Name of process	Process of formation	Mineral example
hydrothermal concentration	minerals in hot solution (brine) crystallise when flowing through cracks in rocks	hydrothermal ores, e.g. zinc
contact metamorphism	sedimentary rock is changed through the temperature and the pressure of hot molten material (magma) being forced between the rock	silver
magmatic segregation	as magma cools, different constituents solidify out at different times (fractional crystallisation), and so will become separated	chromite – a chromium ore
residual deposits	ores become concentrated in minerals through the removal of another substance by weathering processes	bauxite (through the removal of silica by weathering of silicate rocks)
secondary enrichment	due to weathering and transportation, minerals become concentrated within existing deposits	copper and iron
placer deposits	concentration of heavy minerals deposited out of flowing water	sand, gravel and gold
chemical precipitates	precipitation of minerals out of solution in salt water	manganese and phosphorus
evaporite deposits	deposition of salt minerals previously in solution, after the evaporation of water	halcite (sodium chloride) from salt water; sodium carbonate from fresh water

Important mineral deposits are associated with three geological areas. **Continental shields** are areas of crust (Chapter 3), exposed at the surface, which have been unaltered for a long period. They may be rich in minerals because of the volcanic activity that helped to form them (for example the one-billion-year-old Canadian Shield of North America). **Stable plateaus** are areas within continental shields which have become covered in a thin layer of sediment. Ore deposits can be moved close to the surface and can concentrate in fractures caused by thrust faulting and folding (that is, in **fold and thrust mountains**). These ranges are long arc formations several thousand kilometres across (for example the Himalayas, the Alps, the Appalachians and the Urals).

Mineral exploration

It is likely that the first miners located deposits by simply observing surface rocks, but modern exploration geology is a highly detailed and complex process. Analysing geological structures makes it possible to determine how the deposit was formed. This information allows mining companies to make more efficient use of that deposit and to locate new deposits, because geologists know that if a similar structure is located elsewhere then it is likely that a related mineral deposit will also be present at that location.

Geologists explore for mineral deposits by using various **field survey** and **remote sensing** techniques. Remote sensing is the study of an object using instrumentation placed at a distance from that object (for example aerial photography or satellite imagery). This allows a large area to be analysed far more quickly and easily than through laborious fieldwork but involves a higher cost.

Field survey techniques include simple observations, analysis of the rock composition (**geochemical analysis**) and **geophysics**, which allows a detailed view of the underlying geology through, for example, the interpretation of seismic data (Figure 15.1).

These forms of survey are often used in combination; geochemical analysis of morainal material (Chapter 3) in a mountainous area enables the chemical composition of material further up the glacier to be derived. Viewing aerial photographs allows the source area of material in the moraine to be located by retracing glacier movements.

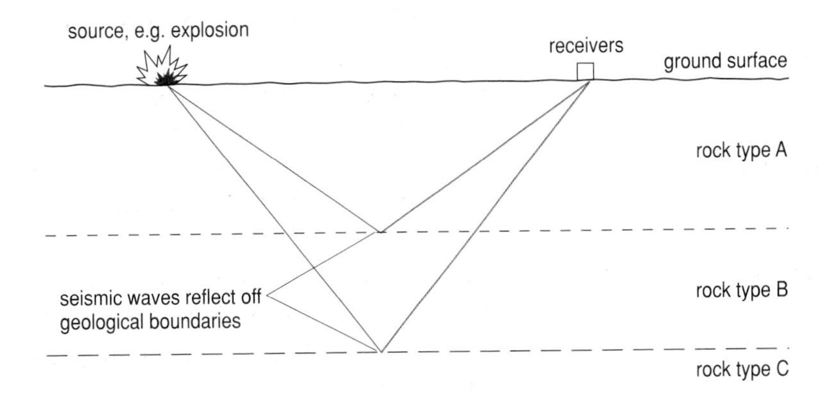

Fig 15.1 Seismic reflections.
The interpretation of seismic data allows the geological structure of the underlying layers to be determined. Seismic waves at the surface will reflect off boundaries between different rock types so that they can be received again at the surface.

Methods of extraction

Extraction of a mineral can occur through mining or dredging. There are four main types of mining:

- **Strip or opencast mining**
  When a mineral is situated close to the surface, the soil and overlying layers (overburden) can simply be stripped away and the minerals dug out (Figure 15.2). This is the cheapest form of mining.

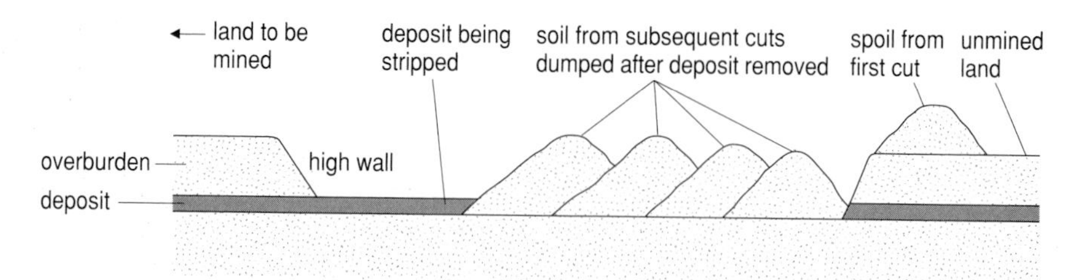

Fig 15.2 Strip mining.

- **Quarrying**
  When the mineral is present at the surface, it can be simply dug and blasted out. The need for building stone has resulted in a lot of small, localised quarries around the country.
- **Adit mining**
  Horizontal galleries may be tunnelled out when a mineral is situated in a horizontal seam.
- **Shaft mining**
  A vertical shaft is sunk deep into the earth, from which a series of galleries are made at different depths (Figure 15.3). This was the original method of exploiting minerals at depth, as proven by the neolithic flint mines which have been discovered around the country. However, modern techniques often allow the mineral to be extracted by the removal of thick layers of overburden, that is, strip mining.

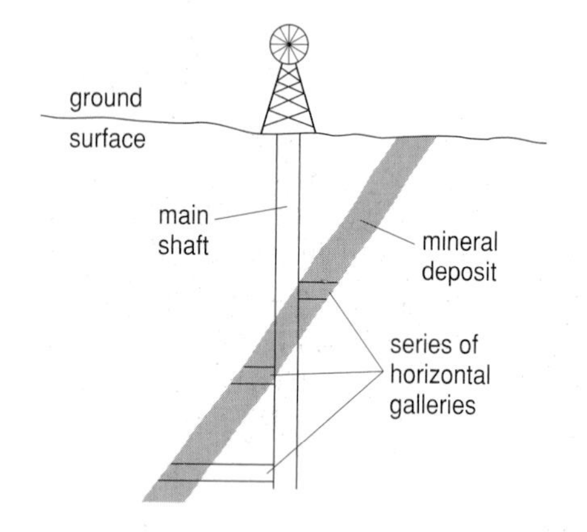

Fig 15.3 Shaft mining.

Fig 15.4 An opencast copper mine.
Opencast mining is cheaper and easier but has a greater environmental impact than deep mines (Chapter 17). In recent years there has been a large swing from deep-mined coal to opencast coal in this country, largely because it is cheaper.

QUESTION

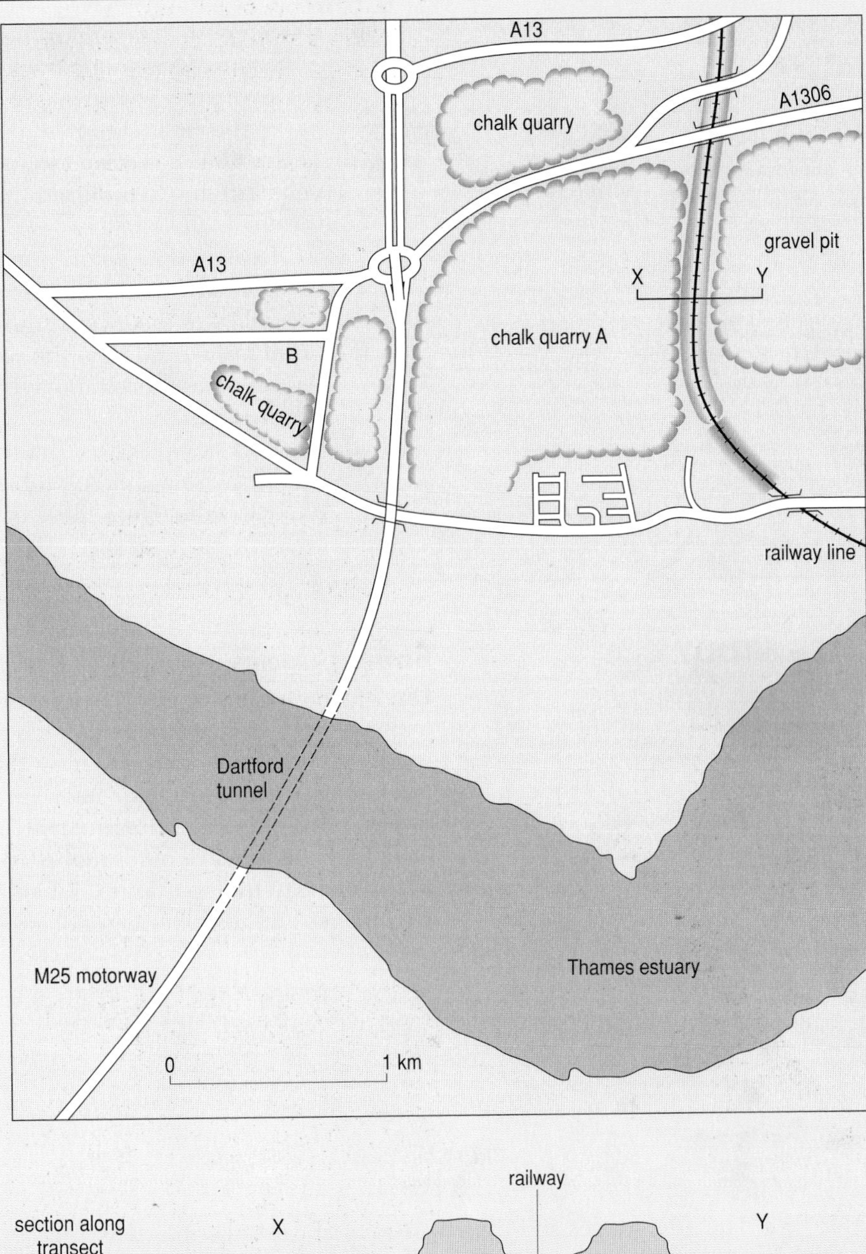

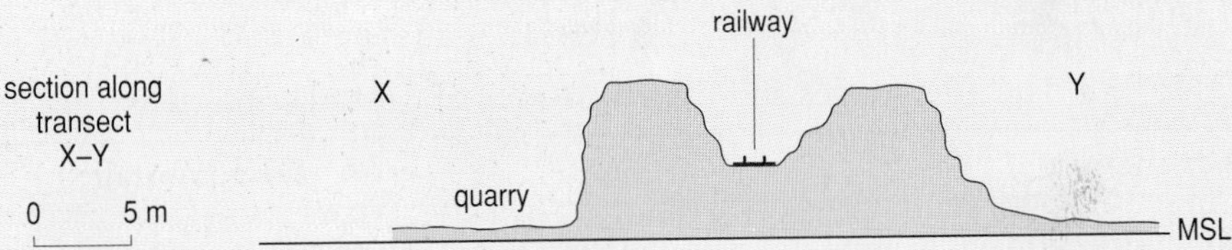

Fig 15.5 Chalk pits near the Thames estuary.

15.1 The map shows sites adjacent to the Thames estuary made derelict by extraction of chalk for the cement industry. Extraction ceased when at mean sea level (MSL).

(a) (i) Give **two** possible reasons why continued extraction below MSL proved uneconomic.

(ii) What evidence from the map indicates that extraction has had to cease in some quarries?

(b) (i) What is the *approximate* area of chalk quarry A? If it is on average 20 metres deep calculate the likely volume of material needed to infill to the original level.

(ii) Give **two** reasons why it is unlikely that infill is a suitable method of reclamation for *these* sites.

(c) (i) Suggest a reason for the nature of the section **X–Y** across the route of the single track railway line.
(ii) Explain why the role of the railway is likely to be minimal in the redevelopment of the derelict sites.

(d) The road at **B** is closed to four-wheeled vehicles (except farm vehicles) due to undermining of the road by solution holes in the chalk quarry face.
(i) Explain how solution holes develop in chalk.
(ii) Give **one** possible reason why closure of the road was the policy of the local authority.

(e) What influence might the recent construction of the M25 (London outer orbital) have had on the redevelopment of the derelict sites?

(f) Disused clay pits exist in the area as a result of the cement industry. How are the problems and methods of reclamation likely to differ from those of the disused chalk quarries?

AEB June 1987

CASE STUDY

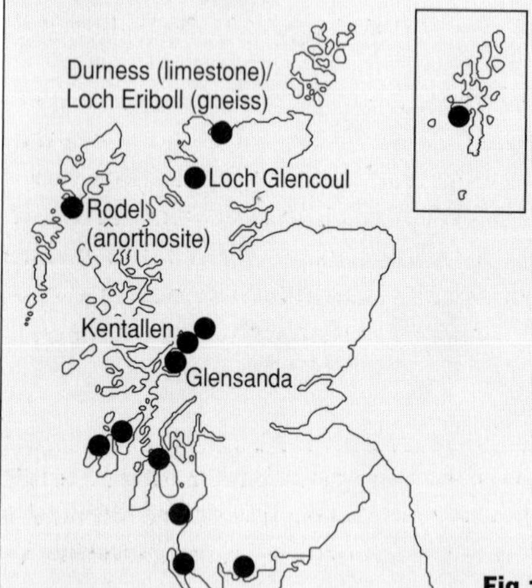

Fig 15.6 Location of Scottish superquarries.

Superquarries

Due largely to new roads, the demand for aggregate is increasing. The majority of this is used in areas such as south-east England, where there is not enough raw material to meet the demand. Superquarries in Scotland, such as Glensanda near Fort William, are seen to be one way of meeting this demand. Superquarries are usually situated in quiet areas, have huge volumes of high-quality rock and have a deep harbour nearby to allow cheap transportation. Table 15.4 outlines some of the many advantages and disadvantages of superquarries.

Table 15.4 The advantages and disadvantages of coastal superquarries

Advantages	Disadvantages
they cause less environmental damage than numerous smaller quarries because they have the maximum impact in the minimum area, as opposed to a lesser impact in a larger area; six Scottish superquarries represent 45 large inland quarries or 450 sand and gravel pits	if the rock was exported overseas, then the ships would discharge ballast which was taken on board in these foreign waters; this would introduce alien species into a delicate marine ecosystem
greatly improves the economy of rural areas	shipping increases the risk of oil pollution Scottish Nationalists argue that Scottish mountains are being destroyed for the good of England, not Scotland
low disturbance to humans because they have remote locations	very high-quality wildlife areas which have previously been largely untouched will be disturbed; for example, huge quantities of dust are produced, which settles on the sea floor and affects shellfish stocks
other countries have superquarries, and therefore Britain must develop them in order to remain economically competitive	if we conserve mineral resources, then there would be no need for so many superquarries

QUESTION

15.2 (a) Compare and contrast the environmental benefits and problems associated with superquarries.
(b) How would an increase in public transport use alter the need for superquarries?
(c) Outline the reasons why there is particular pressure for mining in the Peak District National Park.

Mineral resources and reserves

The **crustal abundance** of a mineral is the amount that is estimated to exist within the entire earth's crust. An **abundant** mineral has a concentration higher than 0.01% (for example aluminium has 8%) and a **scarce** mineral has a concentration of less than 0.01% (for example gold has 0.000 000 4%). When exploiting an ore, the concentration of a mineral within a rock (the **grade**) is an important factor. To produce a given amount of mineral, a low-grade ore will require much larger volumes and therefore cause higher costs and more environmental degeradation than a high-grade ore. For example, in order to produce 12 million tonnes of copper, it is necessary to exploit 60 million tonnes of 2% grade ore or over 600 million tonnes of 0.6% grade ore. The exploitation of the latter would clearly have a much greater environmental impact.

It is possible to calculate the grade of ore which is necessary to produce a profit (the **cut-off grade**). The cut-off grade is very different for each mineral, and will vary through time with the factors which affect the resource base. For example, uranium is mined in quantities of a fraction of a percent so that 250 tonnes of ore are needed to produce around one tonne of usable uranium fuel (Chapter 12), whereas iron ores which are exploited have concentrations of around 60% iron. Table 15.5 shows the cut-off grades for several metals.

Table 15.5 The cut-off grades for metals

Metal	Average % of metal in crust	Cut-off grade /% metal	Number of times metal must be concentrated above average to reach cut-off grade
aluminium	8.1	30	3.75
copper	0.005 5	0.4	73
gold	0.000 000 4	0.000 01	25
iron	5.0	25	5
lead	0.001 3	4	3077
nickel	0.007 5	0.5	67
mercury	0.000 008	0.2	25 000
tin	0.002	0.5	2500
uranium	0.000 18	0.1	556
zinc	0.007	4.0	571

The total **resource** of a mineral is the volume within the earth's crust which is estimated to be economically exploitable now or in the future. The resource base includes those minerals which are exploitable now (demonstrated), those that are known to exist and may become exploitable in the future (inferred) and those which are speculated to exist (undiscovered). A mineral **reserve** is that part of the resource base which is economically exploitable

Fig 15.7 Tin mine in Cornwall.

under the present socio-economic conditions. There are several important socio-economic factors which will affect the mineral reserve.

Demand

The demand for any particular mineral fluctuates through time. This may be because of changes in lifestyles and technology, and also because another mineral can be substituted for the original. For example, as plastic replaced tin in food packaging, demand for tin fell and, consequently, so did its price. In turn, this affected the cut-off grade and led to the widespread closure of tin mines in, for example, Cornwall.

Technology

As technology improves, industry can extract more mineral from an ore, make more efficient use of the raw material and exploit ore bodies that were not possible in the past.

Supply

New reserves are still being found throughout the world. If a particularly good reserve is located, the supply will increase and less economical sites may close. A site which was close to the cut-off grade may no longer be a profitable reserve.

Political factors

A government may wish to subsidise mining in a region which has little other industry. Conversely, mining may be discouraged, perhaps to encourage the use of alternatives or because of the environmental problems which might result from extraction.

Environmental factors

The environmental impacts of mineral extraction play an important part in the mining industry. For example, new, large-scale mining operations may not be allowed in national parks (Chapter 18).

Although there are vast quantities of minerals within the earth's crust, every mineral is finite because it can only be mined once. When the amount of reserve has been established, the figure can be divided by an estimate of annual consumption to produce a **life index**.

QUESTION

15.3 (a) Explain the difference between a reserve and a resource.

(b) It is interesting that before the nuclear age uranium had no mineral value. Describe what effects the use of uranium may have had on the economic reserve of coal.

Processing of minerals

Every ore will contain within it waste minerals of no economic value which must be removed through **processing**. The dumping of this waste material can cause severe environmental problems, for example silver is extracted from silver sulphide, resulting in the production of acidic sulphide waste.

Iron is separated from iron ore in the blast furnace. The ore – impure Fe_2O_3 – is heated with limestone and coke. The coke, which is basically carbon, reduces the iron oxide to iron, while the limestone combines with some of the acidic impurities present. Remaining impurities in the resulting 'pig iron' – which will include carbon – are 'burnt out' by blowing oxygen through the molten iron. This produces a mixture of waste gases including carbon dioxide.

Bauxite is the ore from which aluminium is obtained. There are three stages involved in bauxite processing:

1. Fine clay and quartz particles are removed by crushing, washing and screening.

2. Aluminium hydroxide is removed by dissolving the sediment in sodium hydroxide. This is called the Bayer process and results in the precipitation of alumina (Al_2O_3).

3. Pure aluminium is removed through electrolysis of a mixture of molten alumina (Al_2O_3) and cryolite (Na_3AlF_6). The cryolite is added to reduce the temperature necessary to melt the alumina and hence reduce the energy input required. The spoil produced from bauxite processing is often dumped in huge lagoons which become highly alkaline.

15.2 THE EFFECTS OF MINING, QUARRYING AND REFINING

Dereliction of land

Mining activity causes a considerable loss of land and often permanent scarring of the land surface (Chapter 17). **Scarification** is the creation of surface pits through the mechanical extraction of minerals. Disused sand and gravel pits have substantial areas damaged by scarification.

Subsidence

The presence of old, deep mines may cause the ground surface to **subside** in a vertical or horizontal direction. This may severely damage buildings, roads and farmland, as well as alter surface drainage patterns.

Pollution

Mining operations often pollute the atmosphere, surface waters and ground water. Huge volumes of **dust** generated by explosions, transportation and processing may kill surrounding vegetation. Water is sometimes sprayed over opencast mines in an attempt to minimise the dust released. Rainwater seeping through spoil heaps may become heavily contaminated, acidic or turbid, with potentially devastating effects on nearby streams and rivers (Chapter 10).

Kaolinite is sometimes mined using high pressure water jets which wash away the softer kaolinite from the granite bedrock. This method pollutes water courses, has a high demand on water resources and is highly erosive both on the site and downstream.

Noise

Blasting and transport cause **noise disturbance** to local residents and to wildlife; as well as rail traffic, there are around 4000 lorry movements per day from the Mendips, which is the most heavily mined area in Europe.

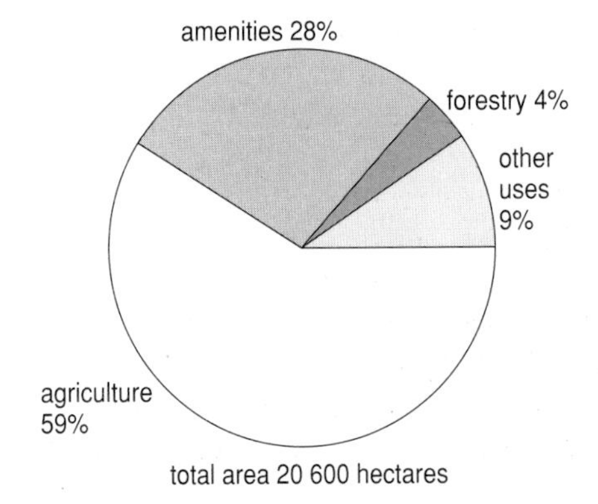

Fig 15.8 Usage of land reclaimed after mineral workings in England, 1982–88.

Energy

Extraction and transportation requires huge amounts of energy (Chapter 12), which adds to impacts such as acid rain and the enhanced greenhouse effect (Chapter 1).

Most modern mining permissions contain guarantees of restoration, although many old permissions which do not contain such guarantees are still valid. Restoration is expensive, takes a long time and is often hampered by the massive environmental problems on the site. Despite the problems involved, many successful reclamation schemes have taken place. Between 1982 and 1988, 20 600 hectares of land (approximately 10 000 football pitches) were reclaimed from mineral workings. Much of this reclaimed land is then used for agriculture.

QUESTION

15.4 Figure 15.9 shows an abandoned deep-shaft coal mine in the South Yorkshire coalfield. The site is now to be converted into a Country Park. The top table shows the physical and chemical characteristics of an opencast mined area and colliery spoil. The lower table shows the typical nutrient content of wastes that can be used in reclamation schemes.

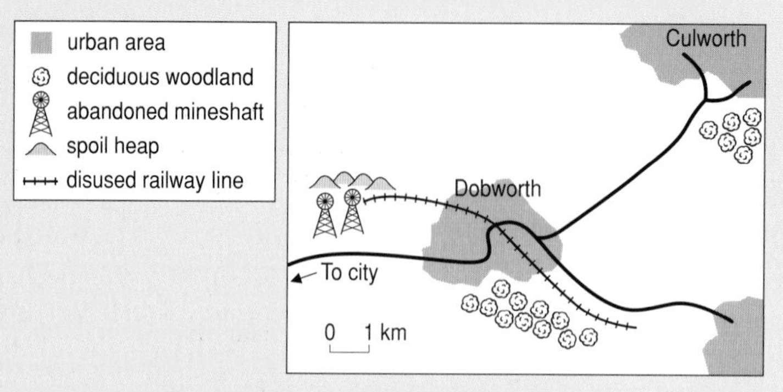

Fig 15.9 An abandoned coal mine in South Yorkshire.

material	texture and structure	stability	nitrogen	phosphorus	pH
opencast mined area	◎	◎	◎	○	slightly acidic
colliery spoil	●	●	●	●	highly acidic

● extreme deficiency or problem ◎ moderate deficiency or problem ○ slight deficiency

material	percentage composition			special problems
	nitrogen	phosphorus	organic matter	
sewage sludge	2.0	0.3	45	may contain toxic metals
farmyard manure	0.6	0.6	24	toxic if applied directly to plants

(a) How might the physical characteristics of the site be improved?

(b) Using information in the first table and your own knowledge, explain why it is difficult to establish vegetation on coal spoil heaps.

(c) State **one** advantage and **one** disadvantage of opencast mining.

(d) Explain:

(i) how leguminous species can be of value in the reclamation of mined areas

(ii) why sewage sludge is more useful than farmyard manure on colliery sites.

(e) Explain why pulverised fuel ash:

(i) has a variable composition

(ii) does not contain nitrogen.

(f) What is a Country Park?

(g) Name **two** types of countryside interpretation facility which would be appropriate at the park.

(h) Suggest why some residents of Dobworth opposed the idea of establishing a Country Park on the site.

AEB June 1995

CASE STUDY

Mining in national parks

National parks (Chapter 18) contain over 3% by area of the UK's working mines (Table 15.6), which yield around 16 million tonnes per year of aggregate (limestone, sandstone and igneous rocks (Chapter 3), used for road building and construction), as well as important specialised minerals such as potash, fluorspar and china clay.

Table 15.6 The number and area of active mineral workings in Britain's national parks

National park	Surface workings		Waste disposal sites	
	Number	Area /hectares	Number	Area /hectares
Brecon Beacons	27	611	2	5
Dartmoor	9	510	7	166
Exmoor	–	–	–	–
Lake District	20	172	21	87
Northumberland	5	50	-	2
North York Moors	21	266	4	23
Peak District	121	1353	26	122
Pembrokeshire Coast	10	133	1	2
Snowdonia	92	117	168	241
Yorkshire Dales	13	386	1	3
total	318	3318	230	631

The Council for National Parks state that 'a clear and significant environmental problem is being caused under current mineral policies'. Mining especially affects the landscape beauty and recreational use of areas, which is precisely why national parks were set up. The view of individual national parks is that mining in or near national parks should only take place if a pressing necessity can be proved.

The impact of quarrying is likely to increase because crushed rock is replacing sand and gravel deposits for some end uses, and old mining agreements contain inadequate restoration agreements (these old agreements also mean that mining will continue for some time – at least another 47 years in the Peak District). The demand for aggregate is increasing owing to new road developments and new uses such as flue-gas desulphurisation in power stations (Chapter 10).

However, mines provide much-needed jobs for rural communities where there is little other industry – 10% of the population of the Peak District, for example, work in the mining industry. Such employment boosts the local economy and helps persuade young people not to leave the area.

QUESTION

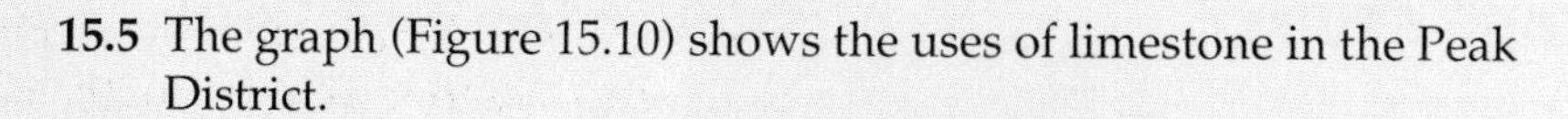

15.5 The graph (Figure 15.10) shows the uses of limestone in the Peak District.

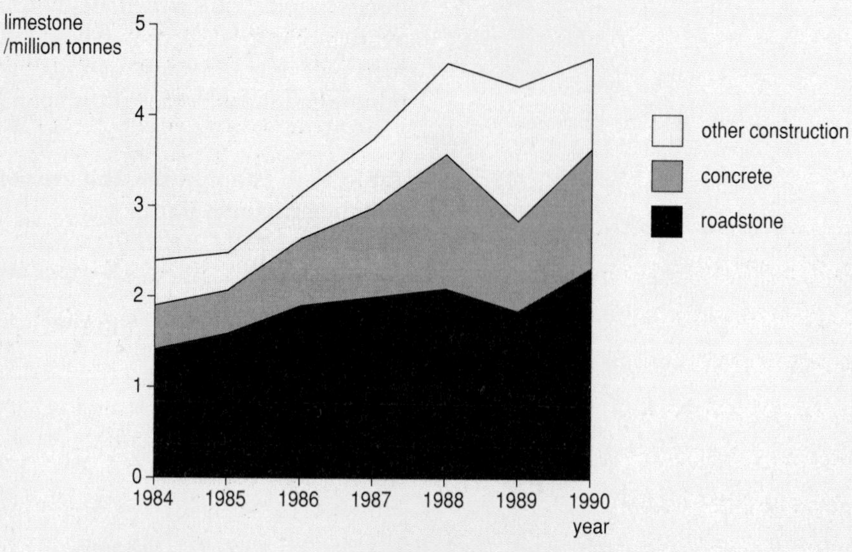

Fig 15.10 Uses of limestone in the Peak District.

(a) To which major group of rocks do limestones belong?
(b) From the graph calculate the percentage of limestone which was used for road construction in 1990.
(c) It is expected that demand for limestone will increase dramatically over the next 15 years. Suggest **one** advantage and **one** disadvantage to the local community of an increase in quarrying in the Peak District.

AEB June 1995 ES

15.3 CONSERVATION OF RESOURCES

Predicting demand and rates of exhaustion

Minerals are finite (they can only be mined once) and will, therefore, at some point in the future run out. However, it is not easy to predict precisely how long a particular mineral will last. It is likely that not all of the geological resources have been mapped and, because of changes in lifestyles, technology, population growth and mineral substitution, future demand may be difficult to establish.

Mineral conservation

Conservation of mineral resources can be achieved by reducing the amount of packaging used, reusing and recycling materials, substituting some non-recyclable materials for alternatives which can be recycled, or manufacturing items that will last longer. The efficiency of mineral use could also be increased (for example at present around 10% of aggregate material is wasted on construction sites) and alternative mineral sources, such as coal and china clay waste, could be used (this would also solve many derelict land problems) (Chapter 17). In addition, it is possible to impose a tax on new resources used, which would promote recycling, reusing and using less of the resources.

An important factor in predicting the rate of exhaustion is the amount that will be recycled and reused. Modern society has something of a 'throw away' mentality, in which items are only used once and simply discarded when something goes wrong with them or when the latest model is created. By reusing items, demand for new resources and energy is reduced, less waste is produced, and the lifetime of the mineral is prolonged.

Recycling is not a modern idea; in the past, all drinks bottles were made from glass and a refund was given if the bottle was returned. It is only in recent times that plastic, which can be recycled, has become widely used as a packaging material. Recycling schemes for such materials as glass, paper, cardboard, plastic, clothing fabric, ferrous metals (for example steel) and non-ferrous metals (for example aluminium, copper, lead and tin) have started all over the UK (Chapter 16).

SUMMARY ASSIGNMENT

1. What are the advantages of conserving mineral resources?
2. Metallic elements have great economic importance. The UK has a rich variety of mineral deposits, although most of the ore fields of England and Wales now contain only sub-economic reserves and large amounts of metal are imported. The figure shows the UK consumption of metals and their crustal abundance.

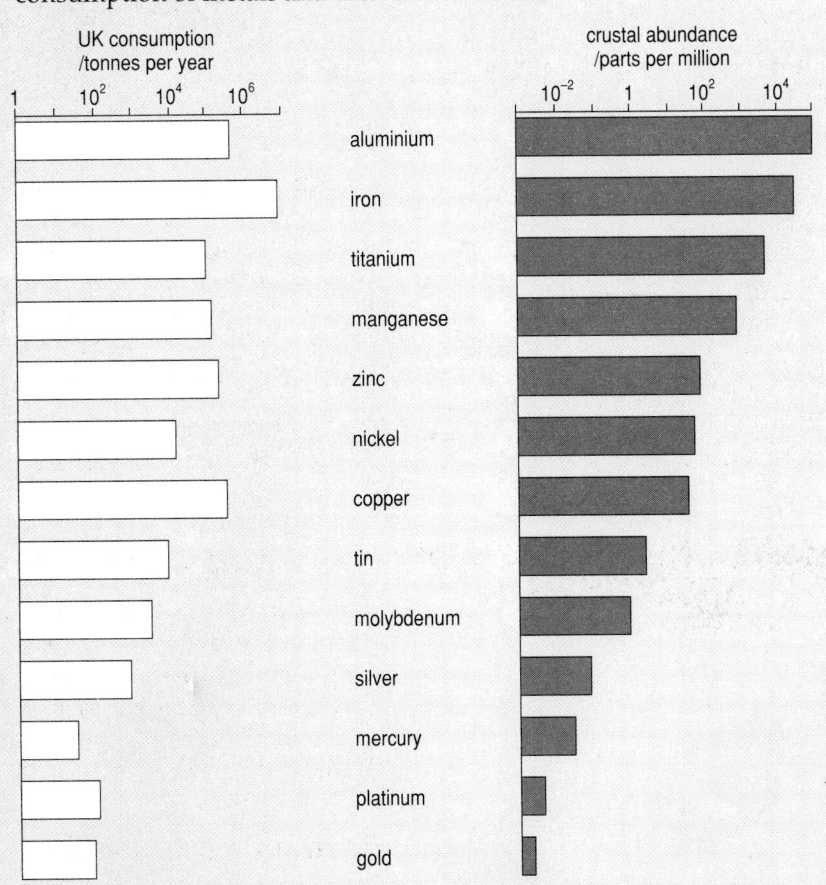

(a) With reference to the figure:

(i) ignoring copper, describe the relationship between UK consumption and crustal abundance

(ii) suggest **one** reason why copper does not fit well into this trend

(iii) what are the advantages of using a logarithmic scale as shown in the figure?

(b) The world reserves of copper are currently estimated at about 6×10^8 tonnes, whereas current world production is about 9×10^6 tonnes per year.

(i) Calculate the reserves/production ratio.

(ii) Explain the significance of your answer.

(c) The figure below shows how the energy for mining and milling copper ore varies with the copper content of the rock.

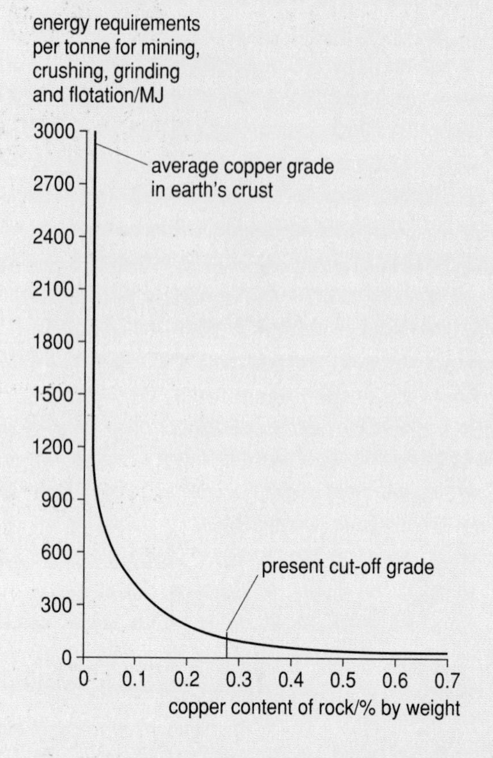

(i) Using the figure, discuss the implications for copper mining of possible changes in the world economy and the price of energy.

(ii) Discuss the environmental consequences of these changes.

NEAB June 1995

Chapter 16

WASTE

Every year, developed countries produce hundreds of millions of tonnes of new waste, much of which is harmful to humans or other plants and animals. Disposing of this waste in an environmentally safe way is becoming increasingly difficult and expensive, and many people are beginning to realise that the best approach is not to produce so much waste in the first place. This chapter will begin by describing the sources, nature and disposal of waste before considering some of the strategies by which we might reduce the amount of waste which we produce.

LEARNING OBJECTIVES

After completing the work in this chapter you will be able to:

1. identify what is meant by the term 'waste' and the problems associated with waste
2. describe the different types of waste
3. explain the growth in the amount of waste produced today compared with the past
4. outline the different types of waste disposal
5. analyse the benefits of recycling compared to alternative forms of waste disposal.

16.1 WHAT IS WASTE?

Waste is material which has no direct value to the producer and so must be disposed of. Waste can be solid, liquid or gaseous material and arises from a wide variety of sources including the generation and use of energy, extraction of raw (virgin) materials, industrial processes, transport and agricultural activities. It also arises from construction, selling of goods or products, using products and waste disposal. This continual production of waste material is known as the **waste stream** and includes the whole variety of waste produced through domestic, industrial, construction and commercial processes.

Waste production

The generation of waste products is not only inevitable but also enormous; food results in sewage, food packaging becomes unwanted paper and plastic, and so on. For example, the European Union countries generate around two billion tonnes of waste per year, and in the United States waste generation is equivalent to 850 kg per person per year. The generation of large quantities of waste can cause a wide range of problems.

The waste problems

- Disposal of this waste is an increasing problem. Traditional disposal sites are rapidly becoming filled (Figure 16.1) and the volume produced continues to grow.

Fig 16.1 Disposal site in the USA.

- Rising waste disposal exacts costs from local government and so too from the public. Due to the volume of waste generated and its potential environmental effects, safe and acceptable disposal of waste is becoming increasingly expensive.
- Environmental impact of disposal can be high in many areas. The main disposal techniques all have associated environmental problems. Pollution can occur to soil, water and air, along with the possible loss of large areas of land used for disposal.
- Hazardous and toxic material in the waste stream, such as paints, solvents and dry-cell batteries, may represent a danger to human health and cause significant damage to the environment.
- Raw material and energy consumption is high. Consequently there are the associated problems of mining and energy production, particularly the burning of fossil fuels (Chapter 12, Chapter 15).

Types of waste

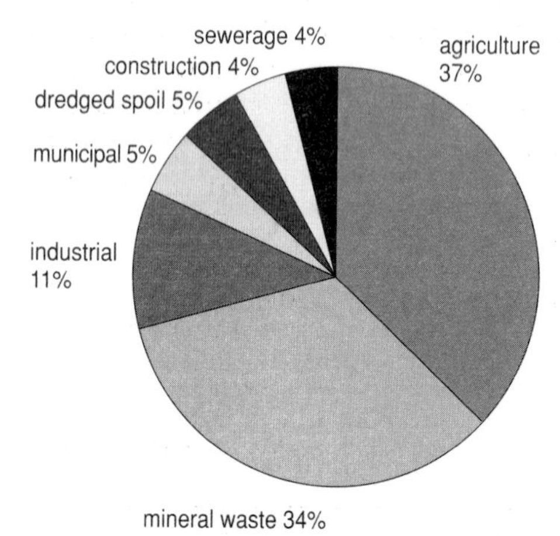

Fig 16.2 Components of UK waste production.

Figure 16.2 shows the components of the 400 million tonnes of waste (excluding air pollution emissions) that the UK produced in the late 1980s. It classifies waste into a wide variety of components. For convenient analysis, it can be divided into the following four main classifications, each of which will then be considered in further detail:

- municipal waste
- industrial waste
- agricultural, mining and construction waste
- hazardous and toxic waste.

The particular problems of waste generated during production of electricity in power stations and through transport are covered in Chapter 12.

Municipal waste

Municipal waste is the waste generated from urban areas, particularly houses and shops. Although Figure 16.2 shows that municipal waste accounts for only a relatively small fraction of total waste production, it is the form of waste that most people can personally control. The composition of municipal waste varies from country to country, but an average composition is shown in Figure 16.3.

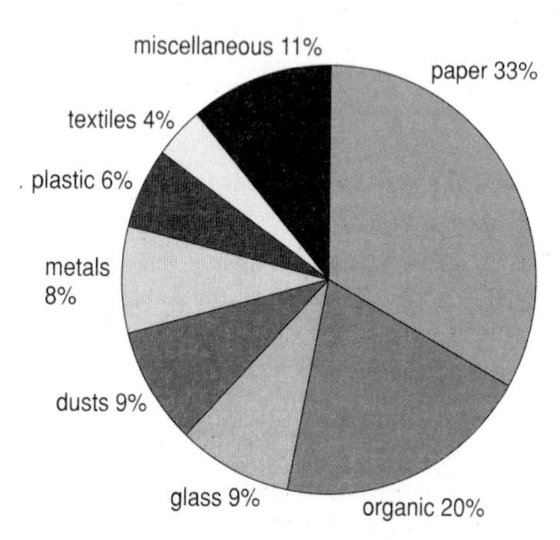

Fig 16.3 Typical components of municipal waste in developed countries.

Organic waste, ranging from garden waste to food scraps, is still the main component. However, since the 1950s, organic waste has been declining as a proportion of municipal waste. This is mainly due to increasing amounts of paper, cardboard and plastics and can clearly be associated with a change in people's lifestyle.

Industrial waste

This is the unwanted material from industrial processes, of varying composition from country to country, depending on their particular industrial base. For example, the major components of industrial waste in the UK are blast furnace and steel slag and power station ash. In addition, a wide variety of industrial activities, including food manufacture and horticulture, produce significant volumes of waste. However, many modern industrial activities make a considerable effort to minimise or reuse waste. This includes selling unwanted materials to other industries, since it makes economic sense to produce as little waste and as much saleable product as possible. For example, tree bark produced by sawmills is sold on to be used as garden mulch.

Agricultural, mining and construction

This can be sub-divided into three different sources.

Agricultural waste includes organic waste from livestock, both housed and grazed, along with materials such as straw and plastics. The organic forms of agricultural waste (made up of faeces and urea) can be recycled back to the land as fertiliser, as long as the farmer has arable crops as well as livestock. However, this process must involve considerable care because of the pollution problems associated with such wastes entering water courses, particularly eutrophication (Chapter 10). Animal waste is up to 100 times and silage effluent up to 200 times more polluting to water courses than raw sewage.

Mining and quarrying waste (mineral waste), such as soil and rock, forms a major component of overall annual waste production. In the UK, major mining wastes include colliery spoil, china clay and slate.

Demolition and construction wastes include a wide range of materials, predominantly asphalt from roads, along with concrete, masonry, steel, timber and cement from various demolition activities.

Hazardous and toxic waste

Hazardous and toxic waste is present in all the previous three waste classifications. This category includes any discarded material which contains substances known to be **fatal** to humans or animals in low doses, such as cyanide. It also includes **toxic** waste, for example polychlorinated biphenols (PCBs) which were widely used to produce insulators for the electrical industry. The extent of toxicity of these compounds in humans is unknown.

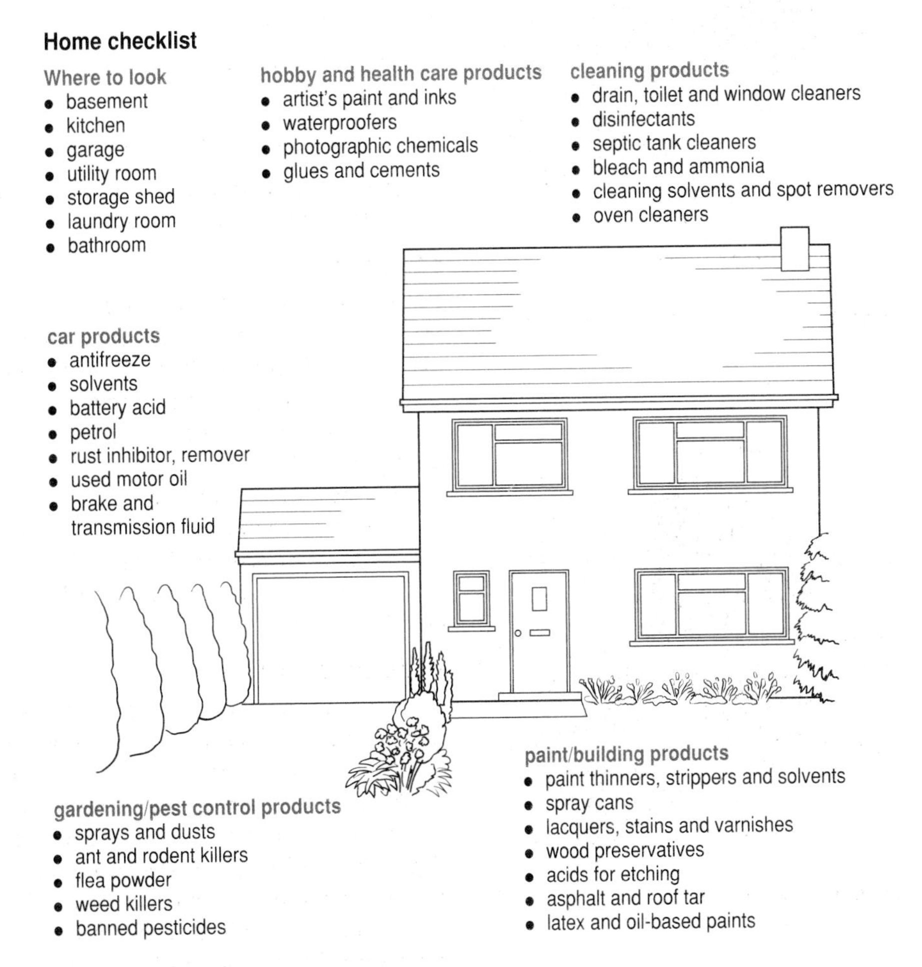

Fig 16.4 Hazardous waste in the home.

Carcinogenic waste can cause cancer. Asbestos causes mesothelioma, a cancer of the chest and abdominal lining. The silvery white metallic element, cadmium, is an example of **mutagenic waste**. It is widely used in industrial processes, in pigments and in nickel-cadmium batteries. **Teratogenic waste** can cause foetal abnormalities to humans or other life forms. In 1955, an outbreak of congenital cerebral palsy in Minimata, Japan, was found to be caused by pregnant women's consumption of fish contaminated with the industrial chemical methyl mercury. Another well-known teratogen is agent orange, a chemical spray containing dioxin used as a defoliant by the US army in the Vietnam war.

Hazardous and toxic waste can be corrosive and explosive or highly reactive (undergoes violent chemical reactions on its own or with other substances).

In the UK, the amount of waste that is considered 'dangerous to life' increased to 2.5 million tonnes by 1991. The greatest volume of all such waste comes from the chemical and petroleum industries, accounting for 71% of all hazardous or toxic waste in the US, for example. Significant producers are companies which make chlorinated hydrocarbons for pesticides, along with the production of materials such as cyanide, waste oils, arsenic, asbestos and heavy metals.

Most individual households do not accumulate much hazardous waste, although many everyday household materials such as solvents, cleaners and fuels are potentially as dangerous as some industrial products. Consequently, accumulation from many households can make up significant quantities of hazardous waste (Figure 16.4). The main problems associated with hazardous waste are summarised in Table 16.1.

Table 16.1 Problems associated with hazardous waste

Type of hazard	Hazard
water contamination	pollution of lakes, rivers and groundwater supplies
soil contamination	disposal-site contamination may make large amounts of land unfit for use and expensive to clean up
habitat destruction	waste entering lakes, estuaries and rivers can destroy whole ecosystems, for example through eutrophication
damage to human health	exposure to many forms of wastes may affect health, from localised poisoning to bioaccumulation and biomagnification of materials such as DDT and PCBs
damage to infrastructure	acids, caustics, solvents and highly reactive compounds cause fires, explosions and corrosion

CASE STUDY

The problems of nuclear waste disposal

High-level and intermediate-level nuclear wastes possess considerable problems of disposal. The essential problem is that the waste is dangerously radioactive for extremely long periods of time, and consequently any permanent storage facility must prevent release of radioactivity for at least 250 000 years. This means that finding suitable sites for **waste repositories** (disposal sites) is extremely difficult. Considerations involve the effects of potential disturbances, including unpredictable events like ice ages, and strong public opposition. Examples in the UK and the US illustrate two different approaches to the problem.

NIREX, the UK national nuclear waste disposal company, plans to build a deep underground rock depository for high and intermediate nuclear wastes near Sellafield, Cumbria in north-west England. The site selection is aided by British Nuclear Fuels which already owns the site, making planning and development easier. However, local people and environmental groups have voiced strong objections to the site, due to potential health hazards. For example, the rocks of the site are linked to a surface-exposed outcrop a few miles inland in the Lake District, and rain entering the rock could wash nuclear waste from the underground store. A study by Her Majesty's Inspectorate of Pollution (HMIP) concluded that plutonium in the depository could contaminate drinking water, potentially exposing local people to radiation doses 10 000 times the legal limit.

In the US, with the absence of suitable land owned by the nuclear industry, a consortium of 33 nuclear utilities offered the Mescalero Apache tribe a deal worth 250 million dollars to build a nuclear waste repository on their land, specifically in the Yucca Mountain of Nevada. An estimated 20 000 tons of spent fuel rods would be stored for 40 years as a transitional arrangement until a permanent plan for disposing the waste is produced. The money derived was proposed to help alleviate the tribe's severe social problems, with 30% unemployment and 50% poverty rates. The proposal was initially rejected by a tribal vote early in 1995. However, subsequent to this, the tribe voted in favour of the dump proposal.

QUESTIONS

16.1 Using the information in Section 16.1, discuss how the content of the waste stream will vary
  (a) between different geographical regions of the UK
  (b) between more-developed and less-developed countries.

16.2 The production of large volumes of waste material will involve consumption of large amounts of raw materials and energy. Outline the harmful environmental effects which will be associated with this.

16.3 Explain how the application of organic animal waste from farming may lead to the process of eutrophication and what may be done to avoid this problem.

16.2 THE GROWTH OF WASTE

The amount of waste generated in the world's economically developed countries is enormous and has been dramatically increasing during the twentieth century, particularly since the Second World War. It has been estimated that between 1920 and 1970, the volume of US domestic waste increased five times as quickly as the population. This increased quantity of waste can be directly linked with increasing economic affluence of a country. A variety of reasons can be identified why economic growth leads to more waste.

Increased consumption of raw materials and energy and the increased manufacture, transport, sale and use of a wide variety of products all cause the production of more waste.

Products with **built-in obsolescence** – products such as toys, electrical and sporting goods – are not designed to last very long and often the simplest problems are difficult to repair. This results in a constant demand for the products since replacement is often cheaper than repair. Many industries rely on this built-in obsolescence to maintain sales.

There is an increase in the amount of packaging used. The development of refrigeration and rapid transport networks has allowed products to be sent around the world, which requires considerable packaging. In addition, huge amounts of packaging are also used simply to make goods more attractive to the consumer.

There has been an enormous rise in demand for convenience products, particularly for disposable consumer goods such as supermarket ready-made meals and disposable razors, pens, nappies, and so on.

The composition of waste has changed considerably, with an increasing proportion of inorganic non-biodegradable waste such as plastics, metal alloys and new chemicals. Consequently, a larger proportion of the waste stream is not recycled naturally, and so greater accumulation of waste occurs.

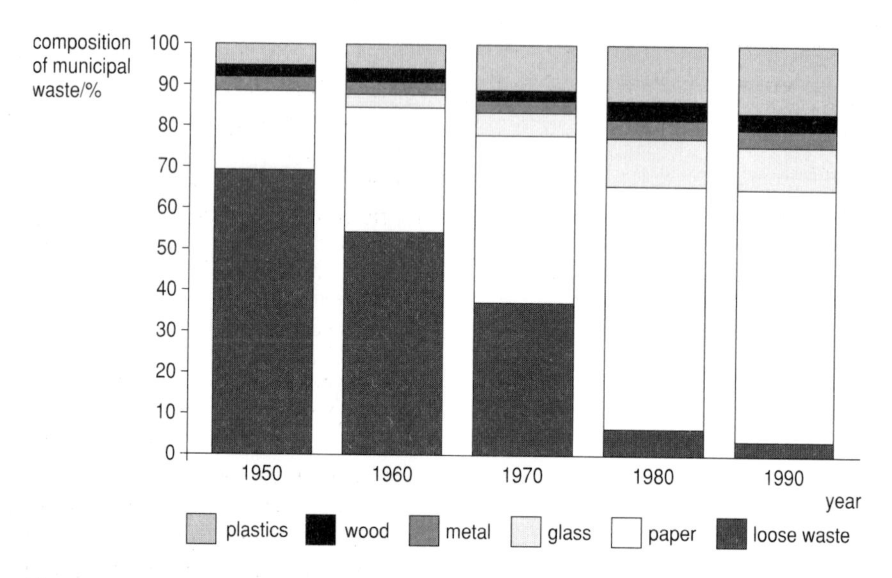

Fig 16.5 Changes in composition of household waste.

QUESTIONS

16.4 With reference to Figure 16.5, describe the change in the volume and composition of municipal waste which occurred between 1950 and 1990.

16.5 How can changes in people's lifestyles be used to explain the increase in production of paper and plastic waste and the decrease in loose waste (defined as ash residue from coal and wood burning fires)?

16.3 WASTE DISPOSAL

The control of the enormous amount of waste continually being generated is a major problem in almost every town and city of the world.

Control of waste disposal and management

Historically, waste disposal simply involved dumping waste materials either into open-air tips, rivers or the sea. However, the increasing volume of waste generated, the changes in its composition and greater understanding of the effects of uncontrolled dumping on the environment mean that such methods are no longer acceptable. Consequently, waste management is becoming increasingly regulated by strict legislative controls.

In the UK, anyone wishing to operate disposal, storage and treatment facilities has to have a licence from the appropriate Waste Disposal Authority

(WDA). Furthermore, under the **Environmental Protection Act** (1990), licence holders cannot simply cancel the licence at their own convenience. The surrender of the licence must be accepted by the Waste Regulation Authority, after it has confirmed that no further pollution is likely to occur from the disposal site. Similar regulations are in existence in the rest of Europe and the United States.

Methods of waste disposal

Table 16.2 The number of the different types of waste-disposal facility in the UK

Disposal method	Number of disposal licences
landfill	4196
incineration	212
treatment (physical, chemical, biological)	122
other (including recycling)	366

Table 16.2 indicates different waste disposal facility in the UK. It is important to realise that the number of licences shown in the table is not necessarily an indication of the volume of waste being disposed of via that particular method.

In the UK the major methods of waste disposal are **landfill burial, incineration** and **recycling and reuse**. In addition, there are a variety of other disposal methods, including disposal at sea, discharge into water courses and tidal waters, and use in construction and road building. In terms of volumes of waste disposed of, these methods are less significant than landfill, incineration and recycling, although they may be important on local scales or for specific wastes, such as the disposal of sewage or radioactive material into water courses.

Landfill

Landfill techniques are the main method for disposal of waste. In the UK, landfill accounts for the disposal of 85% of all waste, excluding mining and agricultural material, and in Germany, 67.6% of municipal waste. Modern landfill sites are an extension of traditional tips and rubbish dumps, with improvements in design to regulate and control pollution, smells, litter and insect and rodent populations (Figure 16.6). The widespread use of landfill is mainly due to the variety of associated economic benefits.

Advantages of landfill

Landfill is a convenient technique and initial cost is low. This is because the basic requirements are only land for the site and arrangements to transport waste to the site.

Landfill is considered the best method for hazardous waste disposal. For example, many industries have on-site large-scale burial schemes for their wastes.

Sites produce combustible gas, particularly methane, through the anaerobic decomposition of organic material within the waste. This can be used for heating or to generate electricity.

During the 1980s significant economic and environmental problems have emerged over the use of landfill.

Disadvantages of landfill

Land that is suitable for waste disposal near major urban areas is becoming increasingly scarce and expensive. Landfill now has to compete with many other land uses which may be able to pay more for the site.

The cost of transporting waste to the landfill is increasing, as sites are being forced to move further from urban areas.

Landfill sites produce hazardous fumes, solids and liquids from the buried waste. Rain percolating down through the site may drain out as a polluted **leachate** contaminating soil and local ground and surface water supplies. This has resulted in an accumulation of contaminated sites, many of which have been built on or landscaped without appropriate cleaning.

The build-up of combustible gas within the landfill causes the problem of rubbish fires. Landfill gases also act as significant greenhouse gases within the atmosphere, with methane an estimated 60 times more powerful contributor to global warming than carbon dioxide. In fact, landfill gas emissions account for approximately 21% of UK methane gas emissions.

Such environmental problems have resulted in tougher legislation to control landfill sites. Under EU Landfill Directives, site operators are financially liable for all future environmental impacts of sites. This means that operators pay the costs of monitoring and controlling sites until they are no longer considered a risk to human health and the environment. The result of the potential environmental problems associated with landfill is that modern landfill sites are increasingly complex (Figure 16.6).

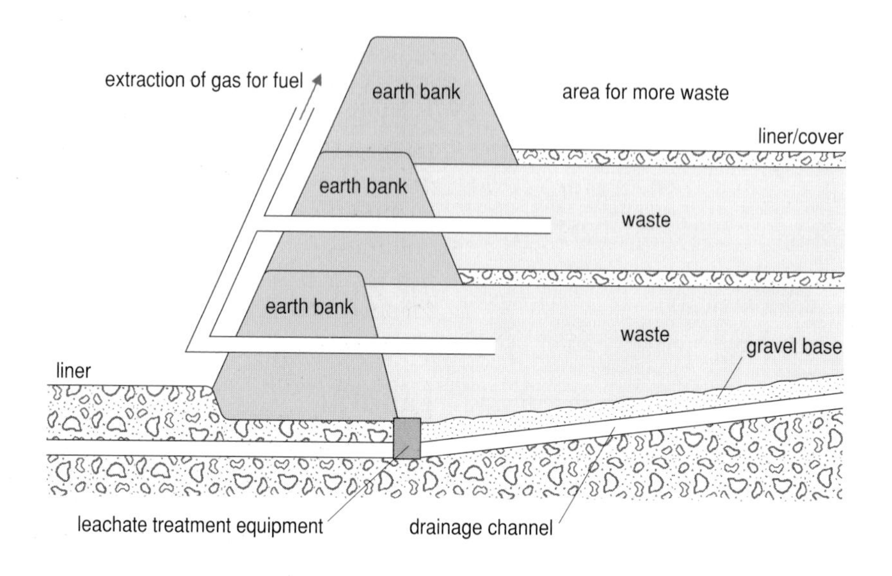

Fig 16.6 Cross-section of a landfill site.

The requirements of a modern landfill site are:

- The site should be as far as possible from centres of population, permeable or faulted rock, water courses and aquifers to minimise potential disturbances and pollution to the local environment.
- An impermeable lining of natural clay or synthetic material such as plastic membranes must be used to prevent leachate seepage.
- Leachate must be controlled using pipes installed at the bottom of the site to collect toxic substances. This leachate is sometimes drained off to a separate treatment plant.
- Landfill gas must be controlled through a pipe network. The gas may be simply vented into the atmosphere, flared (burnt) off or, in more advanced systems, captured and used to generate electricity.
- The waste must be compacted and covered every day with a top layer of soil to control airborne pollution and pest populations, particularly flies and gulls. Unfortunately, this soil layer may occupy up to 20% of the landfill space.

Consequently, obtaining and operating a landfill site is becoming increasingly expensive. It is estimated that worldwide landfill costs will rise between 37% and 135% by the year 2000. This, along with increasing public resistance to the presence of a landfill site, has reduced the appeal of landfill for waste disposal in Europe and the US. However, landfill is still likely to continue as the main method of waste disposal in many countries for at least the near future.

CASE STUDY

The future of landfill in the UK

One of the major methods proposed to reduce the volume of the waste stream is to reduce the amount of waste going to landfill. In 1995, the UK Department of the Environment proposed that it would reduce this volume by 10% in ten years. Methods for achieving this include encouraging composting of organic waste, easier access to recycling centres for eight out of ten UK households and taxing the volume of landfill.

By the year 2000 it is proposed that 25% of all household waste should be recycled and 60% of packaging waste incinerated. However, many environmentalists express concern over this technique for reducing the waste stream since, they argue, it will simply encourage increased incineration of waste, which could have more serious environmental consequences. For example, in 1995 incinerators accounted for two-thirds of UK atmospheric dioxins (carcinogenic chemicals even in minute quantities).

Incineration

Incinerators are specially designed plants which burn thousands of tonnes of waste per day. There are two basic types of plant. The simplest means are **mass burn** plants where all the waste is shredded into small pieces and then as much as possible is burnt. The alternative is to sort the waste to remove most of the incombustible or recyclable waste before burning. These are known as **refuse-derived fuel sources**. Although sorting the waste is expensive and unpleasant, there are fewer problems of air pollution and incinerator corrosion.

The waste is burnt with a flame of between 1500°C and 2000°C, and the combustion gas passes through a series of filters to draw off toxic and particulate material. Some plants use the high temperatures to produce steam which can either directly heat buildings or be used to drive a turbine to generate electricity. Worldwide there are over 1000 of these **waste-to-energy** plants.

There are over 500 large incinerators for domestic waste disposal in the European Union, of which 80% are waste-to-energy plants. In addition, recent improvements in design and profitability, along with demand for non-fossil fuel energy sources, has increased interest in the use of incinerators. However, worldwide there is still disagreement as to whether incinerators are a practical method for large-scale waste disposal, reflected in the varied extent of use between countries (Table 16.3). For example, less than 10% of municipal waste is incinerated in the UK, compared with 72% in Japan.

Table 16.3 The number of waste incinerators in various countries

Country	Number of plants	Municipal waste incinerated / %
Sweden	23	55
Denmark	38	65
Germany	47	30
Netherlands	12	40
France	170	42
Spain	22	6
Italy	94	18
UK	34	8
USA	168	16
Japan	1893	72
Canada	17	19

This variation in the use of incinerators is due to there being significant advantages and disadvantages in using incinerators for waste disposal. These are summarised in Table 16.4.

A 1993 report on waste disposal by the UK Royal Commission of Environmental Pollution concluded that incineration, using the latest burning and pollution control technology, is the best option for waste disposal. Consequently, there are various techniques being introduced at incinerators to tackle some of the environmental problems.

Improving incineration pollution control technology

Filters, screens and combustion conditions are being developed to prevent any hazardous waste being released and to meet tougher emission controls set by the 1990 UK Environmental Protection Act (EPA), which is the main Act of Parliament that controls pollution. This includes strict legislation on the storage, collection, transport, treatment and disposal of waste to protect the environment.

Table 16.4 The advantages and disadvantages of using incinerators for waste disposal

Advantages	Disadvantages
1. reduction in volume of waste for final disposal by 60%–90% (65% by weight) 2. removal of toxic substances and waste sterilisation 3. electricity produced brings in money to pay for incinerator operation and reduces the burning of fossil fuels which generates greenhouse gases 4. occupy less space than landfill, so can be located in urban areas, reducing problems of waste transport 5. ash residue from combustion can be recycled, e.g. for road building and metal recovery	1. production of urban air pollution due to toxic material in ash; emissions of particular concern are: • dioxins and furan molecules from burning chlorine-containing compounds such as plastics and bleached paper; toxic at concentrations as low as 1 ppb • heavy metals such as lead and cadmium, mainly from burning batteries • nitrous oxides produced from burning organic waste 2. residual ash (not released into the atmosphere) containing dioxins, furans and heavy metals in enhanced concentrations 3. waste disposal by landfill is cheaper because incineration requires a large capital investment, up to £100 million 4. a continuous stream is often required due to high initial capital cost and energy costs; this demand for waste does not encourage reduction in waste production

The first part of the EPA contains the '**polluter pays principle**'. This means that industries which cause pollution must buy a licence, which makes the site operator responsible for ensuring safe disposal (known as 'duty of care'). In addition, charges (as taxes) are placed on any industry that causes pollution. As a further incentive to reduce pollution, industries can receive grants or subsidies if they develop anti-pollution equipment.

Removing certain wastes before burning

Removing wastes such as batteries and chlorine-containing plastics before incineration eliminates many of the pollutants discussed in Table 16.4.

Despite these efforts, there are still major concerns over the use of incinerators, with strong public opposition to siting incinerators near residential areas. The greatest concern is over the efficiency of emission control and the ability to effectively monitor incinerators, since waste is burnt 24 hours a day, 365 days a year.

Recycling

Recycling can be defined as the collection and separation of materials from the waste stream and their subsequent reuse or processing to produce a marketable (or saleable) product. Some recycling processes reuse materials for the same product, for example old aluminium cans and glass bottles are recast into new cans and bottles. Other processes turn the waste product into an entirely new product; old tyres become rubberised road surfacing, kitchen waste is turned into fuel pellets, and steel cans are turned into cars.

Recycling can be divided into four distinct areas, known as **the hierarchy of recycling options**:

1. **Preconsumer recycling** is industrial recycling as part of the production process (Section 16.1).
2. **Product reuse** covers a variety of returnable products, such as milk bottles, but also includes repair and refurbishment and the elimination of built-in obsolescence (Section 16.2).
3. **Primary recovery** is the collection of materials and their subsequent use to

Fig 16.7 Bottle bank for recycling glass.

replace **virgin raw materials**, that is, materials produced from extraction processes, notably mining. For example, paper can be collected and processed rather than manufactured from timber, or compost can be produced from organic waste to replace the use of peat and fertilisers.

4. **Secondary recovery** is the use of waste to produce fuel, heat and electricity through waste-to-energy incinerators and collection of landfill gas. It has been estimated that the potential energy value in municipal, industrial and agricultural waste could meet 10% of the UK energy needs, but the present energy derived is a fraction of this.

The main emphasis of recycling policy is to encourage recycling of municipal waste through product reuse and primary recovery. There are three main methods available for the recovery of reusable materials from municipal waste: **collection** or kerbside schemes, **bring systems** based around on-street collection facilities, or **centralised sorting facilities** (Figure 16.7), which involve normal rubbish collection systems followed by sorting of recyclable material at a centralised centre. This technique has declined since the 1970s. The advantages and disadvantages of these methods are summarised in Table 16.5.

Table 16.5 Advantages and disadvantages of the methods of recycling waste

Technique	Advantages	Disadvantages
1. collect scheme: separate municipal waste for kerbside collection	high level of material recovered; significant public participation; low levels of waste contamination	high capital and labour costs to local government relies on public co-operation
2. bring system: individuals take recyclable materials to collection sites, e.g. bottle banks	low capital and labour costs	low levels of recycling local environmental impact of recycling centres, such as congestion of car parks
3. centralised sorting plants: manual and mechanical sorting of mixed waste	low labour costs can be combined with energy generation less reliance on the public	high capital costs only moderate levels of recycling local opposition to siting of plants pollution problems of incinerator

Clearly any form of recycling needs to persuade people to take responsibility for their own waste and to make more effort than simply throwing rubbish away (Figure 16.8).

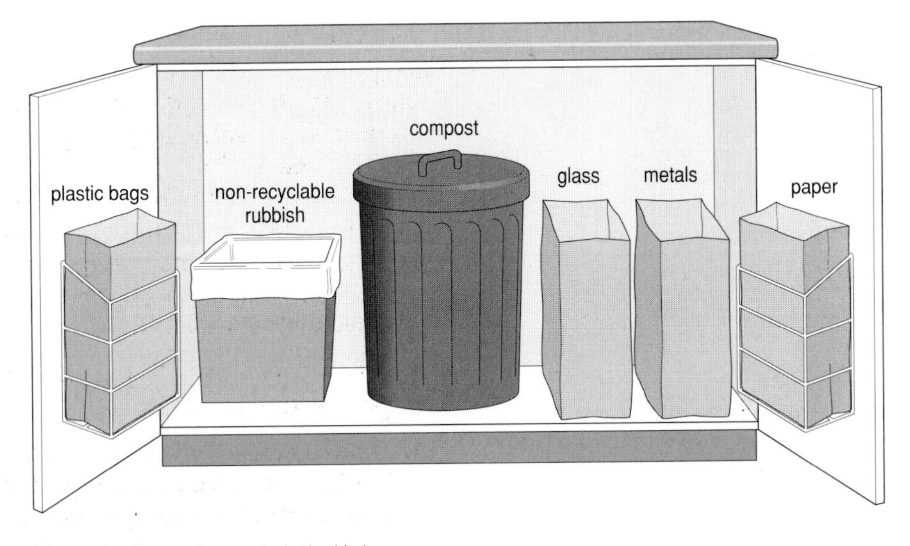

Fig 16.8 Separating waste in the kitchen.

Recycling may involve sorting and washing different types of rubbish, using recycling centres like bottle banks, and thinking more about what we buy.

Advantages of recycling

Most of the advantages of recycling are of benefit to the environment, although some economic benefits can be found.

- Recycling some materials, such as paper and aluminium drinks cans, helps to conserve finite resources by reducing demand. This also reduces reliance on raw materials from a single country or group of countries. *Recycling one tonne of aluminium saves four tonnes of bauxite (aluminium ore) and 700 kg of petroleum coke which has to be extracted from the ground.*
- Reduction in demand through recycling will reduce production of goods and the associated energy consumption and emissions of greenhouse gases. *Recycling plastic bottles can save 50% to 60% of the energy needed to make new bottles; similarly making new steel from scrap is an estimated 75% energy saving.*
- Recycling results in a reduction of pollution from extraction industries, production processes and waste disposal. *Recycling one tonne of aluminium reduces aluminium fluoride emissions into the air by 35 kg.* In fact, it has been estimated that by doubling aluminium recovery worldwide, one million tonnes of air pollutants would be eliminated annually (Table 16.6).
- Recycling may result in reduced waste disposal costs; this is important as the cost of landfill sites is increasing.
- Recycling brings about job creation, for example in recycling, repairing and renovating industries.
- Recycling can reduce the problems of litter. Discarded paper, glass, metal, foam and plastic packaging produces an unsightly litter problem, which requires expensive collection and cleaning. By encouraging the use of recyclable containers to reduce packaging, this problem will decline.
- Participation in recycling helps to raise our awareness of the environment. This means that we have an increasing awareness and responsibility towards waste production.
- Recycling different materials has different environmental impacts. Table 16.6 summarises the benefits of using recycled materials in terms of savings in resource consumption and pollution.

Table 16.6 Environmental benefits of using recycled materials

Material	Reduction /%				
	Energy wastage	Water usage	Air pollution	Water pollution	Mining waste
aluminium	90–97	–	95	97	–
steel	47–74	40	85	76	97
paper	23–74	58	74	35	–
glass	4–32	50	20	–	80

Disadvantages of recycling

The main disadvantages are economic. At present in terms of profitability, recycling is not able to compete either as a method of waste disposal or as a manufacturing process.

- Recycled products have to compete with 'virgin' raw materials which are established in the market and may be subsidised in price. For example, a plastic recycling scheme in Sheffield between 1989 and 1992 operated by the British Federation of Plastics cost £100 000 a year to run, but resale of the plastics made less than £50 000 per year. However, comparing costs is difficult because the environmental cost of not recycling is hard to estimate. If disposal of products not recycled is taken into account in the costing of new materials, the economics of recycling becomes more favourable.

- The production of recycled materials is not determined by demand but by production of waste. This causes economic problems since the supply of recycled products cannot respond directly to changes in demand. This has occurred significantly in the paper recycling industry, where the amount of recycled paper far exceeded the demand, and resulted in great financial problems to the industry.
- Recycling may require sponsorship to be able to operate. For example, Pepsi-Cola International is one of several companies providing money for a 'collect' scheme in Adur, West Sussex.
- Some material recovery may not be environmentally beneficial, since energy and resource consumption for recycling may be greater than simply manufacturing new materials. It has been estimated that at maximum possible level of paper recycling, emissions of sulphur dioxide, nitrous oxides and carbon dioxide would be greater than if some paper waste was incinerated in waste-to-energy plants.
- There may be a lack of space, both in the domestic and urban environment, for storing material to be recycled. However, this is not really as important as the preceding points. It is important to note that individuals may have to put with some personal inconvenience to recycle, compared to the overall benefit to the whole community.

The success of recycling schemes is extremely varied. Japan has probably the world's most successful recycling programme, with a recycling rate of between 40% and 50% of all municipal wastes. The remaining 50% is therefore split between landfill and incineration. In Europe, Finland manages to recycle around 40% of municipal waste, whereas the UK's present recycling rate is around 3%, with a government recycling target of 50% by the year 2000 (equivalent to 25% of domestic waste). Table 16.7 clearly shows the success of recycling of some materials in European countries.

Table 16.7 Recycling rates of selected materials in European countries

Country	Glass / %	Aluminium / %	Steel cans / %
France	34	*	24
Italy	40	29	**
Netherlands	53	*	45
Norway	6	60	**
Sweden	22	82	**
Switzerland	55	26	**
UK	20	10	10
West Germany	39	*	45

* negligible percentage
** data unavailable

CASE STUDY

The case against recycling

The arguments for recycling are clear and the recycling of some materials has more environmental benefits than others. For example, the recycling of some metals such as aluminium uses 95% less energy than manufacturing aluminium from bauxite (aluminium ore).

However, recycling of some materials may in fact be detrimental to the environment. Plastics require large amounts of energy and money for recycling. In addition, the availability of recycling options may be used as an excuse for continuing to produce large amounts of waste. A 1995 UK government study on recycling concluded that shifting away from products with a short lifespan and making products last longer is more important than recycling. In Sweden these principles have been implemented. For example, Swedish cars are on average designed to last 17 years, as opposed to 12 years in the UK. The argument for waste reduction and reuse can be illustrated by the fact that in the UK, two million cars, six million large kitchen appliances and three million vacuum cleaners enter the waste stream each year, much of which is not recyclable.

QUESTIONS

16.6 **(a)** Explain how legislation, specifically the UK Environmental Protection Act (1990), can put pressure on landfill site operators to ensure that their sites have as little effect on the environment over a long period of time as possible.

(b) In Russia, in excess of 20 000 cubic metres of radioactive coolant water from nuclear submarines must be disposed of every year. Discuss the main problems encountered in disposing of such radioactive material using landfill-based techniques.

(c) Between 1950 and the late 1980s the number of people living in cities worldwide rose from 600 million to two billion, a process known as **urbanisation**. What problems does this present to the use of landfill for waste disposal?

16.7 **(a)** The proposed siting of an incinerator generally provokes considerable local opposition due to possible effects on public health. What other objections could local residents have about the siting of an incinerator?

(b) Incinerators have been associated with the release of dioxins, heavy metals and nitrous oxides into the atmosphere. State the potential environmental effects of release of these materials.

(c) Japan has 1893 incinerators burning 72% of its municipal waste. In addition to the direct advantages and disadvantages discussed concerning incinerators, suggest what other factors may encourage such a high level of waste incineration.

16.8 Although most people in the UK have heard of waste recycling, very little recycling actually takes place. A national survey of Local Authority waste recycling schemes by Friends of the Earth found that only 2.6% of domestic waste is recycled in the UK.

(a) Explain briefly what is meant by 'waste recycling'.

(b) State **two** general factors which affect whether any waste product is likely to be recycled.

Glass recycling is one scheme which does seem to have public support. The figure shows how the number of bottle banks in the UK changed between 1977 and 1990.

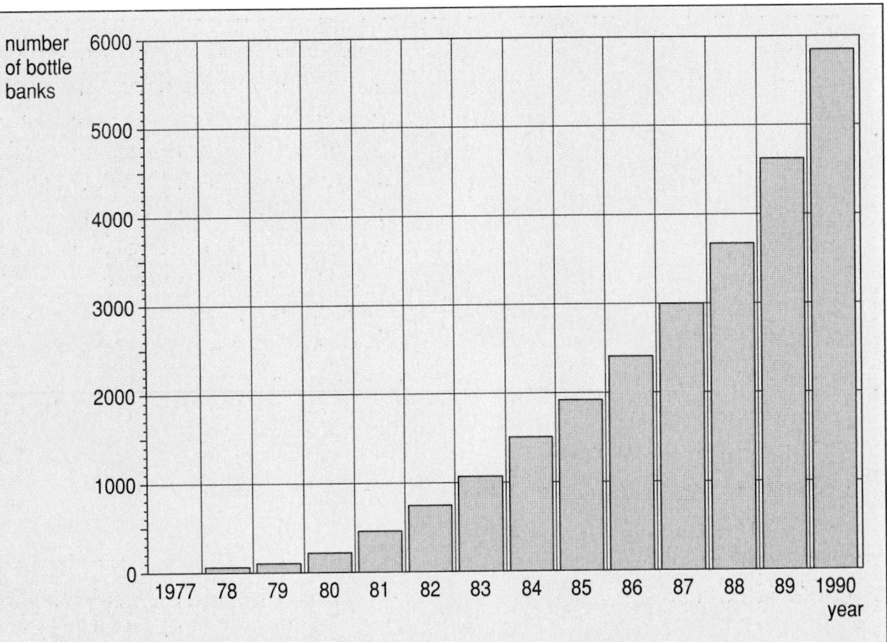

(c) (i) Describe the change in the number of bottle banks in the UK between 1977 and 1990.

(ii) Outline **two** important reasons for the change in the number of bottle banks in the UK between 1977 and 1990.

(iii) State **two** major factors which affect the extent to which people will use bottle banks to recycle their waste glass.

(d) The figure below shows how much glass is recycled by European countries.

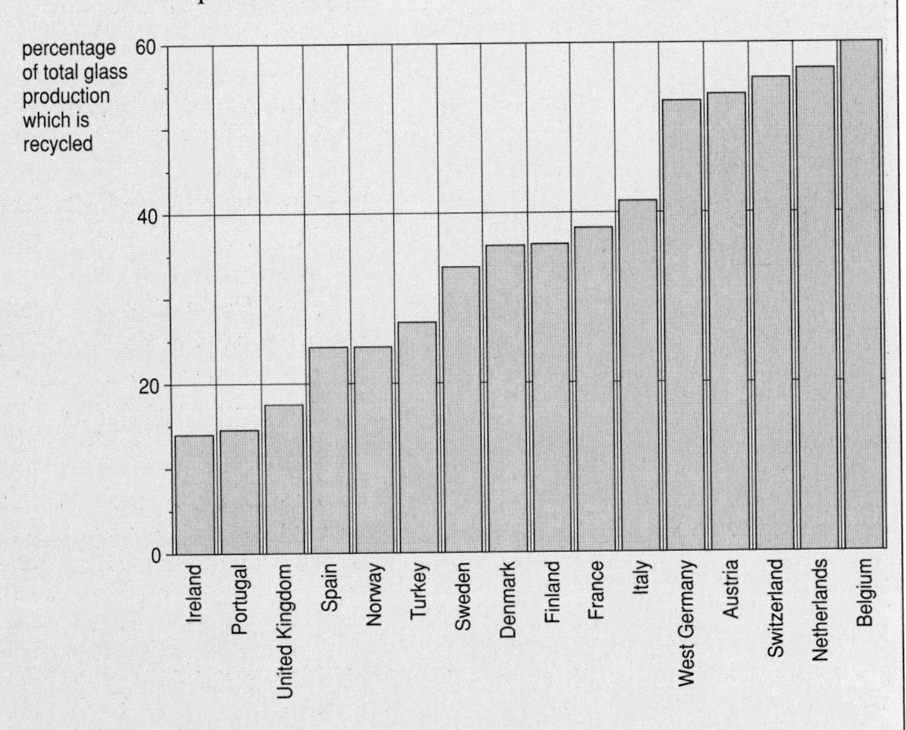

(i) Explain why the data in the figure are shown as a bar chart rather than a line graph.

(ii) Comment briefly on the data given in the figure.

(e) The figure below shows the total amount of glass wasted each year by European countries.

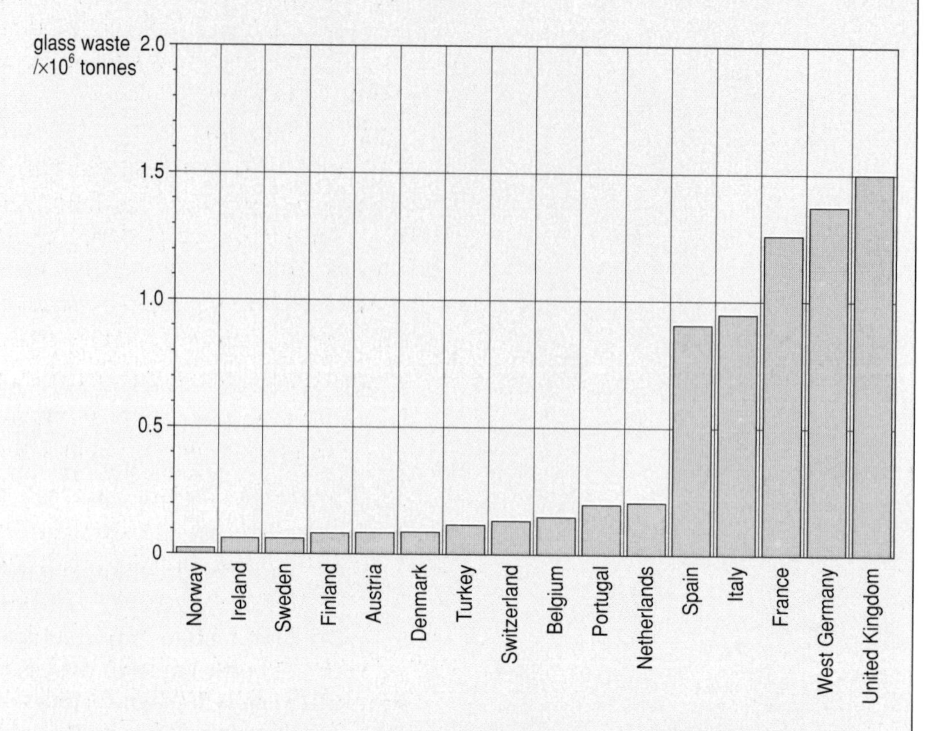

Comment briefly on the data given in the figure.

(f) Use the data in the figures in (d) and (e) to calculate UK total glass production each year.

(g) Glass collected from bottle banks is delivered to a processing plant, treated and crushed to make cullet and then delivered to the furnace. Making glass is an energy-intensive process. To produce 1 tonne of glass **without using cullet** requires 19 000 MJ of energy. The data below show the energy balance involved in making 1 tonne of glass using cullet.

Energy balance details	**MJ tonne^{-1}**
Energy saved in raw material manufacture	4060
Energy saved in delivering raw materials to the furnace	215
Energy saved in the glass furnace	1650
Energy saved in not having to dispose of waste glass	87
Energy needed to collect and deliver cullet to the processing plant	313
Energy needed to treat cullet and make it usable in the furnace	97
Energy needed to deliver cullet to the furnace	49

(i) State which **one** of the above would give the greatest reduction in energy usage in glass production.

(ii) Calculate the net energy saved per tonne of glass made through the use of cullet.

(h) Outline feasible ways in which the UK Government could encourage more glass recycling.

NEAB June 1995

16.4 THE FUTURE FOR WASTE

The management of waste is a complex subject involving environmental and economic issues. No single disposal technique is ever likely to be solely sufficient, yet we should try to minimise the amount of waste being produced in the first place. This will have the added benefits of conserving energy and reducing the volume of waste produced. There are a number of ways in which this can be done.

- Encourage recycling.
- Reduce the excess packaging on food and consumer products. A Friends of the Earth report in 1993 found that a major German retailer considered 98% of all secondary packaging, such as boxes around toothpaste and shampoo, unnecessary.
- Produce easily repairable products to change the principle that a replacement is cheaper than repair.
- Increase the use of washable and refillable food and beverage containers.
- Use easier-to-recycle materials such as glass or metal wherever possible and encourage greater use of biodegradable plastic products.
- Convert municipal organic waste to compost rather than dispose of it.

This reduction in waste management will involve a certain amount of change away from the widespread human behaviour which favours convenient throw-away materials, from food containers to packaging. Such a change by manufacturers is only likely to occur through a combination of legal requirements and economic incentives; in other words, it must be made financially worthwhile not to produce waste. This might come about through implementation of the 'polluter pays' principle (Chapter 10).

CASE STUDY

The problem of the plastic cup

It seems to be more 'environmentally friendly' to use a reusable ceramic mug for hot drinks than a disposable polystyrene foam cup, since this will reduce the amount of polystyrene entering the waste stream. However, since energy consumption for manufacture and for cleaning is much greater for a ceramic mug than for plastic cups, and this has a wide range of environmental implications, this assumption may not be true (Chapter 12). In fact, calculations have suggested that the mug needs to be used more than 1000 times to become as energy efficient as a disposable cup.

The plastic cup example illustrates that when considering any environmental impact, it can be extremely misleading to just look at one factor. In this case, the disposable cups have a significant impact on waste production. However, their relative environmental impact may be less than the ceramic mugs when manufacturing is also considered. This approach of analysing every possible environmental effect of an action is known as a holistic approach.

SUMMARY ASSIGNMENT

1. Using the information from this chapter and also from previous chapters, explain how the extraction of raw materials and the generation and use of energy can result in material entering the waste stream.
2. For each of the problems associated with waste production (pages 225–226, explain which ones have an environmental impact and which have an economic impact. State whether any of the problems of waste have be considered to have both an environmental and economic impact.

3. Discuss one technique which could be used to reduce the amount of waste produced by industrial activities, agriculture and mining activities.
4. List the main features which can be said to have caused an increase in the volume of the waste stream, since the end of the Second World War.
5. Explain the advantages and disadvantages that are associated with using either incineration or landfill as techniques for waste disposal on a local scale.
6. The implementation of a recycling scheme can be said to cause problems to the local environment and the general public. Contrast these with the widespread environmental benefits which are provided by such schemes.
7. List the different efforts that can be made by manufacturing industry and the general public to reduce the volume of waste from entering the waste stream.

Chapter 17

LAND RESOURCES

With the possible exception of remote sea cliffs, practically every square metre of the UK has been affected by human activity. The landscape has been sculpted by landforming processes that alter the environment at such a slow rate, they are unnoticeable in human lifetimes (Chapter 3). However, the landscape is constantly changing as a consequence of human society.

LEARNING OBJECTIVES

After completing this chapter you will be able to:

1. appreciate the conflicts of multiple land use
2. account for the formation, distribution and reclamation of derelict land
3. describe and suggest possible solutions to the social and environmental consequences of urban life.

17.1 LAND USE

The term **land use** describes the type of use of the land, such as agriculture, forestry, industry, residential areas, mining, derelict land and conservation areas. The term **land cover** describes the appearance, features and characteristics of that land in more detail. Land cover therefore describes characteristics such as roads, buildings, woodlands and crops.

Land which is in use or has the potential for use is surveyed on the ground or by remote sensing. Remote sensing is the collection of data at a distance from the site, such as aerial photography or satellite imagery (for example Landsat). Remote sensing involves a high initial cost and requires expert knowledge, but enables large areas to be studied quickly and more cheaply than laborious field work.

17.2 CONFLICTS OF MULTIPLE LAND USE

There are many instances when different land uses conflict with each other. For example, Chapter 12 highlighted the conflict between wind power and landscape conservation in national parks. Some of the most well publicised problems are outlined below.

National parks

National parks, which are designed to protect valuable landscapes, are under threat from activities such as energy generation, mining, forestry, agriculture and the impacts of increased recreation. Many conservation areas are being damaged, because the flood of tourists means increased trampling, fires, disturbance and increased demand for car parks.

Water bodies and catchments

The summer drought of 1995 highlighted the need to protect water resources. To do this, it is vital that the surrounding land is used appropriately. Water authorities buy the catchment area around reservoirs and groundwater bore holes to control land uses that may affect water quality (Chapter 11). In this way, land around reservoirs is often forested. This reduces soil erosion and silting up of the reservoir and evens out the flow of water into the basin by acting as a water store. However, forests may dramatically reduce the amount of water that enters reservoirs or that is able to reach groundwater reserves.

Forestry

There is continued pressure on lowland areas, and so afforested areas are largely limited to upland regions (Chapter 13). There are many benefits of afforestation, although the planting of fast-growing, hardy coniferous species (for example Sitka spruce) also causes many problems within the UK: soils and watercourses become acidified; the native mosaic of colour is replaced with a single tone of green; contour and feature changes are masked; habitats such as peat bogs are destroyed for ever.

Transport policies

It is widely predicted that the use of cars will increase in future years, with wide-ranging consequences. Habitats and agricultural land will be destroyed to make way for more roads. There will be more respiratory complaints due to increased air pollutants (for example carbon monoxide and PM10s, that is, particles less than 10 nm in diameter) and greater negative consequences of oil exploration (for example spills). There might be an increased risk of accidents during oil transportation. More traffic also means an increase in noise disturbance.

An increase in traffic will mean greater pressure on national parks and other areas due to easier transport and the need for more mining of the aggregates which are used in road construction (Chapter 15). More cars will increase pollution of water reservoirs and lead to increased mineral exploitation and energy use.

Agricultural land

Changes in agriculture affect the entire landscape but are not subject to many of the planning restrictions which affect other developments. Since 1945 there have been dramatic changes in agriculture, as summarised in Table 17.1. Such changes have dramatically affected the rural landscape (Table 17.2).

Table 17.1 UK agricultural policy and changes in farming

Time period	Agricultural policy	Impact on farming and land use
1945–73	aim to increase agricultural production through guaranteed prices, grants and subsidies for machinery and improvements	fewer farms and the promotion of farm amalgamation (farms combined into larger units), intensification (use of machines, fertilisers, pesticides, herbicides, new varieties, monocultures), and alterations to farm environments (draining wetlands, cropping rough grazing land, removal of hedges)
1973–84	implementation of the Common Agricultural Policy (CAP) after joining the European Union; aim to increase output by funding for farmers in rural areas and intervention buying	further intensification and alteration to farm environments; farm products have now become much more specialised, resulting in limited crop rotations; increased wheat and oil-seed rape production (owing to a favourable intervention price for oil-seed rape when the UK entered into the CAP)
1984–96	restriction of output, due to surplus stocks, by production controls, e.g. stock is set aside; diversification grants	increased organic and low-output farming; increased part-time farming and on-farm recreational facilities

Table 17.2 Habitat loss in recent years

Habitat	% lost 1951–86
lowland herb-rich grasslands	95
chalk and limestone grasslands	80
lowland heaths	60
limestone pavements	45
ancient woodlands	50
lowland fens and marshes	50
lowland raised bogs	over 60
upland grasslands, heaths, mires	over 30

QUESTIONS

17.1 Name three land uses that should be restricted near a drinking water borehole.

17.2 The definition of a national park is as follows:

'extensive tracts of country which, by virtue of their natural beauty, the opportunities which they afford for open air recreation, their character and their position in relation to centres of population, should be designated National Parks'.

Explain why each of the following activities may lead to conflict if they are practised within a national park:

(a) monoculture forestry using densely planted Norway spruce

(b) quarrying of limestone

(c) water-skiing

(d) mountain-biking.

The role of protest movements within land use conflicts cannot be understated. Protest organisations can halt, alter, or at the very least bring such conflicts into public view, as illustrated by recent protests over road developments (for example Newbury, Twyford Down).

Environmental Impact Assessments

An **Environmental Impact Assessment** (EIA) is a management tool which attempts to put environmental factors on an equal footing with economic ones. EIAs have been a part of the UK development control process since 1988, when they were introduced as part of a European Community Directive. An EIA involves the developing company producing an environmental statement which includes a background on the project, an analysis of significant effects of the project, a description of measures to reduce the impacts, and an assessment of how the project compares to regulations and statutory plans.

The communication of information gathered by ecological and social surveys is conveyed by several different methods, the most popular of which are outlined as follows:

Overlays – an overlain map for each impact clearly shows any potential conflicts. A map which shows the proposed route of a by-pass may have one overlay showing the projected impact on quarrying and another on increased emissions of exhaust gases, for example.

Flow charts – allow the consequences of a particular aspect of impact to be visualised.

Models – for example, a computer model may be used to predict how the burial of radioactive waste might lead to contamination of water supplies at increasing distances from the burial site.

Checklists – a systematically arranged checklist will ensure that all components are accounted for.

Matrices (for example the Leopold matrix) – this is traditionally the most common method in which the magnitude and importance is given for each action of a proposal. The data is presented in a matrix grid with environmental characteristics on the left and the actions across the top (Figure 17.1).

Instructions

1. Identify all actions (located across the top of the matrix) that are part of the proposed project.
2. Under each of the proposed actions, place a slash at the intersection with each item on the side of the matrix if an impact is possible.
3. Having completed the matrix, in the upper left-hand corner of each box with a slash, place a number from 1 to 10 which indicates the *magnitude* of the possible impact; 10 represents the greatest magnitude of impact and 1, the least (no zeros). Before each number place + (if the impact would be beneficial). In the lower right-hand corner of the box place a number from 1 to 10 which indicates the *importance* of the possible impact (e.g. regional vs local); 10 represents the greatest importance and 1, the least (no zeros).
4. The text which accompanies the matrix should be a discussion of the significant impacts – those columns and rows with large numbers of boxes marked and individual boxes with the large numbers.

sample matrix

	a	b	c	d	e
a		2/1			8/5
b		7/2	8/8	3/1	9/7

		A Modification of regime													B Land transformation and construction																			A Resource extraction						
	proposed actions	a exotic flora or fauna introduction	b biological controls	c modification of habitat	d alteration of ground cover	e alteration of groundwater hydrology	f alteration of drainage	g river control and flow modification	h canalisation	i irrigation	j weather modification	k burning	l surface or paving	m noise and vibration	a urbanisation	b industrial sites and buildings	c airports	d highways and bridges	e roads and trails	f railroads	g cables and lifts	h transmission lines, pipelines and corridors	i barriers including fencing	j channel dredging and straightening	k channel revetments	l canals	m dams and impoundments	n piers, seawalls, marinas and sea terminals	o offshore structures	p recreational structures	q blasting and drilling	r cut and fill	s tunnels and underground structures	a blasting and drilling	b surface excavation	c subsurface excavation and retorting	d well drilling and fluid removal	e dredging	f clear cutting and other lumbering	g commercial fishing and hunting
Chemical characteristics – 1 earth	a mineral resources																																							
	b construction material																																							
	c soils																																							
	d landform																																							
	e force fields and background radiation																																							
	f unique physical features																																							
Chemical characteristics – 2 water	a surface																																							
	b ocean																																							
	c underground																																							
	d quality																																							
	e temperature																																							
	f recharge																																							
	g snow, ice and permafrost																																							

Fig 17.1 A section of the Leopold matrix.

QUESTION

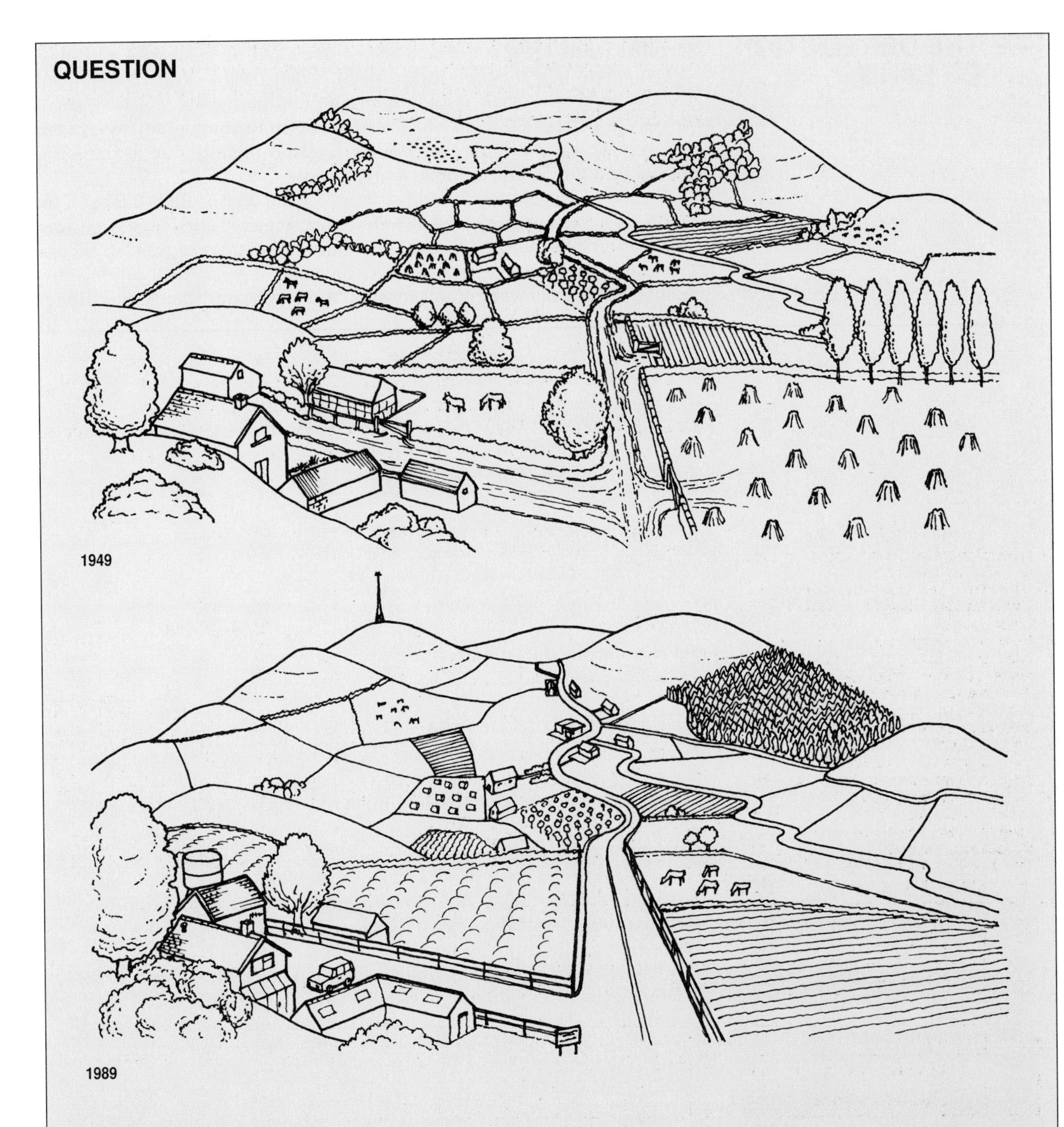

Fig 17.2 A rural area, 1949 and 1989.

17.3 Study the two diagrams above, which represent the same rural area in 1949 and 1989.

(a) List five changes in the two diagrams between 1949 and 1989.

(b) Explain the reason for each of the changes outlined in (a).

(c) Give evidence provided in the diagrams for (i) habitat destruction (ii) advances in technology (iii) a potential increase in soil erosion.

17.3 THE DERELICTION OF LAND

Derelict lands are those which are so damaged by industrial and other development that they are incapable of beneficial use without treatment. Derelict land is a major issue in Great Britain because it is a very small, densely populated island in which land is a precious commodity. Today there are over 41 000 ha of derelict land in England, mainly due to industrial decline and the closure of mines and quarries.

Figure 17.3 indicates that most derelict land lies in the Midlands, the south-west and northern England. However, some counties have large areas due to particular circumstances. There are large areas of derelict land in Cornwall, due to the former tin and china clay mines which produce five million tonnes per year of silica sand, and in Greater Manchester due to industrial decline (for example old mills).

Fig 17.3 Derelict land in England by region, 1988.

CASE STUDY

The greening of Greenham

Greenham Common was known as the place where nuclear missiles were stored; its public image was of a high-security US airbase surrounded by protesting women. Today, the cruise missiles have gone, and three years after the final departure of the American airforce, Greenham is going back to nature.

The airbase is surrounded by a mixture of woodland, grassland and heathland of SSSI quality. English Nature have argued that as the area was originally heathland, it would be logical to turn back the clock and restore the site to its pre-Second World War state.

This argument has found favour with Newbury District Council, the MOD and the local residents. Part of the site is to be restored to heathland. Last year, work started on removing over a billion tonnes of runway concrete; the former runways will then be covered with the underlying local gravel. This will allow the indigenous lowland heath plants to recolonise naturally.

Lowland heath is one of Britain's most endangered habitats, so this move has been greeted with enthusiasm by conservationists. The airbase residents kept the site free of the birch scrub which is frequently endemic in neglected heathlands, allowing the full heathland plant community to flourish. Calcium-loving plants grow near the former runways, thriving on the minerals from the concrete. Birds have also settled the area; wintering flocks of golden plovers and lapwings have joined nightjars on the heath.

Newbury District Council hope to purchase the reclaimed area. They plan to maintain the open spaces by traditional grazing. If successful, they may go on to acquire the remainder of the site. This could either be returned to heathland or used for leisure activities such as walking or riding.

Fig 17.4 Heathland on MOD land at Hankley Common, Surrey.

Table 17.3 illustrates some of the physical and chemical problems of derelict land along with possible solutions. The benefits of reclamation include the reduction of visual pollution, and the regeneration of inner city areas reduces pressure on rural areas. Reclamation also removes dangerous sites. If left alone, these sites will undergo a succession, eventually becoming woodland, although this takes a very long time. Ironically, some sites are now preserved because they developed a specialised plant community due to their high levels of pollution.

Table 17.3 Problems of derelict sites and possible solutions

Physical		Chemical	
Problem	**Solution**	**Problem**	**Solution**
poor soil structure due to compaction, e.g. by machinery	plough, add organic matter, replace or add topsoil	generally low in nutrients, e.g. nitrogen and phosphorus, owing to low organic-matter content	fertilise with sewage sludge and plant leguminous species
instability, e.g. due to steep slopes and poor construction	lower gradients, vegetate	toxic concentration of one or more elements, e.g. Pb, Cu, Zn	plant resistant species
waterlogging, due to compaction and impermeable matter, e.g. clay	create drainage channels	extremes of pH, e.g. china clay wastes are highly alkaline and colliery spoil wastes are acidic	plant resistant species, neutralise pH
subsidence from deep mines	avoid heavy machinery and construction developments	high salinity	plant resistant species (halophytes)

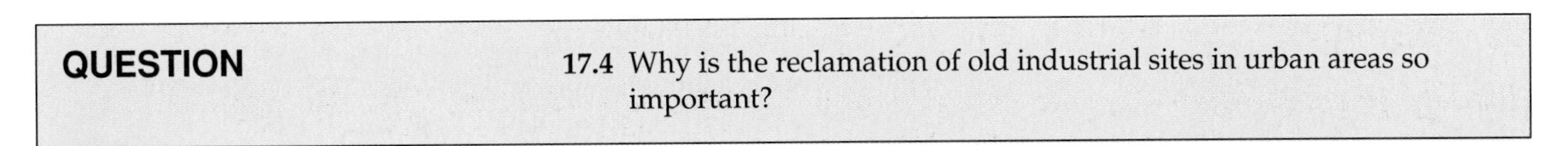

QUESTION

17.4 Why is the reclamation of old industrial sites in urban areas so important?

It is impossible to reclaim a site without change of some sort. For example, after a quarry is filled with refuse and topsoil returned, the underground drainage will have altered considerably, even if revegetation and landscaping succeeds. However, many reclamation schemes have been a great success (Figure 17.5), with some providing valuable recreation or conservation amenities.

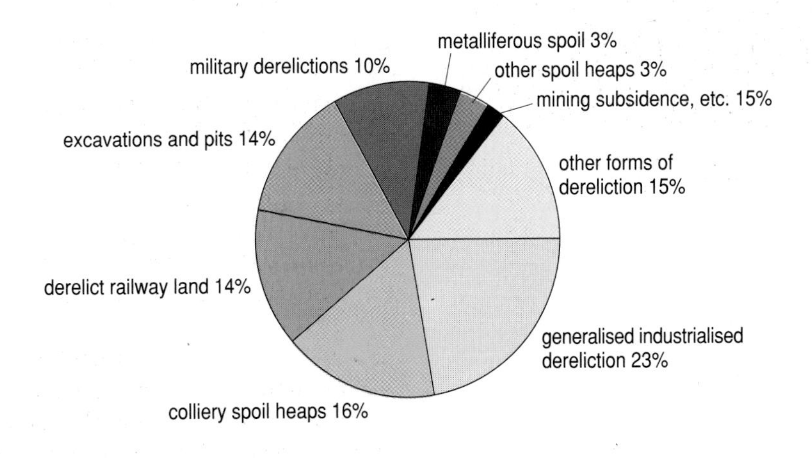

Fig 17.5 Derelict land reclaimed after mineral workings, 1982 to 1988.

QUESTION

17.5 The maps show an area called Dinton Pastures near Reading, Berkshire. Map **A** shows the boundaries between the various gravel workings in 1969. Map **B** shows the developing Dinton Pastures country park in 1974.

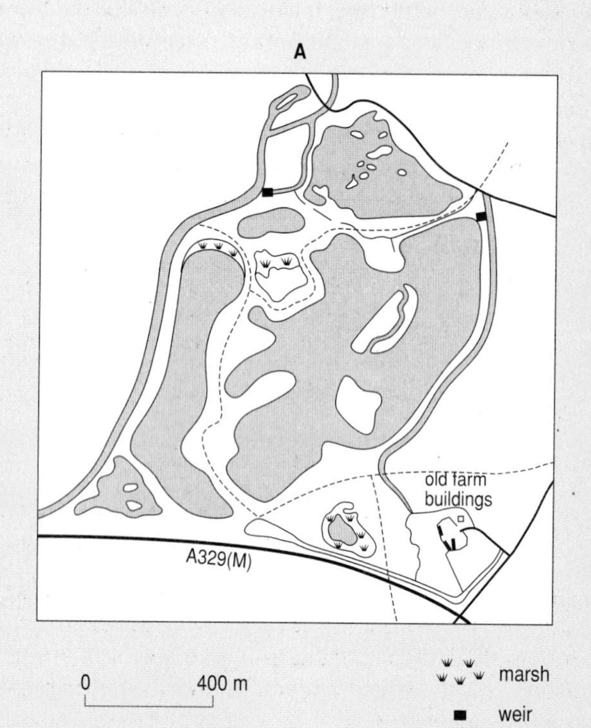

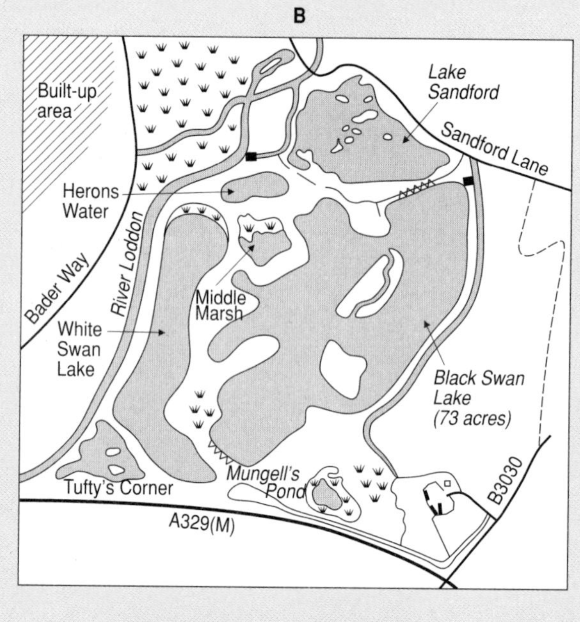

marsh
weir
bank liable to erosion
boundary of land owned by Wokingham District Council

Berkshire County Council's 1970 plan for the area had the aim of allowing 'the maximum reasonable extraction of gravel consistent with leaving the land to shapes and contours which can then play their part in the evolution of a valley park'.

(a) Describe **three** decisions which the gravel extraction company should have taken so that their work was consistent with the council's aims.

(b) State **three** possible leisure pursuits encouraged within the developed Dinton Pastures country park.

(c) Describe **two** ways in which the park management could conserve wildlife within the area.

(d) Consider the use of open water areas in the country park and state

(i) **two** conflicting uses of water

(ii) **two** non-conflicting uses of water.

(e) It was suggested at one stage that the marsh, adjacent to the built-up area, should be drained. State **one** advantage and **one** disadvantage of doing this.

(f) Explain the need for each of the following in the area, and state how each might be achieved.

(i) noise screening

(ii) bank stabilisation

(iii) interpretative information.

AEB June 1991

Desertification

Desertification is the conversion of productive land into desert-like conditions. The environmental, social and economic consequences of desertification have affected millions of people in both developed and developing countries. For example, poorer quality soils have led to declining agricultural yields which have, in turn, led to overuse of artificial fertilisers, cultivation of marginal lands and, finally, the movement of people away from rural areas into cities.

Desertification may occur naturally as a result of drought, outbreaks of fire or epidemics of pests, such as locusts. Human activities have accelerated desert formation. All of the following factors are intensified when population pressure increases.

- Poor irrigation schemes cause a rise in the water table (the level below which soils are saturated), bringing up salts from depth which then concentrate at the surface. These salts can form a hard impermeable salt crust, which reduces infiltration, increases runoff and increases soil erosion (Chapter 14).
- Over-cultivation reduces the fertility of the soil.
- Overgrazing removes vegetation and compacts the soil, which decreases infiltration, increases runoff and therefore enhances soil erosion.
- Deforestation (Chapter 13) increases the amount and speed of runoff, causes an increase in surface wind speeds and reduces the organic content of soils (Chapter 4). When there is no wood to burn, dung and other organic wastes, which would otherwise act as an important organic fertiliser, are often burnt instead.
- In many developing countries the growing of cash crops (crops for money, such as cotton and cocoa) forces food production onto more marginal land, which has a greater risk of desertification.

Combating desertification involves improving affected land, such as re-establishing vegetation and improving fertility. However, the real aim must be to ensure that sustainable forms of agriculture are practised (Chapter 14).

17.4 URBANISATION

The creation of cities has always been viewed as a sign of development. However, there are many social and environmental problems associated with urban areas. **Urbanisation** is the process by which an increasing proportion of a country's population come to live in urban areas. In more developed countries urbanisation started as a result of industrialisation, when people flooded into cities to take the newly created jobs. Urbanisation in developing countries is occurring at a much faster rate and on a much larger scale than was experienced in the more developed world. In developing countries, migrants move to cities because they believe they will have a better standard of living, although in reality many suffer horrendous conditions in the cities; in this form of migration it is the migrant's perception which is important.

The reasons for migration can be categorised by those which 'push' people away from an area and those which 'pull' people towards somewhere else (that is, **push and pull factors**). Push factors (negative factors) in rural areas include poor services (for example health care, education, transport, entertainment) and low employment. Pull factors (positive factors) in the city include higher employment, better services and an overall better standard of living. However, as we shall see, people are also pushed away from cities and pulled towards the countryside.

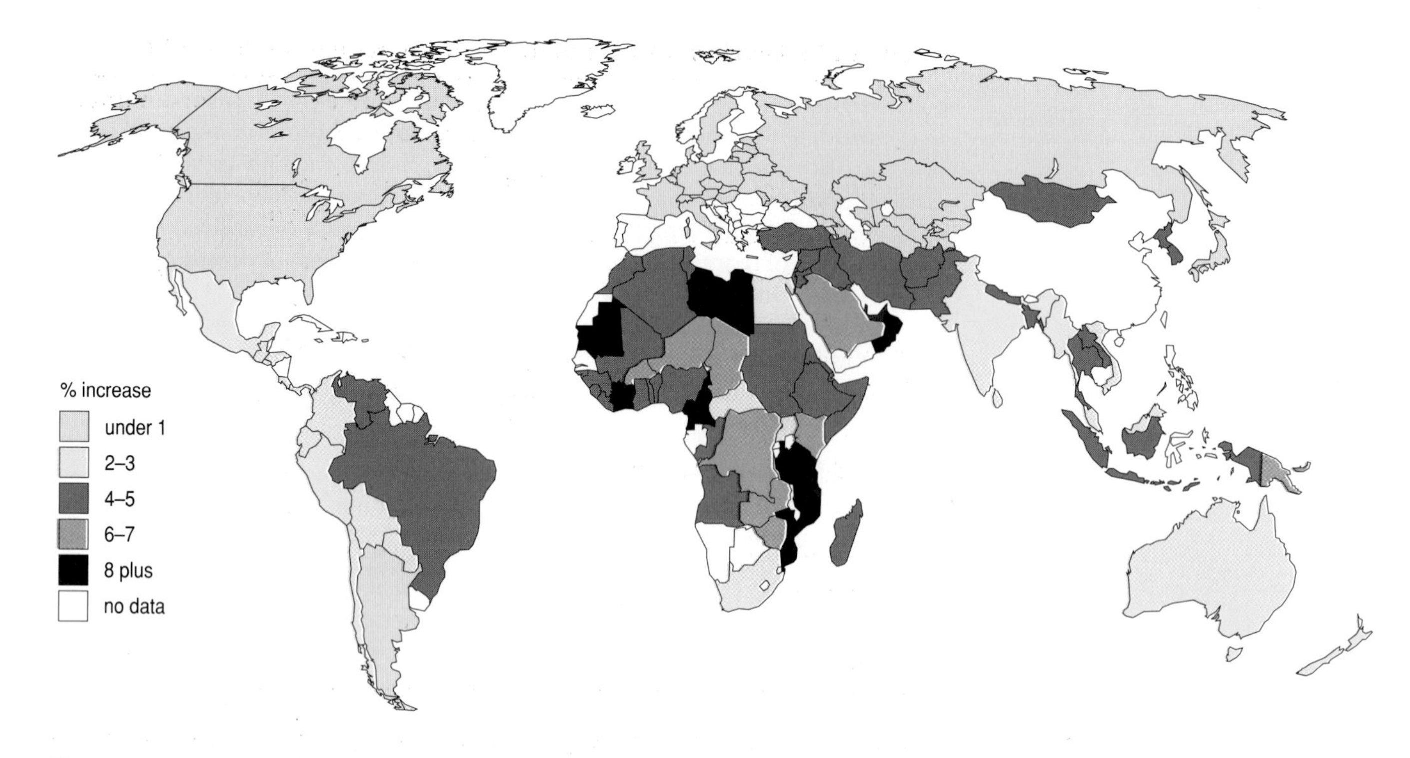

Fig 17.6 Increasing urban population per year.

Examine Table 17.4, which shows the growth of the world's largest cities, and Table 17.5, which shows the increase in urbanisation throughout this century. By the year 2000 it is estimated that 3090 million people, or 51% of the world's 6112 million population, will live in urban areas. The prediction of 1010 million people who will be urbanised in more developed countries accounts for 83% of those countries' population. However, the remaining 2080 million people in developing nations represent only 43% of the population. This rapidly increasing urban population and the increasing extent of urban areas, especially in developing countries, is causing many problems.

Table 17.4 The largest cities in the world

1950	Millions of people	1985	Millions of people	Predicted 2000	Millions of people*
1. New York	12.3	Tokyo	18.8	Mexico City	24.4–31.0
2. London	10.4	Mexico City	17.3	São Paulo	23.6–25.8
3. Rhine-Ruhr	6.9	São Paulo	15.9	Shanghai	14.7–23.7
4. Tokyo	6.7	New York	15.6	Tokyo	23.7
5. Shanghai	5.8	Shanghai	12.0	New York	22.4
6. Paris	5.5	Calcutta	11.0	Beijing	11.5–20.9
7. Buenos Aires	5.3	Buenos Aires	10.9	Rio de Janeiro	13.0–19.0
8. Chicago	4.9	Rio de Janeiro	10.4	Bombay	16.8
9. Moscow	4.8	London	10.4	Calcutta	15.9–16.4
10. Calcutta	4.6	Seoul	10.2	Jakarta	13.2–15.7

* Ranges illustrate the wide variance of predictions.

In the past, the world's largest cities were within more developed nations. However, growth in developing nations has been so great that by the year 2000 it is predicted that only two of the world's ten largest cities will be in developed nations.

Table 17.5 Levels of urbanisation in selected regions of the world

World region	Population living in urban areas / % 1920	1980	Predicted 2000	Total number / million
less developed countries	10	32	43	2080
East Asia	9	31	40	
South Asia	9	25	35	
Latin America	22	60	80	
Africa	7	28	39	
more developed countries	40	70	80	1010
Europe	46	46	71	
North America	52	65	87	
Soviet Union	15	81	85	
Oceania	47	68	80	
world	19	46	51	3090

Urbanisation means that people have to develop a new way of life which often causes social and economic problems, for example the increased cost of housing may mean housing becomes too expensive for lower income groups. Dense housing may lead to stress and tension which can cause delinquency, vandalism and crime.

The environmental effects are felt by all citizens. City expansion destroys habitats and agricultural land, causes traffic congestion, unsightly buildings and a lack of greenery (visual pollution), noise pollution and pollution of water, land (for example waste disposal) and air (for example particulates, ozone, carbon monoxide).

Since the 1960s, some countries such as the UK and the USA have shown a new trend, called **counter-urbanisation**, where particular age groups (commuters and retired people) are migrating from cities to the surrounding

Table 17.6 Consequences of counter-urbanisation

Social	Economic	Environmental
an older population in the rural areas needs special services, e.g. health, nursing homes	general revitalisation of the rural economy by increased investment	destruction of habitats and agricultural land
the children of commuters need a change in local services, e.g. education	inflation of rural house prices often means local people cannot compete for housing	increased dependence on transport and therefore fossil fuels
the dispersal of services makes services unavailable to poorer inner-city dwellers	newcomers still dependon urban areas for many services, e.g. education	increased consequences of recreation, e.g. trampling, litter, noise disturbance
a change in the nature of the rural community life	general improvement of most rural services	deterioration of many inner-city areas
loss of skilled workers from inner-city areas	less industrial investment in inner-city areas	

countryside. This migration has been made possible by the increased number of cars and an improved transport network that allows people to work in the city and live elsewhere. In this case, people are being pushed from the city by factors such as expensive housing, traffic congestion, pollution, a lack of open space and the movement of retail to out-of-town shopping centres. Although they have to pay higher transport costs, these migrants are pulled to the countryside by the opposite of the above factors, as well as the attractions of a slower, healthier pace of life and the opportunities for outdoor recreation.

Table 17.7 Solving the problems of inner-city life

Problem	Possible solution
transport	improve public transport; park-and-ride schemes; reduce city-centre car parking; pedestrianise centres; create bus and multiple-occupancy car lanes; impose restrictions on all-week car usage; increase fuel costs; restrict speed; cycle lanes; promote clean vehicles, e.g. electric
energy	use of local resources; enhance renewable energy, energy conservation and energy efficiency (Chapter 10)
pollution and waste	recycling schemes; reduce packaging; reuse organic wastes
economic	investment in rural and inner-city areas
land use	create open spaces and plant trees to reduce pollution and noise and promote wildlife in urban areas; protect natural areas from development, e.g. by greenbelts; implement self-help housing schemes

There is a great need to ensure the long-term viability of urban areas for future generations through the sustainable development of cities. Table 17.7 illustrates some possible solutions to some of the problems of city life. Not all of these have been a success, for example many motorists still drive in the bus lanes.

Greenbelts

Greenbelts, first introduced in 1955, are areas of land on the outskirts of urban areas in which development can be controlled and shaped. They were implemented to:

- check city sprawl
- protect the countryside
- prevent neighbouring towns merging into large conurbations
- preserve historic towns
- assist in urban regeneration by focusing development within urban boundaries.

By 1989 there were 1.5 million hectares of greenbelt in the UK, which represents around 12% of the country. The largest is nearly 500 000 ha around London, with the second largest in the north-west (Figure 17.7). Permission for development in these areas is not normally granted, but there are exceptions.

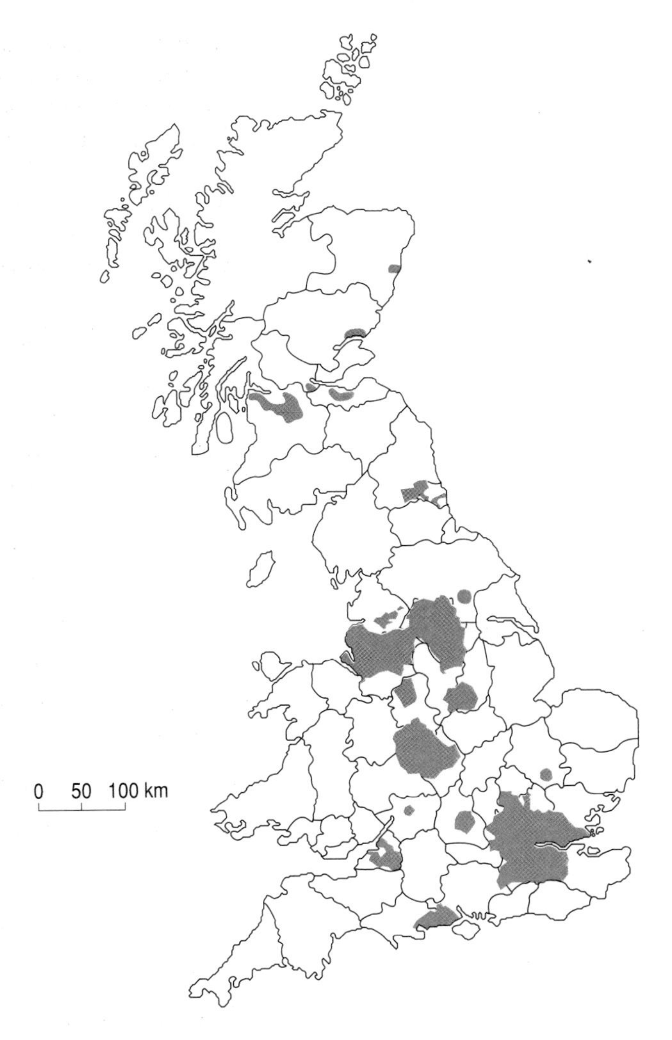

Fig 17.7 The greenbelts of England and Scotland.

QUESTION

17.6 (a) Suggest why greenbelts are often unsuccessful in halting urban expansion.

(b) Why is counter-urbanisation largely restricted to particular age groups?

SUMMARY ASSIGNMENT

Natural and man-made hazards are a feature of life in many of the rapidly expanding cities of less developed countries (LDCs). One such place is Mexico City, situated on the sediments of a former lake bed. Some 19 million people live in Mexico City today, compared to 14 million people ten years ago. Continued migration and a high rate of natural increase make it likely that many millions more will be added during the rest of the 1990s.

Such a large city, located in a relatively poor country, faces many seemingly complex problems. Housing conditions for many are difficult. Service provisions, though not disastrous, are inadequate. Most houses do have access to electricity and to water by tap or standpipe. Much of the water is obtained by over-extracting groundwater from beneath the city. However, the homes of over 10% of the population lack mains drainage and the garbage system is deficient.

Transport is a massive problem; the number of road vehicles has increased ten-fold in 30 years to reach a figure of 2.7 million. The result is large-scale congestion, long journey times to work and high levels of atmospheric pollution. In times of intense sunlight, especially when temperature inversions occur, the hazard is greatly increased.

Other hazards experienced by the inhabitant include the location of oil depots in residential areas, earthquakes (122 in 150 years), floods after heavy rain, and dust storms made worse by soil erosion and deforestation on the urban fringe.

1. Describe some of the reasons why large urban areas in LDCs experience a rapid rate of population growth.

2. For each of the problems listed below describe how it can result in a major hazard for the citizens of the LDCs.
(a) limited access to potable water.
(b) lack of mains drainage.
(c) poor refuse disposal.

3. Name four gaseous pollutants likely to be in the atmosphere of Mexico City as a result of the large number of motor vehicles. Outline how bright sunlight and temperature inversions increase the number and concentration of atmospheric pollutants.

4. Suggest reasons why:
(a) Mexico City suffers from rapid flooding after heavy rain.
(b) deforestation and soil erosion are features of the urban fringe.

5. Suggest the main ways in which it might be possible in Mexico City to reduce the impact or extent of each of the following hazards.
(a) earthquakes.
(b) floods.
(c) subsidence due to groundwater extraction.

NEAB June 1994

Chapter 18

CONSERVATION

The Large Blue butterfly only lays its eggs on very closely grazed grass, usually on the leaves of wild thyme.

Caterpillar feeds on thyme and then flicks itself to the ground. By inflating the skin behind its head, the caterpillar mimics a red ant grub. A secretion is released which attracts the ants.

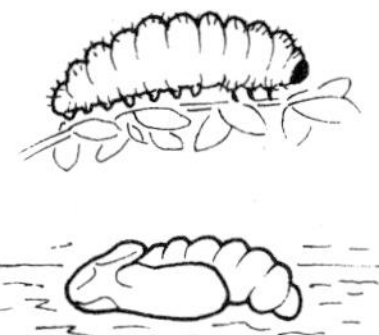

The ants carry the caterpillar into the ant nest where the caterpillar feeds on the ant grubs, hibernates, feeds on the grubs once again and emerges as a butterfly the next June.

Conservation of the Large Blue butterfly has meant artificially recreating their habitat; grasslands have been closely grazed, wild thyme has been established and red ants have been persuaded to build nests nearby!

Fig 18.1 Maintenance of habitats of the Large Blue butterfly.

There are millions of species of plants and animals on earth today. However, many species are becoming extinct. There are many reasons why we might want to maintain or conserve the huge diversity of plant and animal life that still exists. This chapter will review some of the problems and the solutions.

LEARNING OBJECTIVES

After completing the work in this chapter you will be able to:

1. outline the justifications for conservation
2. describe different approaches to conservation
3. appreciate some of the conflicts associated with conservation measures
4. outline the roles of governmental and non-governmental conservation agencies.

Conservation has been defined in many different ways, for example, as:

- protecting the environment from human activities
- preventing animals and plants from becoming extinct
- creating new habitats
- managing natural resources and ecosystems
- keeping the balance of nature
- preserving the natural environment.

QUESTION

18.1 Rearrange the list in what you consider to be their order of importance, with the most important first and the least important last. Are there any which you do not think count as conservation?

There are different levels of conservation. **Biological conservation** refers to maintenance of species, habitats, communities and ecosystems. These are often interconnected; **species** (such as the Large Blue butterfly) may only be maintained if their **habitats** are conserved or deliberately recreated through conservation management (Figure 18.1). As we shall see, conservation includes preservation, management, reclamation and creation of new habitats.

The utilitarian argument

Utility means 'use', and this justification of conservation usually refers to the economic benefit of conserving plants, animals and their habitats.

Food

Plant species need to be maintained as a potential source of food. Out of a known total of at least 250 000 species, we use about 30 on a large scale. As discussed in Chapter 14, monocultures used in intensive agriculture are susceptible to pest attack, and wild varieties, which may possess natural genetic resistance to these pests, therefore represent a form of biological insurance. This is one of the main arguments for **genetic conservation**. Most of the wild relatives of our crop plants live in the tropics, and unfortunately some parts of the tropics are under great threat so these species may be lost as a source of potential crop plants for the future.

Medicine

American scientists have recently estimated that the value to society of the as yet undiscovered drugs from tropical rainforests is around $150 billion (Table 18.1). In other words, the rainforests may be worth more alive, as a source of valuable drugs, than they are dead, when they are used as a source of timber.

Already, 47 major drugs, including quinine and codeine, come from tropical plants, but only a small fraction of the total number of plants have been investigated. Indigenous people, whose ancestors have lived in the forests for thousands of years, often have a tremendous knowledge of the potential uses of many plants, but as yet developed countries have learned little from such people.

Table 18.1 Medicinal plants

Name of plant	Name of compound	Traditional use	Potential medicinal use
Atropa belladonna	atropine	pupil dilation	heart functioning
Cinchona ledgeriana	quinine	anti-malarial	anti-malarial
Datura matel	scopolamine	sedative	sedative
Silybum marianum	silymarin	liver disorders	liver disorders

Animals, too, continue to be a source of drugs, and have an important role in medical research, particularly in cytological research (analysis and study of cells) and physiological research (study of the function of living organisms) (Table 18.2).

Table 18.2 Medicinal animals

Animal	Product	Use
leeches	saliva contains anticoagulants	microsurgery of several body parts
horseshoe crab	blood clots rapidly when in contact with a particular type of bacteria	test for presence of this type of bacteria
snakes	venom of some species contains anticoagulants, neurotoxins and cytotoxins	anticoagulants prevent clotting, neurotoxins may be used as anaesthetics, cytotoxins in cancer research

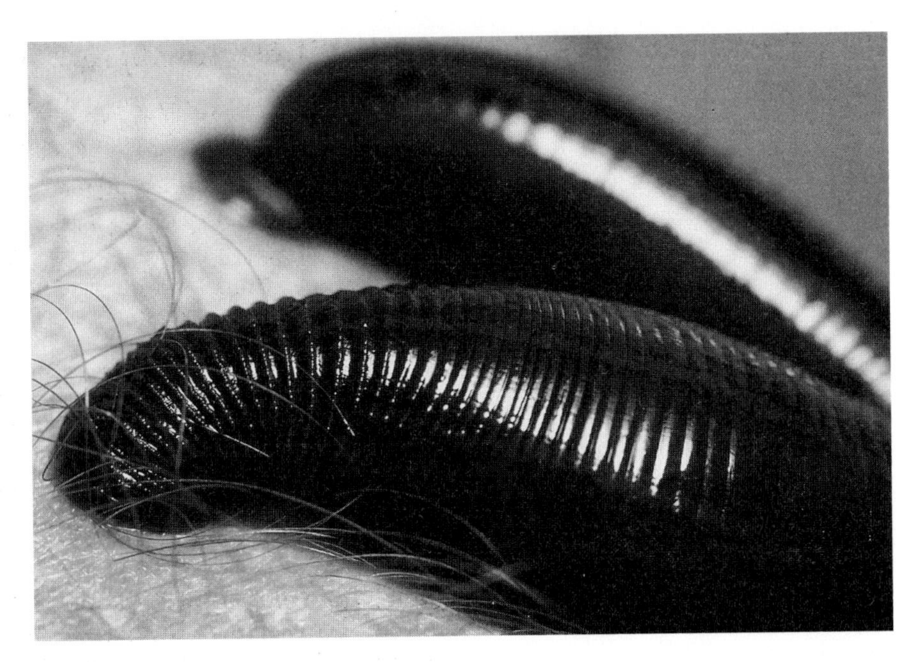

Fig 18.2 Leeches are sometimes used in surgery. These two leeches are sucking blood from a human arm.

Conserving genetic material

Wild populations may be faced with constantly changing environmental conditions; disease, drought and frosts may come and go, and only the fittest plants and animals will survive. Such individuals may possess genes that make them resistant to these pressures. Conversely, modern varieties of crops tend to have lost these genes because they were selected on the basis of productivity rather than for their resistance to adverse conditions or to disease.

Modern agriculture has made up for a lack of disease-resistance by using pesticides, and for a lack of drought-resistance by irrigation. Problems occur when the conditions in which the modern crop is growing change. With minimum genetic variability, modern crops may not be able to tolerate environmental change, productivity may decline or the crop may die. Plant breeders can transfer the desired genes from the wild variety into the modern crop by cross breeding (an example of artificial selection) or genetic engineering, as long as the wild variety still exists!

CASE STUDY

Biodiversity saved by the lottery!

The Royal Botanic Gardens at Kew have been given £21.5 million to launch a Millennium Seed Bank which will preserve the seeds of every British plant, along with those of thousands of endangered species from the desert areas of Africa and Asia.

Kew scientists have already identified 6000 plants vital to people who live in arid lands – plants used for hedges, food, medicines and even as toothbrushes. By drying and freezing the seeds at –20°C, it is hoped that the seeds will keep for 200 years.

[from 'The green shoots of recovery', *The Guardian* Online, 28/12/95]

Industrial products

The list of essential products that we obtain from plants and animals, such as timber, paper, fuel, gums, dyes and oils, seems endless. To continue to enjoy them, we need to conserve the plants and animals from which these products are derived.

Ethical reasons

Everything has a right to live and we ought to respect this. Different cultures and religions have very different moralities; an animal which is sacred in one culture may be insignificant in another.

It appears that extinction is a natural process. The fossil record suggests that the vast majority of species that have ever existed are now extinct, and that humans had nothing to do with most of these extinctions. Do all organisms have an equal right to live? Is a ladybird more important than the bacterium which causes leprosy? These kinds of issues can be quite subjective. Many people believe that we have the responsibility to maintain the present species diversity for future generations to enjoy.

Aesthetic reasons

An aesthete is a pleasure-seeker, and the natural environment, whether we are experiencing it on the television or in actuality, is a great source of pleasure. City-workers may flock to the parks at lunchtime, urban-dwellers may escape to the countryside at the weekend, and wildlife-watching holidays and ecotourism are big business. To enjoy it, we must conserve it.

CASE STUDY

The rise of ecotourism

People are attracted to areas of natural beauty, from the landscapes of the Grand Canyon in America and Ayres Rock in Australia, to the unique wildlife of Africa and the Great Barrier Reef. Ironically, the presence of large numbers of people can easily have a detrimental effect on the areas visited. Since tourism is such a big business, generating over $195 billion per year (the second largest industry in the world), it clearly benefits tourist operators to conserve natural resources for their economic benefits.

Ecotourism can be defined as travel that provides tourists with access to natural resources but preserves the ecosystem visited. It also produces economic benefits for the area visited, enabling further conservation.

The Central American country of Belize, for example, experienced the benefits of ecotourism. In 1980 the country was essentially virgin territory, with an undeveloped infrastructure and few accommodation facilities. However, the attraction of coral reefs, rainforests, wildlife and historical ruins has produced a rapidly growing tourist industry, with Belize becoming recognised as an international paradise, with a Ministry of Tourism and the Environment. The benefits to the country have included an improved infrastructure and capital available for conservation.

Ecological reasons

Plants are an essential part of the biosphere; they help to regulate the atmosphere and the water cycle and are essential components of the carbon and nitrogen cycles. They help to form the soil which they then hold in place, and they form the basis of all food chains.

The life cycle of many plants is entirely dependent upon the activities of insects – pollinators such as bees, for example – and we are only just beginning to understand the complexities of food webs. Our attempts to manipulate food chains in pest-control programmes have often had totally unforeseen effects because our understanding of food chains and webs has been superficial. For example, the cane toad was introduced into areas of Northern Australia to control insect crop pests. Unfortunately, this introduction has had many significant unforeseen effects; the cane toad is an aggressive predator and not only preyed on the crop pests but also on many important local species, such as native frogs and other reptiles, whose numbers have significantly reduced. In many areas, the cane toad has now become totally dominant over all the other species.

Similarly, organisms such as lichens have proved to be very sensitive indicators of pollution, and by studying such organisms we have been able to identify and control harmful human activities. We have been able to use their presence or absence from a particular area as an indication of sulphur dioxide levels. By conserving and studying the natural environment we can reduce our harmful effect upon it.

ANALYSIS

Is conservation the same as preservation?

Table 18.3 describes some conservation management techniques. Copy and complete the table by filling in the boxes.

Table 18.3 Conservation management techniques

Area	Action	Effect	Conservation value
overgrown hazelwood	chop 70% of trees to just above ground level (called coppicing)	multiple shoots spring up from stumps; varied canopy height and light levels result	
tropical rainforest	ban import of tropical logs into the UK		
North Sea fishery	restriction on minimum allowable net mesh size		
mangrove swamp	restrict number of power boats allowed on swamp	reduces swash	
beech woodland	grey squirrels shot and/or poisoned	competition with red squirrels reduced	

Preservation is perhaps the simplest form of conservation. As the human population has increased, demand for land and resources has also increased. This has often made simple preservation of land and habitats impossible.

In many countries management is essential if areas or resources are to be conserved. In the UK, for example, grasslands and ponds would simply disappear through succession if they were not actively maintained; coppice woodlands, with their great variety of microhabitats, would turn into high forest without periodic cutting, and fisheries worldwide would be exploited to extinction without agreements on how much fish individual countries are allowed to catch.

Conservation can also be achieved by the creation of new areas or the re-creation of habitats that have been lost. By planting native broad-leaved woodlands across 180 square miles of derelict and polluted ex-mining land across central England, the National Forest is creating many new valuable wildlife habitats (see Chapter 13).

18.2 CONSERVATION ORGANISATIONS

Statutory agencies

In the UK the statutory agencies include **English Nature** (formerly the **Nature Conservancy Council**), the **Countryside Commission**, the **Countryside Council for Wales**, **Scottish Natural Heritage** and the **Environment Service of the Department of the Environment Northern Ireland**. Generally, these agencies have five major functions, to

- advise the government on nature conservation policies
- notify sites of great wildlife, geological or natural value
- establish and manage national nature reserves
- organise and carry out research
- disseminate the results of this research.

English Nature and the Countryside Commission are both involved in the conservation of the countryside. English Nature has its origins in the 1949 National Parks and Access to the Countryside Act, with a range of duties from scientific research and site management to producing educational literature on the environment. The Countryside Commission was formed from the 1968 Countryside Act as an extension of the National Parks Commission. It is responsible for enhancing the beauty of the countryside and improving outdoor recreation in the countryside. This includes managing country parks. These parks are usually situated near to centres of population to reduce recreational pressure on national parks.

Voluntary organisations

Voluntary or non-government organisations (NGOs) play a key role in nature conservation in the UK. Besides owning or managing thousands of important sites across the UK, they act as **pressure groups**, organising campaigns and protests against developments which threaten wildlife. Many have an important educational role, producing everything from leaflets to scientific reports on specialist environmental issues.

Royal Society for Nature Conservation Wildlife Trusts have a membership of 250 000 and manage 2000 nature reserves. There are 47 County Wildlife Trusts and 50 Urban Wildlife Trusts in the UK.

The **Royal Society for the Protection of Birds** is the largest wildlife conservation organisation in Europe. With a membership of 870 000, the RSPB employs 12 full-time scientists and manages over 120 reserves.

The **National Trust** is a multimillion-pound operation and the third largest landowner (after the State and the Crown) in the UK. It has over 1.5 million members, and its aims include preservation of buildings and landscape, nature conservation and providing public access.

The **British Trust for Conservation Volunteers** organises conservation working weekends, holidays and long-term placements all over the country. Volunteers learn practical conservation skills such as coppicing, dry stone walling, hedgelaying, bridge-building and pond-making.

Friends of the Earth is an international pressure group active in 52 countries. With 250 local groups throughout England, Wales and Scotland, they are the biggest pressure group in the UK. Relying for 80% of their income from individual donations and subscriptions, FoE's successful campaigns have addressed:

- the protection of wildlife habitats, for example Oxleas Wood
- the banning of CFCs
- reduction in the trade in rainforest timber
- the strengthening of pollution-control legislation.

QUESTION

18.2 What is meant by the following terms:
(a) non-government organisations
(b) aesthetic justification for conservation?

18.3 WILDLIFE AND LANDSCAPE CONSERVATION IN THE UK

Wildlife conservation

Wildlife conservation is mainly achieved through the designation of important sites, but it is also achieved through restrictions on the import or release of non-native (exotic) species and by **captive breeding** and **reintroduction** (the breeding of species in research centres or zoos and then their release into the wild, for example the North American condor). Designated areas for wildlife conservation include nature reserves (local, national and marine) and areas of special scientific interest.

Local Nature Reserves are controlled by local planning authorities, with the help of many voluntary organisations and English Nature.

There are 270 **National Nature Reserves** in the UK, designated to protect areas of natural or semi-natural vegetation and their fauna. NNRs are controlled by local authorities but, again, many are managed by voluntary organisations.

Fig 18.3 Lundy island, Devon.

Marine Nature Reserves include the island of Lundy, off the Devon coast, and Skomer, off the coast of Dyfed. These are protected by a wide range of national and local by-laws and by a voluntary code of conduct.

Sites of Special Scientific Interest are areas containing biological, geological or physiographic features of scientific interest. SSSIs are almost always found on privately owned land. Many have been damaged or completely destroyed by the owners who thought that designation would reduce the value of the land or the output from it. Under the 1981 Wildlife and Countryside Act, the statutory bodies of the UK (English Nature etc.) notified the owners of all SSSIs, negotiating management agreements with the owners of the land and advising them against carrying out any potentially damaging operations (PDOs), such as ploughing up a field containing orchids, for example.

However, if landowners are prevented from making money, for example by draining or ploughing a water meadow, then English Nature are likely to compensate the landowners for any profits lost. Such compensation payments have proved extremely expensive for English Nature and there have been several public inquiries into compensation claims.

Landscape conservation

National parks were set up following the 1949 National Parks and Access to the Countryside Act. The aim of national parks is to maintain valuable landscapes and to promote quiet recreational activities within them. Tight planning restrictions operate within the parks so as to protect their individual characters. However, these restrictions do not apply to many agricultural, forestry or quarrying operations.

Areas of Outstanding Natural Beauty include areas such as the Mendips and Malvern hills. Local Planning Authorities are responsible for ensuring that the AONBs are not damaged by development.

There are 45 **Heritage Coasts** in England and Wales, which cover 35% of the total coastline. There are 26 equivalent designations in Scotland (**Preferred Coastal Conservation Zones**), covering 74% of the coastline of the mainland and islands.

Greenbelts are areas of open countryside surrounding major urban areas. The main aim of greenbelts is to limit urban sprawl and to stop adjacent urban areas from merging. Development is supposed to be strictly controlled within the greenbelt.

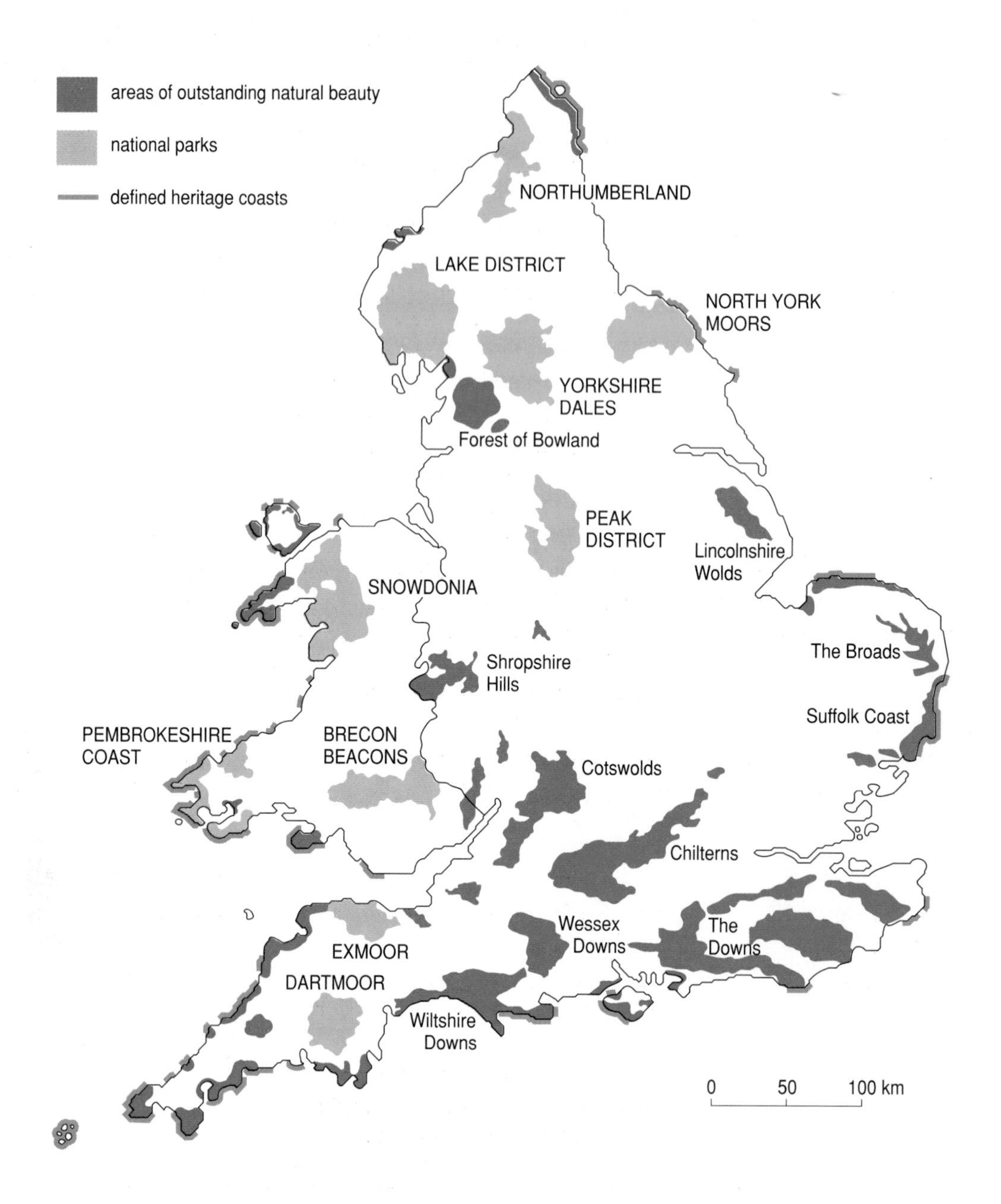

Fig 18.4 Protected areas in the UK. Landscape conservation is achieved through a variety of designations.

Environmentally Sensitive Areas are areas which have high landscape or wildlife value and which are vulnerable to the effects of intensive farming. Farmers are therefore compensated for adopting traditional and less harmful farming practices in a ten-year management plan. ESAs and areas under Countryside Stewardship are discussed in Chapter 14.

Charitable organisations such as the National Trust and Woodland Trust are major landowners and therefore play an important role in both biological and landscape conservation.

Some of the most important conservation designations are summarised in Table 18.4.

Table 18.4 Examples of conservation designations

Designation	Aim	Problems
zoos	breed endangered species; release into natural habitat; public education	fraction of total threatened species bred; breeding often unsuccessful; animal welfare concerns
national/marine/ local nature reserve	protection and management of areas to allow plants/animals to live and breed; also used for research demonstration, education and as an amenity	large reserves, e.g. marine nature reserves, have proved very difficult to protect
field gene banks (also botanic gardens)	maintain large numbers of growing plants of one species, e.g. trees or tropical crops, to maintain the widest practicable range of genetic diversity	lack of space
biosphere reserve	areas of land or coast designated by IUCN which have unique importance for research, monitoring, training and demonstration, as well as conservation	
Site of Special Scientific Interest (SSSI) – area of special scientific interest in N. Ireland	protection of sites of great floral, faunal, geological or physiographic interest; SSSIs will form the basis of Special Protection Areas (birds) and Special Areas of Conservation (species and habitats) required under the EU Habitats Directive	legislation has not prevented extensive damage and destruction; management agreements may be very expensive
national parks (the UK)	conserve areas of wild landscape beauty and encourage quiet outdoor recreation	quarrying and forestry activity may threaten landscape; sheer numbers of visitors and some of their noisy recreational activities threaten the isolation and calm which the parks were meant to preserve
national parks (outside the UK)	conserve large natural areas for wildlife	poachers kill wildlife, population pressure threatens boundaries
seed banks	long-term storage of seeds as a potential source of genetic variability	lack of space; 20% of plant species produce seeds that will not tolerate low temperatures and/or drying
biogenetic reserve	set up by the Council of Europe to conserve representative examples of Europe's fauna, flora and natural areas; there are 19 such reserves in the UK	

Conservation legislation

UK wildlife is protected by many voluntary agreements and local, national and international legislation. The 1981 Wildlife and Countryside Act:

- increased the number of animals and plants that are legally protected
- prohibited the release of non-native animals and plants
- encouraged people to get out into the countryside by making local authorities produce definitive footpath maps
- ordered that all SSSIs be re-inspected and the owners notified of the designation
- stated that conflicts between landowners and nature conservation bodies should be resolved by voluntary agreement. In practice, this would mean management agreements and compensation payments.

Table 18.5 Conservation conventions

Legislation	Description
Convention on International Trade in Endangered Species (CITES)	aims to regulate trade of and products from endangered species; this includes species for which no commercial trade is permitted, e.g. African elephants, Great whales, some orchids; also includes those species which may become threatened with extinction unless trade is strictly regulated, e.g. all owls, all crocodiles, all cacti; also includes species which require protection in certain countries only, e.g. hippos in Ghana
Ramsar Convention	lists wetland sites which are of international conservation importance
International Convention for the Regulation of Whaling	aims to protect whales from over-exploitation but to encourage sustainable exploitation
Convention on Biological Diversity	aims to encourage the conservation of biodiversity *in situ* by encouraging economically sustainable use
World Heritage Convention	designates sites of universal conservation or scientific value

CASE STUDY

Fig 18.5 Giant panda at Wolong Giant Panda Research and Conservation Centre.

No home to go to

The success of China's captive breeding programme for giant pandas – with four cubs born in Autumn 1995 at the Wolong Giant Panda Research Centre in Sichuan – is undoubtedly a triumph for the Chinese and WWF scientists concerned. It comes at the end of a decade of captive breeding success: rare parrots, falcons, mice, marmosets, frogs and snails have all been successfully reared, and nowadays, by and large, we are able to breed most species at will.

But breeding success has, ironically, brought sharp criticism of the programme. While the numbers of captively-bred, endangered species has increased, little has been done to reduce pressure on the animal's living space from agricultural expansion, poaching and deforestation.

In south-east China, dedicated work by Chinese zoologists has resulted in the 'saving' of the endangered Chinese alligator, known to the inhabitants of the Middle Kingdom as the Ice Dragon. It is the only one of the world's crocodiles to hibernate.

By 1992, scientists had a total of 6000 hatchlings, and reintroduction of the alligators was imperative, but analysis of the alligator's habitat using Landsat satellite images and field surveys revealed that there were no suitable sites for reintroduction. The breeding programme has now been halted.

There is a further problem that recurs with species the world over: local human resistance to reintroduction, as in the case of the Manchurian (Siberian) tiger, which is extinct in the wild in China and is down to 300 animals divided between populations in Siberia and northern Korea. By contrast, the captive breeding programme, established with just 22 tigers in 1986, now has a total of 300 animals. The problem, once again, is that no one wants a tiger released anywhere near their village, town or city, and certainly not in their backyard.

Even when reintroduction has been sanctioned by the locals, the problems do not disappear. Many mammals and some birds simply cannot cope with translocation. Once a giant panda has become used to living in an area, the animal seems unable to accommodate a change of scene: of two wild pandas caught and released into new areas, one returned to its old haunts immediately, and the second, apparently completely disoriented, despite a move of only 20 km, set off on its own 'long march' in search of home and was later found dead in very steep terrain, many kilometres from the panda's release point.

Captive-breeding programmes have been a success, but attempts at reintroduction have not. Many scientists are now calling for a halt or at least a huge reduction in breeding efforts until the real, but infinitely less glamorous and more difficult, task of protecting or recreating habitats has been addressed.

QUESTION

18.3 Summarise the major obstacles to successful reintroduction of captive-bred species to the wild.

18.4 WILDLIFE CONSERVATION IN THE UK – SUCCESS OR FAILURE?

Sites of Special Scientific Interest, of which there are 6103 covering 20 000 km^2, are supposed to be one of the most important designations for wildlife conservation in the UK. However, many have been extensively damaged or destroyed by agriculture and forestry.

Landowners who intend to carry out any operation that would damage the conservation of a site must give four months' notice, during which time the statutory body can negotiate with the owner to try to persuade them not

Table 18.6 Native species at risk or extinct (selected species)

Species group	No. of native species	No. extinct	No. endangered
land mammals	44	1	–
birds	523	6	–
amphibians	6	0	1
freshwater fish	41	2	2
flies	6000	12	188
bees/wasps/ants	542	17	41
beetles	4000	68	164
butterflies/moths	2600	35	126
caddis flies	199	2	7
mayflies	48	0	1
spiders	687	0	22
leeches	16	0	0
seed plants	2230	19	48
ferns	70	0	3
liverworts/mosses	1000	16	31
lichens	1700	33	44
fungi	20 000+	n/a	72

– = not known

to carry out the operations. Compensation payments usually help. If agreement cannot be reached within four months, the Secretary of State for the Environment can enforce a Nature Conservation Order to extend the period of notice so that negotiations can continue. However, if no agreement can be reached, the only alternative may be compulsory purchase of the site.

In the last ten years, over 500 SSSIs have been damaged and hundreds of sites have had to be denotified. However, the rate of damage and destruction by agricultural activities has been dramatically reduced in recent years as a result of successful management agreements.

Since 1900, over 200 of our native animal or plant species have become extinct (Table 18.6), with habitat-loss the most common cause; since the 1940s over 90% of flower-rich meadows, 50% of fens and wetlands and 368 000 km of hedgerow have been lost. However, recent initiatives to make agriculture less intensive and the government's decision to reduce the number of new road schemes are encouraging signs that conservation is at last being taken more seriously.

CASE STUDY

Raiders of the lost bark

Light-fingered local groups have been raiding timber yards this summer, seizing planks of Brazilian mahogany and handing them over to the police as goods stolen from the Amazon Indians … the campaign aimed to draw attention to the illegal logging of mahogany in Indian reserves in Brazil.

The Independent, 18 February 1995

National Trust gets tough

The UK National Trust is to change its conservative image and become a campaigning organisation prepared to 'engage in political debate'. The Trust owns 1% of the British Isles and has already threatened to invoke rarely-used parliamentary powers to protect land threatened by road schemes.

The Guardian, 29 September 1995

On yer bike!

The Millennium Commission has given a £42.5 million grant to help construct a 6500 mile network of cycle routes across the UK. The award has been given to Sustrans, a voluntary transport organisation which expects half of the route to be along derelict railways, canal paths and bridleways with the other half along minor roads.

The Independent, 11 September 1995

RSPB want new marine protection

The RSPB has called for new laws to protect marine wildlife. After surveying 58 of the most important seabird sites around the UK, it found that almost all of them lacked proper management or protection. It has called upon the government to uphold their commitments under the EU Birds Directive and the EU Habitats and Species Directive.

The Independent, 4 May 1994

WWF launches CITES airport campaign

The World Wide Fund for Nature has launched a campaign at British airports to educate people about how to avoid buying illegal products made from endangered species. Last year customs officials at London airports confiscated 12 853 wildlife items, including chameleons and souvenirs made from ivory, reptile skins, coral and stuffed turtles.

The New Scientist, 30 April 1994 © *New Scientist*

Fig 18.6 Ivory collection.

ANALYSIS

Declining fish stocks

Worldwide, fish stocks are in decline, and two students decided to carry out a library research investigation to find out why.

The students had a sneaking suspicion that declining stocks were due either to overfishing or pollution, and so they concentrated their research on these two aspects. Some of the first data they collected are shown below.

Table 18.7 International fish catch / thousand tonnes

Year	Cod	Plaice
1982	273	185
1984	205	138
1986	163	169
1988	150	109
1990	105	54
1992	97	81

Table 18.8 Average concentrations of contamination in fish caught in UK fisheries/ mg kg^{-1} wet weight

Year	PCBs		Mercury	
	Cod	Haddock	Cod	Haddock
1982	6.6	–	0.09	–
1984	–	0.6	0.08	0.06
1986	2.1	–	0.08	0.04
1988	2.9	0.3	0.08	–
1990	2.1	0.1	0.09	0.05
1992	–	0.1	–	0.05

– = no available data

1. **(a)** By selecting data from Tables 18.7 and 18.8, plot a graph to illustrate any possible trend between fish stocks and pollutant contamination.
   (b) In terms of the students' aim, what are the limitations of this data?

The students then discovered the following graph.

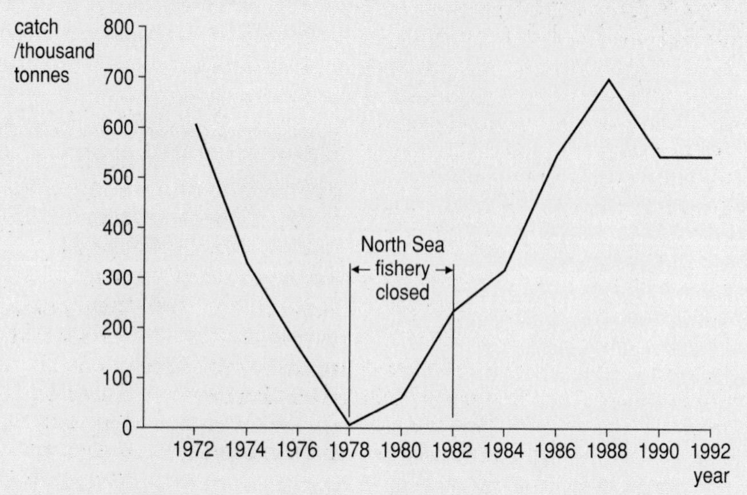

Fig 18.7 International herring catch.

The students concluded that overfishing had been the cause of declining catches of herring, and that this was also probably the reason for declining catches of other species.

(c) What other data would have been useful in this exercise?
(d) What data are needed in order for fisheries' scientists to be able to calculate the maximum sustained yield of cod in the North Sea fisheries?

SUMMARY ASSIGNMENT

1. In 1984 the Council of Europe published a list of 368 animal and 108 plant species in imminent danger of extinction in Europe.
   (a) List **three** major human activities which may have contributed to these species becoming endangered.

   Various arguments have been put forward to justify the conservation of animals and plants faced with extinction.
   (b) Outline briefly the arguments for wildlife conservation using the following headings.
   (i) ethical/moral argument
   (ii) aesthetic argument
   (iii) scientific argument.

   NEAB June 1995

2. Explain why it is important that we do not let
   (a) tigers
   (b) mosquitoes, become extinct.

3. Conduct a library research assignment into the work of an environmental non-governmental organisation (NGO) in your area.

Appendix A

GLOSSARY

A

Abiotic factors – factors resulting from the non-living components within the environment.
Absorption spectrum – a graph showing the relative absorption of different wavelengths of light by a pigment.
Acid rain – a general term encompassing dry deposition and precipitation which has an increased acidity due to contamination by carbon dioxide, sulphur dioxide and nitrogen dioxide.
Action spectrum – a graph which shows the relative effectiveness of different wavelengths of light in stimulating photosynthesis.
Adit mining – the extraction of minerals through a horizontal gallery.
Aerobic respiration – the breakdown of molecules such as glucose to release energy in the presence of oxygen.
Afforestation – the planting of trees on land which did not previously have trees.
Air mass – a large body of air in which the properties at any one horizontal level are similar. The characteristics of an air mass are determined by the source area.
Albedo – the percentage of solar radiation which is reflected back from a surface.
Ambient – the surrounding environmental conditions.
Anabatic and katabatic winds – winds caused by differential heating of the valley sides and floor, respectively.
Anaerobic respiration – respiration without the presence of oxygen, causing the incomplete breakdown of organic compounds. See Respiration and Aerobic respiration.
Anthropocentric – a way of viewing nature primarily from the human point of view.
Anticyclone – a large weather system in which air is sinking; associated with calm, clear weather conditions.
Aquifer – a layer of water-bearing, porous rock below the earth's surface.
Area of Outstanding Natural Beauty (AONB) – an area of land designated for high scenic, amenity and scientific value, which is smaller and has fewer planning restrictions than a national park (e.g. the Cotswolds).
Autotroph – an organism which synthesises organic molecules from simple inorganic molecules, using an external energy source (e.g. the sun: photoautotrophs; inorganic compounds: chemoautotrophs).

B

BAT – a pollution control system which uses the 'best available technology'.
BATNEEC – a pollution control mechanism which uses the 'best available technique not exceeding excessive cost'.
Bedrock – a general term for the local solid rock.
Bioaccumulation – the accumulation of non-biodegradable substances within living tissue. Compare Biomagnification.
Biochemical oxygen demand – the combined measure of the amount of dissolved oxygen within water utilised by micro-organisms and chemicals. The biochemical oxygen demand will include the biological oxygen demand and chemical oxygen demand.
Biodiversity – a measure of species variation in a region.
Biogeochemical cycles – the movement of matter, such as oxygen, nitrogen, carbon and phosphorous, within ecosystems by biotic and abiotic factors.
Biological oxygen demand (BOD) – a measure of the amount of dissolved oxygen utilised by living organisms when organic pollution enters a body of water. See Biochemical oxygen demand and Chemical oxygen demand.
Biomagnification – the increasing accumulation of a non-biodegradable substance (e.g. heavy metal) going up through a food chain. Compare Bioaccumulation.
Biomass – the total mass of living organisms in a community or area. See Pyramid of biomass.
Biome – an ecosystem, covering a broad area, which has distinctive climatic and soil conditions (e.g. tropical rainforest, tundra).
Biosphere – the part of the earth in which all organisms live.
Biota – all the organisms present within an area.
Biotic factors – factors which result from the living or biological components of the environment.
Birth rate – a measure of fertility. Expressed as the number of births per year per thousand people. See Natural increase.
BPT – a pollution control measure which uses the 'best practical technology' at a reasonable cost.
Built-in obsolescence – products designed for a limited life-span and difficult to repair.

C

C_3 plant – a plant which uses RUBP carboxylase to fix carbon dioxide; so called because the first stable product in the light-independent stage is a molecule containing three carbons.
C_4 plant – a plant which uses PEP carboxylase to fix carbon dioxide; the first product of the light-independent stage is a molecule containing four carbons. C_4 plants are more efficient at fixing carbon dioxide and reducing water loss than C_3 plants.
Calcification – a soil-forming process in which calcium ions form in nodules in the topsoil. Produces the very fertile chernozem soil profile.
Cancer – uncontrolled cell growth. See Carcinogen.
Carbohydrate – an organic molecule consisting of carbon, hydrogen and oxygen atoms (e.g. starches or sugars).
Carbon cycle – the natural cycling of carbon atoms between rocks, vegetation, oceans and the atmosphere. See Carbon sink.
Carbon monoxide – an odourless, colourless, toxic gas produced by incomplete combustion of carbon-containing substances. Can become attached to haemoglobin molecules within blood, thus preventing the transport of oxygen, possibly resulting in death.
Carbon sink – part of the carbon cycle where large quantities of carbon accumulate (e.g. calcium carbonate rocks, coal). See

Carbon cycle.
Carcinogen – a cancer-inducing substance. See Cancer.
Carnivore – organism which obtains food by preying solely on other animals.
Carrying capacity – the number of organisms which a particular ecosystem or area can sustain over a long period of time.
Catalytic converter – a device fitted onto car exhausts to reduce pollutant emissions. The following conversions take place: carbon monoxide to carbon dioxide, nitrogen oxide to nitrogen and oxygen, and organic compounds to carbon dioxide and water.
Catchment – an area of land which collects water to feed a river or reservoir.
CFCs – chlorofluorocarbons – manufactured greenhouse gases which are a cause of ozone destruction. Used as aerosol propellants, refrigerants and in packaging.
Chelation – a soil-forming process in which organic acids, produced via decomposition, attack mineral components of soils.
Chemical oxygen demand (COD) – a measure of the amount of dissolved oxygen that can be used by chemical compounds.
Chemoautotroph – bacteria which use the chemical energy in inorganic substances such as ammonia to fix carbon dioxide. Such organisms play an essential role in the nitrogen cycle.
Chemosynthesis – the process whereby some bacteria synthesise organic compounds using the energy from inorganic compounds. See Autotroph.
Chlorophyll – green pigments within the leaves and some stems of green plants, used to trap the energy in sunlight for photosynthesis.
Chloroplast – the site of photosynthesis within the cells of green plants.
Clear felling – the removal of all trees from a particular area at one time.
Climate – the long-term atmospheric conditions over a large area.
Climatology – the study of long-term atmospheric events which take place over a large area.
Climax community – the final community stage in succession, which is determined by the climate and/or soils (edaphic climax).
Coal – a solid, black fossil fuel derived from dead plant matter. There are three forms of coal: lignite, anthracite and bituminous coal which are classified according to their carbon concentration. See Coalification.
Coalification – the process whereby plant matter which accumulates under anaerobic conditions is converted into coal over geological time. See Coal.
Cold front – See Front.
Condensation – the change of state of water vapour to liquid when the saturation point is reached. Cold air can hold less water than warm air so condensation normally occurs due to cooling. Condensation can only occur around hygroscopic nuclei. See Hygroscopic nuclei.
Condensation nuclei – microscopic particles which water vapour condenses around to form water droplets.
Conservation – the management of organisms, ecosystems or the landscape to enable protection, enhancement or restoration.
Continentality – the climatic effects within the centre of a large continent.
Contour ploughing – a soil conservation technique where ploughing occurs across a slope to reduce soil erosion.
Convection – the transfer of heat through the movement of a liquid or gas; the process by which warm air rises.
Convenience product – goods brought ready to use, requiring little preparation and often involving considerable packaging.
Coppicing – a traditional form of woodland management in which new shoots regenerate from a cut stump.
Core – the hot, dense centre of the earth consisting of a solid inner core and a molten outer core.
Coriolis force – a deflecting force caused by the earth's rotation.
Cost-benefit analysis (CBA) – a way of evaluating large-scale developments by giving each element a monetary value.
Counter-urbanisation – the migration of people away from large urban areas to towns and villages in rural areas.
Cover crops – a soil conservation technique in which fast-growing plants protect the soil from exposure to the wind and rain after a commercial crop is removed.
Crust – the solid, brittle, cool uppermost layer of the earth, up to 70 km thick. There are two types of crust: older, thicker, less dense continental crust and denser, thinner, younger oceanic crust.
Cryosphere – the part of the hydrosphere involving ice.

D

Daily temperature variations – the differences in temperature as a result of the rotation of the earth.
DDT (Dichloro-diphenyl-trichloro-ethane) – a non-biodegradable organochloride which has been used as an insecticide.
Death rate – the number who died per year per thousand people. See Natural increase.
Deciduous plants – trees and shrubs which lose leaves annually at the end of the growing season.
Decomposers – bacteria and fungi which carry out decomposition. See Decomposition.
Decomposition – the degradation of organic material into smaller molecules by fungi and bacteria.
Deforestation – the removal of trees from an area.
Demography – the study of human populations.
Denitrification – part of the carbon cycle in which free-living bacteria in the soil convert nitrates to atmospheric nitrogen and nitrous oxide.
Density-dependent factors – factors which lower the birth rate or increase the death rate as a population grows in size (e.g. quantity and quality of food). See Birth rate and Death rate.
Density-independent factors – factors which affect population size irrespective of population density (usually abiotic e.g. weather, climate, fire). See Birth rate, Death rate and Abiotic factors.
Depression – a large weather system associated with the mid-latitude in which air is rising upwards and inwards.
Derelict land – land which has become damaged by human activities so that it can no longer be used beneficially without reclamation.
Desertification – the conversion of fertile land into less fertile land, by natural or human-induced processes.
Detritivore – the organisms within a food chain which consume dung, debris and other organic litter.
Developed country – a country with much wealth, high energy consumption, high-quality services, high literacy rates and generally a high standard of living (e.g. USA, Canada, most European countries, Australia, New Zealand and Japan).
Developing country – a relatively poor country, although there is a huge variance between countries and within countries themselves (e.g. Bangladesh, Brazil).
Dew point – the temperature at which condensation occurs, due to cool air becoming saturated with water vapour.
Discharge – the volume of water passing a point over a particular length of time (e.g. litres per second).

Dissolved oxygen (DO) – the amount of oxygen, expressed in parts per million, dissolved within a volume of water. Will vary with temperature and atmospheric pressure.
Distribution – the way individuals are spread out within a region.
Diversity – describes the number of species within a community.

E

Ecocentric – the moral view that organisms and ecosystems have values.
Ecology – the study of plants and animals and the way they interact between themselves and within their environment.
Ecosystem – a community in which energy and matter are transferred in complex interactions between the environment and organisms.
Ecotone – the boundary between two communities.
Ecotourism – a form of sustainable tourism which has a low impact on the environment. The money received from tourists is used to conserve the environment which attracted them.
Effluent – agricultural or industrial liquid waste.
Electromagnetic spectrum – the range of wavelengths emitted from the sun, including short-wave and long-wave radiation.
Eluviation – the removal of fine soil particles in suspension within the soil profile.
Endangered species – a species which is very close to extinction.
Energy – the ability to do work.
Enhanced greenhouse effect (EGHE) – the intensification of the natural greenhouse effect caused by increased volumes of greenhouse gases released by human activities. See Global warming, Greenhouse effect and Greenhouse gases.
Environment – the conditions which affect a community or organisms.
Environmental Impact Assessment (EIA) – a management tool, using various methodologies, in which the environmental consequences of a development are reviewed.
Environmentally Sensitive Area (ESA)– an area designated for its natural beauty, rarity of habitat and historic interest; compensation is given to farmers who use traditional practices to maintain the character of the area.
Erosion – the loss of soil, sediment or bedrock involving transportation by the wind, glaciers or flowing water.
Eutrophication – the enrichment of water by nutrients such as nitrates and phosphates causing accelerated plant growth, resulting in the disturbance of the ecosystem.
Evaporation – the change of state from a liquid into a gas.
Evapotranspiration – the combined loss of water from water bodies, the land and plants; also describes the combined effects of evaporation and transpiration from plant leaves. See Evaporation and Transpiration.
Evolution – the process whereby species change as a result of random changes in genetic material.
Exotic species – a species which is not native to a country, i.e. one which has been introduced from abroad.
Exponential growth – the rate of increase is proportional to the number of individuals currently in the population. This results in population numbers following a pattern such as 1; 2; 4; 8; 16; 32; . . .
Extinction – the total loss of a species owing to changes in the environment, or one species killing off another.

F

Fauna – the animals present in a community or area.
Fertiliser – a substance added to soils to improve nutrient supply for plant growth.
Flora – the plants present within a community or area.
Flue-gas desulphurisation – a pollution control measure used within coal-fired power stations to reduce carbon dioxide and sulphur dioxide emissions.
Food chain – the transfer of energy and matter in a sequence of trophic levels in which organisms of a lower trophic level become the food of organisms in a higher level. See Trophic level and Food web.
Food web – the interconnection of organisms within several food chains. See Food chain and Trophic level.
Fossil fuels – a stored energy source, originally derived from the sun, created by the conversion of plant and/or animal matter over geological time (e.g. coal, oil, natural gas).
Front – the boundary between two different air masses. A warm front is the leading edge of a depression in which a warm air mass pushes itself above a cold air mass. A cold front is the trailing edge of a depression where a cold air mass pushes itself below a warm air mass forcing the warmer, less dense air to rise. See Air mass.
Frontal rain – the formation of rain due to air in a warm air mass being forced to rise over air in a colder, more dense air mass.
Fungi – one of the five kingdom classifications. Fungi are non-photosynthesising organisms with absorptive nutrition.

G

Gaia – the concept, named after the Greek goddess of the earth, in which the environment and the biota have evolved as a single perpetuating system. In this system the biota ensure that the physical and chemical conditions on earth are suitable for life.
Geomorphology – the study of landforms.
Geothermal energy – energy derived from the earth's internal heat.
Glaciation – an area which has been subjected to erosion and/or deposition by the action of ice and glacial meltwater.
Gleying – the temporary saturation of the soil profile. See Waterlogging.
Global warming – the predicted change in global air temperatures as a result of the enhanced greenhouse effect.
Green revolution – a term used to describe the time at which new techniques dramatically increased agricultural production (e.g. hybrid varieties, pesticides, fertilisers).
Greenbelt – a zone of protected land surrounding cities in which development is controlled in order to prevent neighbouring cities from merging, the growth of urban sprawl, the destruction of habitats and agricultural land, and to preserve historic towns.
Greenhouse effect – the natural 'trapping-in' of heat by greenhouse gases present in the atmosphere.
Greenhouse gases – those gases present in the atmosphere which trap outgoing long-wave radiation, causing a rise in temperatures (e.g. CFCs, carbon dioxide, methane).
Groundwater – water stored within the pores, joints and fissures of rocks within the earth.
Growth rate – the change in population size over time.

H

Habitat – the area in which a species or community lives.
Heavy metal – toxic metals such as lead, cadmium, nickel, zinc, copper, chromium and arsenic.
Herbicide – a chemical which destroys plants (e.g. paraquat).
Herbivore – an organism which obtains food solely by eating plants.
Heterotrophs – all those groups of organisms which need a supply of complex organic material as they cannot synthesise

their own food, for example mammals, earthworms, fungi.
Homeostasis – the maintenance of conditions. It can refer to a particular set of environmental circumstances or to the internal environment of an organism, such as a plant or an animal. See Negative feedback and Positive feedback.
Horizons – distinctive layers within a soil profile which develop due to various soil-forming processes when the soil is in a state of equilibrium.
Humification – the conversion of dead organic matter, collecting as leaf litter, into humus within a soil. See Humus.
Humus – organic matter within soil. Mor humus, produced by incomplete decomposition, is acidic, fibrous and infertile, whereas mull humus is brown, sticky and fertile. Humus combines with clay components to form the clay-humus complex, which acts as a source of nutrients, retains water and binds soil particles together.
Hydrological cycle – the cyclical movement of water through the processes of evapotranspiration, condensation, wind transportation, precipitation, runoff, infiltration and interception.
Hydrosphere – all forms of water present on the earth, incorporating lakes, oceans, streams, groundwater, snow and ice, fresh or saline water.
Hygroscopic nuclei – solid particles suspended in the atmosphere around which condensation can occur (e.g. dust, salt crystals). See Condensation.

I

Igneous rock – intrusive (created within the earth) or extrusive (created on the surface) crystalline rocks formed from the cooling and consolidation of magma.
Impermeable – a surface which does not allow water to infiltrate into it.
Incineration – a waste disposal technique involving combustion of solid waste at high temperatures.
Industrial revolution – the time during which human society progressed considerably as a result of advances in technology.
Infiltration – part of the water cycle in which water percolates into soil and permeable bedrock.
Insecticide – a toxic substance which kills insects.
Insolation – the amount of solar radiation received at a point on the earth's surface.
Intensification – a general term used to describe the dramatic increase in agricultural yield as a result of such actions as mechanisation, increased use of fertilisers and pesticides, improved farming techniques and use of hybrid varieties.
Interspecific competition – competition between members of the different species for resources within a community. See Intraspecific competition.
Intraspecific competition – competition between members of the same species for resources within a community. See Interspecific competition.
Ionosphere – the lower part of the thermosphere.

J

J-curve – a growth curve which illustrates exponential growth; named after the shape of the graph produced. See Exponential growth and Sigmoidal curve.

K

Knock-on effects – the consequences of a change in the environment.

L

Land cover – the appearance, features and characteristics of an area of land.
Land use – the actual use to which land is put (e.g. forestry, housing).
Landfill – a controlled land-based waste disposal technique for solid waste, incorporating pollution control measures.
Lapse rate – the rate at which a body of air cools at it rises.
Latitude – a way of establishing any location on the globe, calculated as the angular distance north or south of the equator. See Longitude.
Lava – magma extruding onto the surface through volcanic events.
Leachate – liquid pollutants produced from degradation of waste substances within a landfill site.
Leaching – the loss of nutrients from the topsoil in solution.
Leguminous plants – plants which have nitrogen-fixing bacteria within root nodules as part of a symbiotic relationship (e.g. peas, clover). The bacteria obtain a habitat and carbohydrate for food and the plants gain an important source of nitrogen. See Nitrogen fixation and Symbiosis.
Life-cycle analysis – the study of the environmental consequences of a product or development during its lifetime (i.e. manufacture, construction, decommissioning, recycling and disposal).
Life expectancy – the average length of time for which an organism will usually survive.
Limiting factors – various factors which limit the distribution or numbers of an organism. See Density-dependent and Density-independent factors.
Lithosphere – the layer of rocks on the surface of the earth.
Longitude – a way of positioning any location on the globe, calculated as the angular distance around the world east or west of the Greenwich meridian.

M

Magma – molten rock produced by high temperatures deep within the earth. See Lava.
Malthusian crisis – the theory created by Thomas Malthus (1896) in which a population which grows exponentially will suffer a population crash caused by a lack of resources and food.
Mantle – a hot, malleable layer of the earth below the lithosphere.
Maritime climate – a mild and wet climate which is produced in coastal regions or islands, because of the proximity to the sea.
Maturation – the formation of oil from microscopic marine life involving increased temperature and pressure over a geological time-scale.
Maximum sustainable yield (MSY) – the largest amount of a resource that can be harvested which does not cause a decline in the basic stock of that resource.
Megawatt (MW) – a unit of electrical energy. One MW = one million watts. See Power.
Metamorphic rock – rock created by the effect of high temperatures and/or pressures on pre-existing rock, over a geological time-scale (e.g. limestone can metamorphose into marble).
Meteorology – the study of short-term, localised atmospheric events. See Weather.
Methane – gas produced through decay of organic material in the absence of oxygen.
Microclimate – small-scale climatic conditions where the regional climate is influenced by localised factors such as vegetation, topography and ground cover.
Mineral – a naturally occurring crystalline element or compound which, when combined, forms rocks.
Mitigation – action taken to compensate for damage to the

environment (e.g. restoration, rehabilitation, or creation of new habitats to replace those destroyed).
Mulch – a substance added to the soil to provide nutrients and protective cover (e.g. manure, wood chippings).
Municipal waste – waste generated from urban areas.

N

National park – a designated area of high landscape value, in which developments are controlled so that the character of the area can be preserved, protected and enhanced.
Natural gas – the least polluting fossil fuel produced either from coalification ('dry natural gas') or from maturation ('wet natural gas'). See Coalification, Fossil fuel and Maturation.
Natural increase – the rate of change in total population calculated by subtracting the crude death rate from the crude birth rate. Both death rate and birth rate are calculated in numbers per thousand people per year.
Negative feedback – a process triggered by an initial change in an environmental variable so that the original 'normal' condition is restored. It can also refer to an internal body process, e.g. temperature regulation of an organism to restore the body temperature. See Positive feedback.
Net primary production – the amount of production passed from the producers onto the next trophic level in a food chain.
Net production – the amount of production after respiration losses are removed.
Neutron – a particle within the nucleus of atoms which has no electromagnetic charge.
Niche – the role, activities and location of an organism within a habitat.
Nitrate Sensitive Area – a designated area of land in which nitrate application is severely restricted in order to prevent nitrate pollution of groundwater (e.g. restricted use of fertilisers).
Nitrification – the conversion of ammonium (NH_4^+) to nitrite (NO_2^-) then nitrate (NO_3^-) by aerobic bacteria (e.g. *Nitrosomas*) as part of the nitrogen cycle.
Nitrogen cycle – the cycling of nitrogen atoms between the atmosphere, rocks, soil and biota. See Nitrification, Denitrification and Nitrogen fixation.
Nitrogen fixation – the conversion of atmospheric nitrogen into ammonia by free-living bacteria, or bacteria within the roots of leguminous plants.
NO_x – a group term which describes the various oxides of nitrogen [nitrogen oxide (NO), nitrogen dioxide (NO_2) and nitrous oxide (N_2O)]
Non-renewable resources – resources which become depleted, because once they are used they are not replenished at a sustainable rate.
Nuclear fission – the splitting of radioactive atoms to produce daughter products and energy. Fission in nuclear power reactors occurs in a chain reaction in which neutrons released from the splitting of atoms are used to split more radioactive atoms in an ongoing process.
Nuclear fusion – a natural process which occurs in the sun under intense heat in which two atoms fuse to form one larger atom, thereby releasing energy.

O

Ocean thermal electricity generation (OTEG) – the generation of electricity using temperature differences within the ocean to drive a gas through turbines.
Omnivore – an animal which obtains energy by eating both other animals and plants.
On-shore and off-shore breezes – winds caused by the differential heating of land and the ocean.
Opencast mining (or strip-mining) – the cheap extraction of minerals close to the surface where, after the removal of overburden, minerals are simply dug out.
Orographic rain – the formation of rain due to air being forced to rise over a mountain range.
Overburden – the layer of rock, sediment and soil which covers a useful raw material, but has no economic value itself.
Ozone – the blue, poisonous, odourless, explosive, triatomic form of oxygen produced naturally in the stratosphere and artificially in the troposphere. Ozone is a greenhouse gas and shields the earth from harmful ultra-violet radiation.
Ozone layer – the bottom layer of the stratosphere (12–25 km) in which there is a high concentration of naturally produced ozone.

P

Parent material – the unweathered source of mineral matter in soils.
Particulate matter – the amount of solid matter suspended in the atmosphere or a water body.
Parts per billion (ppb) – number of parts of a particular element or compound found within one billion parts of a gas, liquid or solid.
Parts per million (ppm) – number of parts of a particular element or compound found within one million parts of a gas, liquid or solid.
Payback time – the time taken for an energy conservation measure to pay for itself by reducing future energy bills.
Peat – a type of soil consisting almost entirely of organic matter.
Permeable – a surface which allows water to infiltrate.
Pesticide – a chemical used to kill animal and plant pests.
Petroleum – liquid and gaseous fossil fuels (e.g. oil, natural gas). See Fossil fuels and Natural gas.
pH – a scale from 0 to 14 which describes the level of acidity or alkalinity of a substance (0 = highly acidic, 7 = neutral, 14 = highly alkaline).
Phloem – a plant tissue containing tubes which run through every root, stem and leaf of a plant. Phloem carries sucrose into which the products of photosynthesis are changed for transportation from the leaves to whichever part of the plant needs it as well as substances such as amino acids, growth substances and some minerals. Flow is bi-directional.
Phosphorus cycle – the cycling of phosphorus atoms between biota, rocks and water.
Photoautotroph – green plants, some bacteria and other organisms which use sunlight energy to fix carbon dioxide.
Photochemical smog – a general term to describe the invisible concoction of pollutants released from urban areas (e.g. hydrocarbons, ozone, NO_x, carbon monoxide). See Smog.
Photosynthesis – autotrophic nutrition carried out in the chloroplasts of green plants in which light energy is converted into chemical energy (carbon dioxide plus water in the presence of light and chlorophyll to give carbohydrates and oxygen).
Photo-voltaic conversion – the conversion of incoming solar radiation into chemical energy, which can then be used to generate electricity.
Phytoplankton – free-floating, microscopic animals and plants.
Pioneer species – the first plant colonisers of a newly created habitat as part of the first stage of succession (e.g. lichens on bare rock, marram grass on sand-dunes).
Plate tectonics – the process whereby the lithosphere, divided into plates, moves horizontally across the surface of the earth.
PM10s – particulate matter of less than 10 microns, associated with car emissions, which is so small it penetrates deep into the lungs.

Podsolisation – an extreme form of leaching, associated with coniferous forests and moorlands, in which the A horizon has become severely depleted in nutrients.
Pollutant – a substance present in harmfully high concentrations.
Pollution – an event which adversely affects the health of organisms or their environment.
Population – the number of organisms of the same species within a given area. See Distribution.
Population pyramid – a model which illustrates both the age and sex structure of a given population.
Positive feedback – a process triggered by an initial change in an environmental variable which causes that variable to deviate further from the original 'normal' condition. It can also refer to the internal environment of an organism where an internal variable, e.g. blood temperature, may exhibit this phenomenon. See Negative feedback.
Power – the rate of energy delivery (units = watts). See Megawatts.
Primary pollutants – substances released directly into the air in a harmful form (e.g. oil).
Primary productivity – the synthesis of organic compounds by photosynthesising plants (or chemosynthetic bacteria).
Primary succession – see Succession.
Producer – an organism which synthesises organic molecules from inorganic molecules. See Autotrophs.
Productivity – the amount of biomass produced in a given time in a given area.
Push and pull factors – the factors which have an effect on migrating human populations. Pull factors are positive factors which cause an individual to migrate towards an area (e.g. better services), whereas push factors are negative factors which cause an individual to migrate away from an area (e.g. drought, lack of jobs).
Pyramid of biomass – a model which illustrates the mass of living matter at different trophic levels.
Pyramid of energy – a model which illustrates the amount of energy present at each trophic level.
Pyramid of numbers – a model which illustrates the number of organisms present at each trophic level.

Q

Quarrying – rock and stone exposed at the surface which is simply dug or blasted out.

R

Radioactive – an unstable atom which constantly emits particles.
Radioactive decay – the emission of radiation from radioactive atoms which gradually changes these atoms into other isotopes or another element.
Rain shadow – an area which has low precipitation due to the sinking of air in the lee of a large mountain range.
Recyclable – material which can be reused, or converted into new materials.
Recycling – reprocessing waste materials into another useful form (e.g. glass). See Reuse.
Renewable resources – resources which are not depleted because they are normally sustainably replaced by natural processes.
Re-radiation – solar radiation that has been absorbed is radiated again from the earth. This re-radiated energy will be largely in the form of long-wave radiation.
Respiration – the breakdown of organic compounds which releases energy and produces carbon dioxide and water. See Aerobic respiration and Anaerobic respiration.
Restoration – the reinstatement of organisms or the environment to previous quality.
Reuse – using materials again for the same purpose without reprocessing. See Recycling.
Rock cycle – the concept that rock material is constantly being created and destroyed in a cycle involving weathering, erosion, transportation, deposition, metamorphosis, melting and cooling.
Rotation – the number of years between the planting and felling of a forest. e.g. If a forest had been planted in 1950 and was felled in 1995, the rotation is 45 years.
Runoff – the flow of water across the surface of the earth. See Hydrological cycle.
Rural area – a relatively low-populated area in which a large proportion of jobs exploit the natural resources.

S

Salinisation – a soil-forming process whereby salts which are brought up in groundwater concentrate in topsoil due to the evaporation of water from the surface.
Salinity – the amount of salts dissolved in a body of water.
Saprophyte – an organism (e.g. fungi) which uses a method of nutrition in which it feeds by digesting dead organic matter before absorbing it.
Seasonal temperature variations – differences in temperature caused by the earth's orbit around the sun.
Secondary pollutant – a substance, non-hazardous when released, which becomes hazardous after chemical reactions in the environment (e.g. tropospheric ozone derives from a complex reaction between hydrocarbons and NO_x in the presence of ultra-violet light).
Secondary succession – See Succession.
Sedimentary rock – a rock formed from the sedimentation and cementation of transported rock fragments or by chemical precipitation.
Set-aside – a farming practice, designed to reduce over-production, whereby farmers are paid compensation not to use an area of land.
Sewage – a mix of domestic and industrial waste in liquid form.
Shaft mining – extraction of minerals in numerous horizontal galleries which stem off a vertical shaft.
Shanty town – a settlement built by the owner using cheap and/or discarded materials.
Short-wave radiation – See Electromagnetic spectrum.
Sigmoidal curve (S-curve) – a population growth curve which shows an initial rapid expansion, and slows down as the carrying capacity is approached. See Exponential growth and J-curve.
Site of Special Scientific Interest (SSSI) – an area which is of national importance because of its flora, fauna, geology or physiography.
Smog – the combination of particulate matter (smoke), gaseous pollutants (e.g. sulphur dioxide) and fog. See Photochemical smog.
Soil – a layer of broken-down mineral and organic matter, which acts as a 'link' between the living world and the non-living environment. See Horizons.
Soil structure – the arrangement of particles within a soil.
Soil texture – the relative abundance of clay, silt, sand and stones within a soil.
Solar radiation – incoming radiation emitted from the sun, mainly in the form of short-wave radiation.
Species – a set of organisms with similar genetic characteristics, which can interbreed to produce fertile offspring. See Diversity.

Stomata – pores in plants to allow molecules to enter or leave the plant including carbon dioxide, and oxygen and water, respectively.
Stratosphere – the layer of the atmosphere above the troposphere ranging from approximately 15 to 50 km, in which temperatures increase with altitude and in which there is a high concentration of naturally produced ozone. See Ozone layer.
Stress – strain placed on an organism by emotional, chemical or physical factors.
Strip cropping – a soil conservation technique in which crops are planted alternately with other plants in rows along the contours to reduce erosion by the wind and rain.
Strip mining – see Opencast mining.
Subsidence – the collapse or settling of the earth's surface to a lower level, due to removal of minerals, water or oil.
Subsoil – mineral horizons below the topsoil with a low humus content and poor structure.
Succession – the natural progressive series in which one plant community replaces the previous community, and which ends in the development of a climax community. Primary succession occurs where there is no soil, organic matter or seeds present (e.g. bare rock after an ice age). Secondary succession occurs where the existing community has been disrupted (e.g. after a fire). See Climax community.
Sustainable development – the implementation of appropriate human activities to ensure development for future generations.
Symbiosis – the relationship between two different species where each species gains from the other.
Synergistic effect – where the combined effect of two substances is greater than the sum of the impacts of each substance.

T

Tectonics – movements and deformation within the earth (e.g. folding, faulting or uplifting).
Terracing – the lowering of a gradient into a series of horizontal levels to provide flat land for farming or to reduce soil erosion.
Thermosphere – the outermost layer in the atmosphere, extending to around 1600 km, where temperatures are extremely high.
Thinning – the selective removal of some trees from a crop either to give some yield or to improve the growing conditions of the remaining trees.
Tidal power – energy obtained from the tides; it can be harnessed by having different levels of water on each side of a barrage; water is then driven through turbines.
Topsoil – the first soil horizon in which organic matter (humus) is mixed with mineral matter.
Toxins – substances which react to kill or alter living cells.
Translocation – the movement of any component within the soil profile. It can also refer to the transport of substances within a plant, e.g. sucrose in the phloem. See Leaching, Eluviation and Podsolisation.
Transpiration – the loss of water to the atmosphere from plants, particularly through stomata.
Transpiration stream – the continuous column of water and minerals which travels up through the plant in the xylem.
Trophic level – a level in a food chain which indicates the status at which organisms feed.
Troposphere – the lowest layer of the atmosphere which has a high concentration of water vapour and in which temperature and pressure decrease with altitude.

U

Ultra-violet radiation – the wavelength of radiation between visible light and X-rays. The amount of UV light which reaches the earth's surface is reduced by the ozone layer.
Urban area – a relatively densely populated built-up area in which the majority of jobs do not exploit natural resources. See Urbanisation.
Urban heat island – the higher temperature of urban areas compared to the surrounding rural land.
Urban smog – a mixture of fog and the increased dust from urban areas.
Urbanisation – the migration of people from surrounding rural regions to urban areas, so that an increasing proportion of the population live in urban areas.

V

Volatile organic compounds (VOCs) – organic compounds which readily evaporate.

W

Warm front – See Front.
Waste stream – the continual production of waste material which contains a variety of waste produced from different sources.
Water cycle – See Hydrological cycle.
Water table – the uppermost level of groundwater saturation.
Waterlogging – the permanent saturation of soil, so that no air exists within soil pore spaces, thus preventing soil organisms and plants from respiring aerobically.
Weather – localised, short-term atmospheric events. The study of the weather is called meteorology.
Weathering – the degradation of rock *in situ* by physical, chemical or biological processes.
Wildlife – non-domesticated organisms which live independently from humans.
Wind farms – a collection of wind turbines used to generate electricity for the national grid.

X

Xylem – a plant tissue containing tubes which run through the root, stem and leaf of a plant, carrying water and minerals such as nitrate. Flow is upwards from the roots (across leaves).

Y

Yield – the agricultural production per unit area (e.g. x tonnes of wheat per hectare).

Z

Zero population growth – occurs when the number of births equals the number of deaths within a population.

Appendix B

INVESTIGATIONS

CHAPTER 1

The earth's atmosphere

1. Measure the effect of factors such as aspect, particulates or sulphur dioxide on lichen distribution.
2. Measure the albedo of different surfaces, for example concrete, soils and vegetation.

CHAPTER 2

Climate and weather

1. What are the microclimate conditions produced by the presence of hedgerows?
2. What are the climatic differences between urban and rural areas (for example temperature variation, rainfall and windspeed)?

CHAPTER 3

Dynamic earth

1. Study the rates of erosion in recreation areas related to human activities.
2. What are the effects of acid rain on different rock types?
3. Analyse the different landforms produced by different rock types.

CHAPTER 4

Soils

1. Compare the physical characteristics between soils from two areas, for example deciduous and coniferous woodlands.
2. Examine the occurrence of different species of flora or fauna related to different soil types.
3. What are the effects of factors such as vegetation cover and gradient on rate of soil erosion?

CHAPTER 5

Autotrophic nutrition and productivity

1. How does light intensity affect the rate of photosynthesis?
2. What are the diurnal temperature variations within and outside a greenhouse and how do they relate to productivity?
3. Analyse the limiting factors which influence the rate of photosynthesis.

CHAPTER 6

Heterotrophic nutrition and respiration

1. What are the factors which affect the rate of decomposition, for example humidity and pH, of litter and the nature of the substrate?
2. Investigate how surface area : volume ratios affect the rate of heat loss.

CHAPTER 7

Ecosystems

1. Show how species diversity of vegetation and fauna changes across a grass verge.
2. Examine how species adapt to environmental conditions, for example freshwater streams.
3. What is the local environmental importance of small freshwater ponds?
4. What is the importance of hedgerows to wildlife species diversity?

CHAPTER 8

Nutrient cycles

1. Measure the relative rates of algal growth produced by nitrates and phosphates.
2. Measure the rate of carbon dioxide uptake by different plant species.
3. Assess the level of eutrophication in lakes located in different areas.
4. Investigate the synergistic effect of fertilisers.

CHAPTER 9

Population

1. Estimate the invertebrate species population using the capture-recapture method.
2. Measure the growth of yeast populations within a nutrient solution.
3. What is the impact of town expansion upon a nature reserve?
4. What is the effect of small town expansion on local land use or wildlife?

CHAPTER 10

Pollution

1. What are the effects of organic matter and temperature on oxygen depletion in fresh water?
2. Examine the possible environmental effects of chemicals (for example herbicides, insecticides and pesticides) used in gardening.
3. What are the effects of heavy metals on the growth of different strains of grass?

CHAPTER 11

Water resources

1. Show the quantitative measurements of variations in dissolved oxygen or BOD concentration along a river course.

2. Examine the concentration of different pollutants (for example aluminium and nitrates) along a stream transect.
3. What are the impacts of recreation activities on canal species diversity?
4. Examine the effect of agricultural activity on river or lake water quality.

CHAPTER 12

Energy resources

1. Conduct an environmental impact assessment of the siting of an energy-generating site, for example a gas pipeline or wind farm in a rural area.
2. What should be the economic and environmental considerations of high-energy electricity supply systems?
3. Examine and analyse public understanding of payback periods.
4. Examine and analyse public perception of the environmental and social implications of different energy-generating techniques.

CHAPTER 13

Forestry

1. Compare species diversity in a coppiced and uncoppiced woodland.
2. Examine the influence of deciduous and coniferous woodland on leaf litter species diversity.
3. Conduct a survey of visitor use in woodlands.

CHAPTER 14

Agriculture

1. Compare crop yields between two different agricultural sites.
2. Examine the effect of pH on seed germination.
3. Examine the effect of aeration/irrigation/salinity/humus on seed germination.

CHAPTER 15

Minerals and metals

1. Examine the effect of pollution from chalk quarries on nearby vegetation.
2. Conduct a cost–benefit analysis on a road-building or mining proposal.

CHAPTER 16

Waste

1. What are the environmental implications of incineration or landfill sites on local areas?
2. Analyse public utilisation of recycling and the environmental benefits.
3. Compare the biodegradability of organic matter.

CHAPTER 17

Land resources

1. Examine the species diversity of vegetation on derelict land.
2. Examine the environmental impact of building a town centre by-pass or a motorway extension scheme.
3. Conduct a Leopold Matrix analysis.

CHAPTER 18

Conservation

1. What are the differences in species diversity between mown and unmown grassland?
2. Examine the management techniques in a selected habitat (for example a national park, nature reserve or country park).
3. Examine the importance of gardens to wildlife conservation, in a particular area.
4. Identify the environmental problems associated with tourist pressure in an area of countryside.

Appendix C

ANSWERS TO ANALYSIS QUESTIONS

Please note: OWTTE means 'or words to that effect'.

CHAPTER 1

The case of the missing CO_2

1. The increased burning of fossil fuels, which contain carbon and release CO_2 upon combustion.
2. It is a place where CO_2 is absorbed or stored.
3. A much higher volume of CO_2 is released in the northern hemisphere, but the concentration of CO_2 is the same in the northern hemisphere as it is in the southern hemisphere.
4. Trees absorb CO_2 through photosynthesis and convert the CO_2 into wood. A tree's growth is shown through its annual rings, with the width of ring indicating the amount of growth. Growth may be proportional to the air's CO_2 concentration in any year. We can age a tree by counting its rings, and the width of the rings can help us to estimate the CO_2 concentrations in the past years.
5. Lower concentrations of CO_2, lower temperatures, and fewer nutrients (such as nitrogen and phosphorus) to build tissues.
6. It is ironic because the pollution from fossil fuels (or acid rain) is often blamed for forest decline.
7. The volume of CO_2 produced by respiration, burning or clearing exceeded the volume of CO_2 absorbed in photosynthesis.
8. The clearing of trees allows for more light/water to enter the forest floor, allowing micro-organisms/bacteria/fungi to break down organic matter faster. Trees store carbon because they are long-lived organisms. Removing trees allows the rapid growth of short-lived organisms which die, are decomposed and so release their carbon, in the form of CO_2, into the atmosphere.
9. Increasing temperature increases the rate of decomposition, raises the incidence of forest fires and kills trees, so increasing the rate of supply of organic carbon to decomposers. All these processes mean that more carbon dioxide enters the atmosphere, which enhances the greenhouse effect, thus raising the temperature.

CHAPTER 10

Vehicle exhaust emissions

1. Commuters begin to drive to work.
2. NO is released from vehicles. NO_2 forms partly from the oxidation of NO, which takes some time.
3. Hydrocarbons accelerate the formation of NO_2 from NO. The NO_2 is then dissociated by light into NO and O. The O can then combine with O_2 to form ozone.
4. NO and NO_2 combine with radicals to form PANs. Radicals begin to form in the morning and then react with the high concentrations of NO and NO_2 (which then decrease).
5. PANs and ozone are still relatively high and will decrease over the next few hours.

The effect of organic pollution

1. (a) The organic material entering the river contains large numbers of bacteria.
   (b) The decrease in numbers is due to increased turbidity, reducing the light available for photosynthesis. There is an increase in algal numbers as turbidity decreases, and a higher availability of phosphate and nitrate.
   (c) The initial increase in numbers is because bacteria/sewage fungi represents a huge food source. There is a decrease in numbers as the food source decreases.
2. (a) The concentration increases rapidly as a result of saprophytic/fungal/bacterial breakdown of organic waste/urea/putrification. It declines through the breakdown of nitrates/nitrification, absorption by *Cladophora*, algae, bacteria and protozoa.
   (b) The concentration gradually increases as it is produced by decomposition/nitrification by decomposers of ammonia. It declines as it is taken up by algae.
3. They are tolerant of low levels of oxygen/high BOD. They are therefore abundant in water with high levels of organic pollutants. They are an indicator species when in high numbers.
4. (a) Water temperature may decrease the recovery time because the rate of respiration of, for example, the bacteria will increase as the temperature increases. The bacteria will therefore break down the organic matter faster. On the other hand, many of the bacteria are aerobic, that is, they require oxygen, and the amount of oxygen which the water can hold will decrease with increasing temperature.
   (b) Water turbulence would decrease the recovery time because turbulence will increase the amount of oxygen in the water, speeding up the activity of aerobic decomposers.

Ecological effects on lakes

1. (a) The pH of all four lakes has decreased since 1700 but the decrease has been greater in Lake Dulyn, Round Loch and Loch Enoch than in Burnmoor Tarn.
   (b) There is an increased deposition of acid rain.
2. (a) The industrial revolution released lead into the atmosphere, as did the use of cars in the twentieth century. This was deposited in precipitation.
   (b) The amount of precipitation/the steepness of the slope of the surrounding land.
3. STUDENT ACTIVITY

CHAPTER 13

Boycott mahogany... save the rainforests

1. High temperatures and moisture levels ensure microbial activity is very high and decomposition rapid. Tree roots are shallow but widespread, speeding up the process of absorption of nutrients.
2. Soils are thinner and contain less humus. Deforestation means that they are exposed to what is often very intensive and erosive rainfall.
3. A rising population means that fallow periods have been reduced. As a consequence, the forest does not have time to recover between cultivations.
4. Deforestation follows the roads, which provide a way into otherwise impenetrable forest areas.
5. Select sample plots using random numbers. Identify co-ordinates by mapping and gridding the area. Set up a circular sample plot of radius $10m^2$ with the centre of the plot at the co-ordinate intersection. Count the number of mahogany trees. Repeat to achieve a sample area of 1%. Multiply up.
6. Reduced demand for tropical timber will lead to a decrease in market value. This will reduce the profitability of extraction and reduce the incentive to deforest.
7. Management of the forest so that the ability of the forest to regenerate is not harmed. This will involve a calculation of the maximum sustainable yield – the maximum volume of timber which can be removed annually without reducing the ability of the forest to replace that volume by new growth.

Conflicts in multi-purpose forests

STUDENT ACTIVITY

CHAPTER 14

Agricultural changes

1. **(a)** $(155/662) \times 100 = 23.4\%$
2. **(a)** More land would be available. There would be less competition for water, light and nutrients close to the hedge. There would be fewer pests living in the hedge. It would be easier to use large machinery.
   (b) There would be no wind barriers, leading to more soil erosion by a stronger wind. Stronger winds mean greater crop damage. Loss of predators living in the hedge.
3. There was an increased use of machinery/fertilisers/pesticides/herbicides/irrigation and drainage systems/hybrid varieties.

Soil erosion

1. All of them.
2. Ploughing breaks up the soil, making it susceptible to erosion. Removing a crop leaves the soil exposed to wind and rain erosion. Compaction of soils by machinery reduces pore spaces, reducing infiltration and so increasing erosion with the increased runoff. Use of inorganic fertilisers means that less organic matter is left in the soil. Ploughing up and down slopes rather than across them leads to high-velocity runoff, which is erosive.

CHAPTER 18

Is conservation the same as preservation?

STUDENT ACTIVITY

Declining fish stocks

1. **(b)** The fish-catch data refer to cod and plaice, whereas the pollution data refer to cod and haddock. International data are used for the catch data but pollution data refer only to UK fisheries.
   (c) Pollution data for herring and plaice, and details of the size of the fishing fleets.
   (d) The present population/the birth rate/the growth rate/the lifespan/the age at which the fish are able to breed.

Appendix D

ANSWERS TO IN-TEXT QUESTIONS

Examination questions are included with permission of the appropriate Examining Boards. Answers are the sole responsibility of the author and have not been provided or approved by the Boards.

CHAPTER 1

1.1 18/25 = 72%

1.2 Global warming will result in higher atmospheric and sea temperatures, which increases rates of evaporation from the water. There will therefore be a decrease in sea-level. In addition, a warmer atmosphere is capable of holding more moisture.

1.3 The precise wavelengths which are absorbed. Different greenhouse gases absorb different parts of the electromagnetic spectrum. Short wavelengths have much greater energy and therefore greater heating effect than longer wavelengths.

1.4 **(a)** Decreased consumption of petrol may reduce CO_2 levels, reducing global warming.
(b) Increased incineration will lead to increased CO_2 levels, but decreased landfill dumping will reduce anaerobic production of methane. The best overall effect would be created by a balance between these two.

1.5 An increasing population leads to an increasing energy use, which is often derived from fossil fuels. This is already the largest anthropogenic source. An increased population may also lead to an increase in the area of paddy and an increased production of all types of waste – increased animal wastes and sewage will further increase methane levels. Clearance of, for example, forests for homes or agriculture will cause increased biomass burning, further increasing methane levels.

CHAPTER 2

2.1 **(a)** In the northern hemisphere, areas with a high latitude have a negative heat budget (that is, they re-radiate more heat than they absorb) and areas near the equator, with a low latitude, have a positive heat budget (that is, they receive more radiation than they re-radiate). The point at which the amount of incoming radiation and outgoing re-radiated energy is the same is likely to be at around 35° latitude. The radiation absorbed decreases with latitude. Radiation lost is greater at the poles than at the equator, but the amount lost is greater than both at latitudes of 30°N and 60°N.
(b) The sun is directly over the Tropic of Capricorn, and so 60°S is closer to the sun than 60°N and receives more diverging solar radiation. At 60°N incoming radiation meets the earth's surface at an oblique angle, thus spreading a unit of energy over a wider area. Solar radiation at 60°N has to pass through a thicker atmospheric layer and therefore more radiation is absorbed, reflected and scattered by the atmosphere. The day is longer and therefore 60°S will receive more sunlight.

2.2 1200 metres

2.3 **(a)** **(i)** As air rises it is put under less pressure and therefore expands.
(ii) The parcel of air loses heat as it rises. Cooler air cannot hold as much water vapour as warm air. When the dew point is reached, condensation occurs around hygroscopic particles.
(b) 2 °C
(c) Clear skies: Insolation high in daytime, so temperature high; infrared radiation lost at night so temperature low, therefore large diurnal change in temperature, e.g. in deserts.
Cloudy skies: Clouds reflect insolation in daytime so temperature lower but clouds also reflect infrared radiation at night back to earth, so temperature does not fall so much, therefore smaller diurnal change in temperature, e.g. in rainforests.

2.4 **(a)** The areas are subject to the same amount of insolation, but the continent loses heat more rapidly during the summer. Rainfall in the UK comes from the rising of moist maritime air over mountains and from depressions, whereas the rain in Kamchatka may come from convection during the summer. Kamchatka is subject to a greater temperature variation than the UK because the maritime climate buffers temperature variations. Major difference is the effect of the North Atlantic Drift on the climate of Great Britain, with a cold current off the coast of Kamchatka.
(b) In Kamchatka the rain falls in the summer, with the need to irrigate to control volumes, and to spread out supplies. There are no data on intensity of rainfall, so it is not possible to speculate on possible effects of soil erosion. There are no winter crops as it is too cold and there is little water. Both have the same amount of light intensity. There is a shorter summer growing season in Kamchatka due to temperatures. In the UK the growing season is restricted in winter by low temperatures and in summer by the availability of water. Irrigation may be necessary in some eastern counties.

2.5 **(a)** A small annual temperature range, wet winters, drier summers, but no drought period.
(b) The shallow seas warm the land in winter and cool it in summer, therefore reducing annual range. This is due to the high specific heat capacity of water acting as a thermal buffer.

2.6 **(a)** Reason 1: Shelter creates a miniature greenhouse which increases the temperature and therefore increases the rate of photosynthesis/other synthetic processes. Reason 2: There is reduced wind/water stress. Reason 3: Shelters reduce damage from herbivores and may prevent the entry of insect pests.

(b) By wilting, water loss from plants is reduced: there is a reduced water need; the surface area of leaves is reduced; there is less chance of heat damage.

CHAPTER 3

3.1 The Pacific Ocean is shrinking and the Atlantic Ocean is enlarging.

3.2 Iceland is largely made up of igneous rocks which have formed from volcanic activity on the constructive plate boundary between the North American plate and the Eurasian plate.

3.3 (a) Salts/gypsum/rock salt (halite)/calcium sulphate are all used for food production and chemical compounds such as soap, detergents, antiseptics, dyes.
(b) Evaporation must exceed precipitation.

3.4 Metamorphic rocks are produced through the effect of intense heat and pressure on existing igneous or sedimentary rocks. This metamorphic rock may become exposed at the surface, it may remain within the crust or it may, through further heat, become molten magma.

3.5 Solar energy is largely responsible for weathering, since it drives climates, and the water and nutrient cycles. Geothermal energy – heat produced by radioactive decay – causes formation of magma.

3.6 Flat, low-lying, often fertile land, food supply, easy transport, aesthetically beautiful, mild climate.

3.7 Groynes: trap moving sediment; boulders and wall protect the base of the cliff.
Drain: removes excess water at the base of the cliff and thus reduces lubrication of the slope.
Metal rods: stabilise the cliff.
Vegetation: binds the soil together and reduces erosion by water and wind.

CHAPTER 4

4.1 Soil texture is the relative amount of inorganic components (sand, silt and clay), whereas soil structure describes the way in which these particles aggregate and also involves components such as water, organic material, biota and the chemistry of the clay.

4.2 The relative proportions of particles of different size present determines to a large extent how the particles will join together.

4.3 It may compact the soil, causing a platey and infertile structure to develop.

4.4. Soils develop slowly. Plant establishment is difficult because there are few nutrients available, drainage is extremely rapid, the dunes may be unstable and are often exposed to wind and wave action. Humus accumulation is therefore slow.

4.5 (a) Clay soils have poor drainage and so may become waterlogged, cold and anaerobic. Plant roots need oxygen and a suitable temperature so that they can develop.
(b) They could add humus to increase the nutrient content and to improve the structure and the water-holding capacity.

4.6 (a) At site C: There may be waterlogging in the valley bottom, for example because of flooding or a high water table.
At site A: There may be a higher amount of precipitation on the top of the hill. The relief is gentle and there will therefore be a slow runoff rate. Peat may also have developed at the top because of the vegetation type and a low rate of organic breakdown. The resulting high organic content will hold a lot of water.

(b) Erosion and mass movement will remove material from the steep slopes and accumulate it at the foot of the slope.

CHAPTER 5

5.1. (a) Pork sausages are produced from pigs which have fed on grains, vegetables, etc.
(b) Fish eat microscopic plants (or they eat other fish that gained their energy from eating such plants).

5.2 They retain their needles all year round and can therefore photosynthesise all year round if the conditions are suitable. They are well adapted to wet and windy conditions of, for example, the UK.

5.3. In a thin leaf CO_2 does not have to diffuse large distances. A leaf with a large surface area allows for maximal light absorption. A leaf with many stomata allows for the entry of CO_2. A leaf with stomata on the underside reduces the evaporation of water. A leaf with a good network of veins allows for the rapid removal of sugars and the entry of water and minerals from the roots, leading to a concentration of chlorophyll towards surface of leaf, which receives the most sunlight.

5.4 (a) C_4
(b) By burning fossil fuels, for example oil burners, in the greenhouse.
(c) Both would show an increased rate of photosynthesis, but the C_3 plants would benefit the most.
(d) By comparing the cost of heating with the extra profit from the extra growth of the crop.

CHAPTER 6

6.1. (a) (i) As the yield increases, digestibility decreases.
(ii) It is the optimum time, as yield and digestibility are maximised.
(b) Wind/rainfall/soil quality/temperature/insolation/diseases/pests/fertiliser application.
(c) Large organic molecules are more difficult to digest than small ones because there are more bonds to break.

CHAPTER 7

7.1 (a) (i) The sun.
(ii) Photosynthesis.
(b) Energy is lost through respiration/faeces/excretion. Not all of an organism at the previous trophic level is consumed.
(c) Number and biomass of organisms decreases at each trophic level.

CHAPTER 8

8.1 The sun provides energy for plants to fix carbon dioxide and convert it, for example into carbohydrates.

8.2. (a) There has been an increased burning of vegetation and fossil fuels, an increased cultivation of soils (more carbon-containing compounds are exposed to air, allowing the carbon to be oxidised), and deforestation.
(b) In summer, photosynthesis is greatly increased, therefore atmospheric levels fall. The reverse occurs in winter.

8.3. (a) $15 \times 20 = 300$ tonnes.
(b) Each hectare will have a fixed 180 tonnes of CO_2. Therefore the farmer needs to plant $300/180 = 1.66$ hectares.
(c) No. They will have produced many tonnes of other greenhouse gases, but they would have dramatically reduced their impact!

8.4. Waterlogging is the major cause of anaerobic conditions, as air spaces in the soil become saturated. Factors causing waterlogging are heavy rainfall, impermeable underlying substrate, low-lying areas or clay soils. Techniques to aerate soils include ploughing, mixing in sand to increase soil air spaces, digging drainage channels, and careful irrigation techniques.

8.5 **(a)** At very low pH values, for example 2–3.
(b) 5–8.
(c) At very high or very low pH values. Litter would accumulate.

8.6 **(a)** Vehicles release NO_x in exhaust fumes.
(b) The leaves may be tougher or unpalatable.
(c) There may be fewer predators. There may be greater survival because of higher-quality food supply. Increased fecundity.

CHAPTER 9

9.1 **(a)** When immigration rate plus birth rate are equal to harvesting plus emigration plus predation plus mortality.
(b) A managed population is one where humans control the population size (e.g. in fish farms). Log phase is desirable since this is the fastest population growth rate. Also, environmental factors are not yet limiting the size of the population (such as food availability, habitat availability and build-up of waste material).
(c) May catch small, immature fish, which have not yet had a chance to breed.

9.2 In the case of rabbits and foxes, an increase in the number of rabbits will mean more food for the foxes. The foxes will then breed, but it will be some months before their cubs are born.

9.3 **(a)** 50%
(b) A: Low infant mortality and a very high mortality after the reproductive period.
B: High infant mortality and a very low mortality until after the reproductive period.
(c) A: Food – when the population reaches old age they cannot catch their own food.
B: Territory – competition occurs to get territory, and once it is obtained there is a high survival rate.

9.4 **(a)** **(i)** The graph should show an initial lag period followed by an exponential (log) phase increase.
(ii) Numbers should stabilise around the carrying capacity of the environment, predominantly determined by the availability of food.
(b) **(i)** To restrict population size, as the deer have no natural predators. An uncontrolled population severely depletes the natural environment (vegetation) eventually leading to large-scale starvation.
(ii) An initial population decline. Long term, in theory, the population will stabilise. A predator-prey relationship develops.

9.5 **(a)** The total number of individuals of a particular species in a given area.
(b) An initial rapid increase because there are no limiting factors. The rapid decrease is because of limiting factors.
(c) Food supply, disease, reproductive inhibition.
(d) Predators, density-independent factors such as hurricanes.

9.6 **(a)** Population = (B+I) – (D+E) or B + I = D + E
(b) The number of organisms born per thousand of the existing population.
(c) Degree of health care/social norms/education/war.

9.7 The population has greatly increased due to lack of contraception/education/social forces, etc. Numbers in the 15–64 category have increased but not by numbers born 30 years ago due to not all individuals having survived/immigration/emigration. A greater survival of old people due to improved health care etc.

9.8 Côte d'Ivoire: Very high birth rate/little availability of contraception and/or education. High mortality rate throughout profile, few very old people, little health care or provision for old people.
Denmark: Declining birth rate, reflecting the fact that women are choosing careers first and children later/excellent education/contraception facilities. Large middle-aged population, reflecting the post-war boom. A much lower mortality rate, reflecting good health care.

9.9 **(a)** **(i)** C **(ii)** B **(iii)** A
(b) **(i)** B
(ii) The birth rate has decreased, so 0–5 is narrower than those immediately above. There is a large middle-aged population because of the post-war boom. There is no sudden increase in mortality at the highest ages.

CHAPTER 10

10.1 Bioaccumulation is the build-up of a pollutant in an organism at trophic level one. Biomagnification is the increasing concentration of a chemical up a food chain as each organism eats the preceding one.

10.2 2512/15 to 13685/15 (167 to 912x).

10.3 Any level of smoke increases the potential harmful effect of SO_2, meaning that synergism may occur.

10.4 The respiration rate of aerobic organisms – and oxygen demand – will increase with temperature.

10.5 **(a)** It increased because of the increased use of nitrate fertilisers, which led to increased leaching.
(b) They were measured in winter as there was no agricultural addition then, so the concentration present reflects additions from previous months.
(c) **(i)** Detergents in sewage. It has increased because of an increased use of appliances such as washing machines.
(ii) The waters initially increased in species and diversity because phosphate is a limiting factor. In the conditions algal bloom forms, algae die and are broken down by aerobic bacteria. Decomposition of large amounts of algae produced during an algal bloom causes an increase in aerobic bacteria. These deplete oxygen in the water, leading to anoxic conditions.
(d) Phosphate because it has point sources.
(e) Boating creates swash. Agricultural machinery/drainage/sewage from boats.
(f) Aerobic organisms have died.
(g) Temperature, numbers of animals, plants, turbulence.

CHAPTER 11

11.1 Per capita ownership of water-hungry appliances such as dishwashers is highest in the USA, as is the ownership and use of cars. Australia uses a huge amount of water for irrigation, while this is rare in the maritime climate of the UK. High industrial demand for water reflects well-developed manufacturing sectors, with Germany having much greater efficiency standards than the USA. Relatively low figures in the UK reflect the decline of manufacturing and/or, in the case of India, general underdevelopment.

11.2 **(a)** Information boards listing, for example, bird species present, and picnic tables.
(b) Time zoning: Prevents visitor access during the nesting season; operates a booking procedure for conflicting activities.
Space zoning: Restricts parking spaces away from sensitive areas; uses fences/wardens to deter access to sensitive areas, etc.

(c) They may be dangerous. They may have a very high conservation importance. The threat of pollution may be too great.

11.3 Chalk, Jurassic limestone, Permo-Triassic sandstone, magnesian limestone.

11.4 Bacteria, fungi and protozoa aerobically break down the organic matter into carbon dioxide and water. Anaerobic breakdown may also occur, when methane will be released.

11.5 We can turn the tap off when it is not in use (for example during brushing our teeth). We can place a brick in the toilet cistern to reduce the volume of water used in toilets (30% of domestic water is flushed down the toilet). We can water plants and gardens with used dirty water (for example after washing up). We can have showers and not baths. We can wash the car less frequently.

11.6 (a) Porous, permeable.

(b) Sandstone, limestone.

(c) (i) Confined aquifers are sandwiched between impermeable rock/are not exposed to the surface. Unconfined aquifers are exposed to the surface.

(ii) Perched aquifers.

(d) (i) Artificially refilling the aquifer with water at times of surplus for use at times of drought.

(ii) Advantage: It is relatively cheap/has a low environmental impact.
Disadvantage: A limited number of sites are suitable.

11.7 (b) Rainfall intensity/the duration of precipitation/the type of precipitation (for example snow will have little immediate effect on stream flow).

11.8 (a) The vertical banks will reduce breaching. Minimise the use of topsoil to reduce the chance of erosion and silting, which would increase the chance of flooding. Maintain the relief channel to take surges. In addition, straightening of river channels and building-up of flood banks.

(b) The retention of well-adapted species such as alder. The maintenance of habitats for flowers and aquatic plants. Pond dredging to avoid silting.

CHAPTER 12

12.1 (a) Coal is derived from plants which converted the sun's energy into biomass via photosynthesis.

(b) The water cycle is driven by the sun, but the actual electricity is generated by water falling under the action of gravity.

12.2 Most developed countries have access to relatively cheap concentrated sources of energy such as fossil fuels.

12.3 The burning (oxidation) of sulphur yields sulphur dioxide, a major contributor to the production of acid rain.

12.4 There is less mining of uranium ore and less waste is produced.

12.5 The amount of radioactive material which must be disposed of will also increase. It is the safe disposal of such waste which is causing most technical difficulties.

12.6 More energy is released for fewer radioactive waste products.

12.7 (a) (i) Uranium/uranium dioxide.

(ii) It slows down and absorbs neutrons so enabling the nuclear reaction to be controlled.

(b) The recovery of unused radioactive material/uranium/plutonium from spent fuel elements.

12.8 (a) The increasing population (in the developing world) and the increasing global standard of living (OWTTE) has led to an increased demand for coal as fuel and other named by-products.

(b) Destructive distillation. Coal is heated in the absence of air and produces coke/coal gas or coal gasification. Produces hydrogen/methane or coal liquification. Produces coal oil.

(c) Long periods of coalification, high temperature and pressure results in a greater loss of volatile materials (for example methane). The carbon percentage content increases therefore increasing the calorific value/energy value, passing from lignite, via bituminous, to anthracite.

(d) Carbon compounds are used as reducing agents. It generates necessary high temperatures.

12.9 (a) The south-facing windows maximise the solar radiation which enters the building, providing natural heat and light, whereas the north-facing windows are small to minimise energy loss.

(b) Water for washing hands etc. only needs to be lukewarm but is often too hot, therefore wasting energy generated from fossil fuels.

12.10 (a) (i) Pumping improves the efficiency by increasing the difference in tide levels on either side of the barrage.

(ii) Predictable generation sometimes occurs 'out-of-sync' with peak demand and therefore it is possible to use a second impoundment which stores water to be used as and when necessary.

(b) The size of the floating machinery causes high start-up costs and it is inefficient. It is therefore likely that, at the present time, the cost per unit of energy will be high.

12.11 More roads may lead to more traffic, hence the greater use of oil, petrol, diesel, etc.

12.12 (a) The payback time would be reduced (that is, double-glazing will pay for itself quicker than was anticipated).

(b) Heating costs vary with fuel prices. Efficiency of double-glazing depends to some extent on other factors, such as the use of curtains, the aspect of the windows, etc.

(c) Students should provide any appropriate conservation measures, with a justified answer referring to payback times and efficiency of measure.

12.13 Some effects will improve and others will degrade water quality. Some habitats for example will be created and others will be destroyed. Greenhouse gas emissions will be reduced so the global environment may benefit but local ecology may suffer.

12.14 Students will all have different answers to this question.

CHAPTER 13

13.1 For: It reduces raindrop impact velocity, thereby decreasing the erosivity of rainfall, reducing erosion and silting. Roots bind soil, further reducing erosion. Screen reservoirs deter recreational use, which may pollute the supply. By intercepting rainfall and reducing runoff, flooding is decreased.
Against: The interception and subsequent evaporation from trees reduces the volume of water reaching reservoirs and rivers. Trees filter pollution and, through stemflow etc., rainfall may be polluted.

13.2 (a) Between 1940 and 1960 the number of workers increased as the plantation area increased. Since 1960 the plantation area has increased but employment has declined.

(b) Before 1960, most operations were manual, and since then forestry operations have become increasingly mechanised.

(c) It shows that employment is unstable. Employment opportunities fluctuate markedly over the life of a forest and are greatest at the end, for harvesting.

13.3 (a) Deciduous/native/less productive species may be used. Open areas would be left. Dead or diseased trees may be left as habitats but may spread disease within the forest.

(b) As above, and facilities for visitors would decrease the area available for planting. Tree spacing would be greater to allow access, therefore there would be fewer trees per hectare.

13.4 (a) Cattle dung would have provided organic matter/humus, which would have helped to bind soil together. A lack of nutrients reduces plant growth and may decrease soil cover or may lead to the cultivation of other land.

(b) Increasing demand in one factor such as agricultural land accelerates other processes (deforestation/shortening of fallow period/fuelwood substitution) which lead to declining productivity. This leads to positive feedback, that is, further demand for agricultural land.

CHAPTER 14

14.1 If humans were to become contaminated with *Salmonella*, the antibiotics which would normally be given to treat the infection may be ineffective.

14.2 (a) Bacteria within nodules in the roots of leguminous plants fix nitrogen from the atmosphere. Fewer legumes therefore means less nitrogen fixation and less chance of nitrogen-rich crop residues being left in the soil.

(b) Cultivation may lead to increased leaching of nitrates or encourage the oxidation of nitrogen compounds, releasing nitrogen oxides.

(c) Nitrates are highly soluble – winter application is unwise as the rainfall during this period tends to be high and plants have only poorly developed root structures.

CHAPTER 15

15.1 (a) (i) The cost of pumping out water is greater than the value of the extraction. The greater depth requires more safety features, and these have a prohibitive cost (OWTTE).

(ii) The extent of the quarries has been limited by the position of the roads and railways.

(b) (i) quarry A = $1.25 \text{ km} \times 1.25 \text{ km} = 1.6 \text{ km}^2$
volume A = $1.6 \text{ km}^2 \times 0.02 \text{ km} = 0.03 \text{ km}^3$

(ii) Chalk is porous and there is therefore the problem of leachates from the infill.

(c) (i) This is where the railway line route has been excavated, to reduce the noise/visual/dust pollution impact.

(ii) It does not allow for access to the quarries.

(d) (i) Through carbonation, chemical weathering and acidified precipitation.

(ii) For safety reasons.

(e) It might have encouraged redevelopment, increased the value of land, and offered potential for recreation, storage and industry.

(f) Problems: They are impermeable and prone to flooding and waterlogging.
Methods: They are suitable for infill, as they are not prone to loss of leachates from the infill.

15.2 (a) See Table 15.4.

(b) It would reduce the demand for roads and reduce the demand for road-building minerals, therefore there would be less need for superquarries.

(c) There is abundant resource availability and employment potential.

15.3 (a) Resource: This is the total global volume of a mineral economically, now or in the future.
Reserve: This is part of the resource economically exploitable at present.

(b) The amount economically exploitable has declined because the value of coal has decreased due to competition from uranium as an energy store.

15.4 (a) Removal of winding gear; levelling of slopes.

(b) Acidic; may contain heavy metals in toxic concentrations; impermeable; no humus or other organic matter; may be subject to temperature extremes; low in N and P; unstable; poor texture and structure.

(c) Advantage: cheap.
Disadvantage: complete destruction of habitat; visually unattractive; dust; noise problems.

(d) (i) Legumes contain nitrogen-fixing bacteria in nodules on their roots. When the roots die, nitrogen is added to the soil.

(ii) More organic matter, more nitrogen and phosphorus. Manure is toxic if applied directly to plants.

(e) (i) Depends on composition of coal, which will vary.

(ii) Nitrogen oxidised to NO/NO_2 which will have been given off.

(f) An area near a town designated for informal recreation.

(g) Wildlife boards; visitor centre.

(h) May increase traffic/congestion/noise.

15.5 (a) Sedimentary/calcareous.

(b) 48–52%

(c) Advantage: It would provide more jobs and more money.
Disadvantage: It would increase noise levels, traffic and damage to the landscape.

CHAPTER 16

16.1 (a) In urban centres (for example the South East and the West Midlands) there will be predominantly municipal, industrial and construction waste. In East Anglia, the South West, North Wales and Yorkshire there will be agricultural waste. In the East Midlands, South Wales and South Yorkshire there will be large amounts of mining waste.

(b) More developed countries will have more plastic, paper and industrial waste. Less developed countries will have more wood and coal ash, and possibly agricultural and mining waste.

16.2 The increased use of fossil fuels leads to an enhanced greenhouse effect, acid deposition, low-level ozone, the production of derelict land, deteriorating visual appearance and dust.

16.3 Nitrates are very mobile, and so they are readily lost from soil through leaching and surface runoff respectively. They accumulate in the groundwater and in streams, where the excess fertilisers cause the growth of algal blooms. Waste should be applied at times of high demand, during the growing season, at levels appropriate to the uptake.

16.4 In 1950 the most common type of waste was loose waste (around 70%) but this has fallen dramatically to less than 5% in 1990. The second most common waste in 1950 was paper, the importance of which has risen dramatically. Similarly, plastic has risen from 5% to 15%, and glass has increased significantly.

16.5 There is an increased demand for newspapers/paper, an increased use of packaging (in transport and in the merchandising of goods), and an increase in disposable goods (for example drinks containers, pens and razors). There is a decrease in coal- and wood-burning fires in the home due to an increase in central heating systems.

16.6 (a) Approval by Local Authority, safe for lifetimes (for example clay, plastic and sand protection), monitoring.

(b) Danger of transporting the waste. It is very expensive. The waste is highly mobile and thus there is the danger that it will leak into the ground-water. It is difficult to contain.

(c) There is an increased demand for land, which reduces the land available for landfill. There are increasing volumes of waste, which increases the pressure on landfill sites.

16.7 (a) Noise pollution through the incinerator operation; the visual pollution of the site; the disturbance caused by lorries bringing in the waste; a reduction in the value of local housing/land.

(b) Dioxins: These are carcinogenic (in very low doses). Heavy metals: These are toxic, persistent chemicals (which bioaccumulate and biomagnify), with lethal and sublethal effects.
Nitrous oxides: These are involved in the production of acid rain and low-level ozone.

(c) The high population density results in a high demand for land, and therefore land is not available for landfill sites, so incineration is a more practical option.

16.8 (a) The collection and reprocessing of waste material into a useful product.

(b) The financial cost of recycling; market demand of product; availability of virgin raw material; availability of waste; convenience of waste disposal.

(c) (i) It increased (significantly – 5,500 fold). The rate of increase is accelerating.
(ii) Government initiatives; public awareness of recycling; cost efficiency for glass manufacture (OWTTE).
(iii) Their awareness/education in environmental issues. The distance to travel to the bottle bank. The ease of using the disposal facility.

(d) (i) The data do not represent change over time but comparisons between countries.
(ii) Generally, the most industrialised/ technologically advanced countries (for example Belgium) have relatively high levels of recycling, whereas less industrialised/technologically advanced countries have low levels of recycling (for example Ireland)(OWTTE).

(e) The data show two distinct groups, with countries with large populations producing large volumes of glass waste, compared with countries with small populations having small volumes of glass waste.

(f) $1.5 \times 106 = 83\%$ of total glass production
100% of total glass production = 100/83
1.8×106 tonnes per year

(g) (i) The energy saved in raw material manufacture.
(ii) 4060 + 215 + 1650 + 87 = 6012 gross energy saved
313 + 97 + 49 = 459 additional energy needed
5553 mJ tonne^{-1}

(h) Through education of environmental benefits, economic incentives/offering a subsidised price on recycled glass, resource taxation on virgin raw materials.

CHAPTER 17

17.1 Agricultural operations such as nitrate/pesticide spraying, recreational activities, industrial activities involving harmful chemicals.

17.2 (a) It would be visually obtrusive, as the species and planting pattern are not natural. Access would be impossible if the trees were planted close together.

(b) Quarrying creates a huge visual scar, blasting can be noisy, and transport of the limestone using huge lorries may be noisy, dusty and dangerous to visitors.

(c) Water-skiing may disrupt nesting or resting birds which visitors have come to see. Besides being noisy, it may also be incompatible with gentler water pursuits such as canoeing.

(d) Mountain-biking may accelerate footpath erosion and be dangerous to hikers / horse riders.

17.3 (a) & (b) Increased field size: increases the area of land under production and allows for a more efficient use of machinery. Presence of coniferous woodlands: highly productive forestry plantations (see Chapter 13). Decrease in hedgerows: allows increased field size. Decrease in deciduous woodlands: removed to increase field size and plant more productive coniferous woodland. Bare fields: large-scale monocultures of crops that are all cleared at the same time.

(c) (i) Habitat destruction: deforestation of the deciduous woodland, removal of hedgerows and increase in field size.
(ii) Advances in technology: increase in field size.
(iii) A potential increase in soil erosion: monocultures producing exposed fields, greater areas of land under agriculture, increased mechanisation, removal of hedgerows.

17.4 It reduces pressure on the greenbelt and provides natural habitats for development. It provides valuable land in the urban areas. It provides safety. It provides potential for employment/recreation/economic growth.

17.5 (a) Extent/depth of working/route of access roads/ consideration of trees etc./slope angles/location of dumped materials/replacement of topsoil.

(b) Walking/boating/fishing/wildlife conservation/bird watching/picnicking/education.

(c) Restrict access/maintain habitat variety/manage vegetation for successional stages/plant native species/screen banks.

(d) (i) Swimming/power boating, etc.
(ii) Fishing/wildfowl attraction, etc.

(e) Advantage: Dry access by foot to the park from the built-up area.
Disadvantage: The loss of wetland habitat.

(f) (i) The urban area and the motorway are adjacent sources of noise – earth mounds will reduce noise disturbance.
(ii) Boating swash causes bank erosion, which can be reduced by netting boulders/wooden groynes/vegetation.
(iii) Greater public enjoyment is achieved through knowledge/increase in accessibility/increase in appreciation of the area.

17.6 (a) Urban development often leapfrogs the greenbelt. Legislation is not tough enough, as some harmful activities (for example road building) are permitted and then lead to increased pressure for amenities.

(b) Retired people move to the countryside for a change of lifestyle. The 30–60 age group move to commuter towns, and live in the country and work in the city. Commuters must, by definition, have the income to be able to afford to travel daily.

CHAPTER 18

18.1 There is no correct order for the definitions. Which is the most important depends upon the specific problems found in an area.

18.2 (a) Organisations, such as Friends of the Earth, which have no political affiliation.

(b) Where conservation of a resource is justified because of the enjoyment or pleasure which that resource gives us.

18.3 Lack of original habitat, resistance of local people, the inability of animals to adjust to a wild environment.

Appendix E

ANSWERS TO SUMMARY ASSIGNMENTS

Examination questions are included with permission of the appropriate Examining Boards. Answers are the sole responsibility of the author and have not been provided or approved by the Boards.

CHAPTER 1

1. The stratosphere contains a natural ozone layer which absorbs much UV radiation, and this would harm life in the troposphere. The fact that the stratosphere is warmer than the troposphere helps to maintain the atmospheric stability of the troposphere, reducing the incidence of violent atmospheric activity which could threaten life.
2. Particulates may cause reflection and scattering of both incoming and outgoing radiation. The greenhouse gases absorb, reflect and scatter outgoing long-wave radiation, effectively delaying its release and so raising temperatures. CFCs and other chemicals which destroy the stratospheric ozone layer allow more UV radiation to penetrate to ground level.
3. **(a) (i)** Air temperature decreases with altitude to 10 km, then increases to an altitude of 45 km. Above 45 km air temperature begins to fall again.
   (ii) –19 to 20°C, 35 km altitude.
   (b) Oxygen is split/breaks up/dissociates/comes apart by UV radiation. Single atoms/free radicals of O combine with O_2.
   (c) The ozone layer protects life on earth from the harmful effects of UV (OWTTE). It also stabilises the climate.
4. STUDENT ACTIVITY

CHAPTER 2

1. **(a)** $(130/240) \times 100 = 54\%$
   (b) There is more snow and bare rock and less vegetation.
   (c) The atmosphere, which absorbs, reflects and scatters incoming radiation, is approximately the same depth all around the earth.
2. Incoming solar radiation is spread out over a larger area at the poles than at the equator. Incoming radiation has to penetrate through a greater depth of atmosphere. There is therefore more reflection, absorption and scattering.
3. Atmospheric gases are transparent to much incoming short-wave radiation but the gases trap re-radiated or long-wave radiation.
4. The sea has a higher specific heat capacity. During the day the land retains more heat than the sea and so there is a lower pressure over the land (where the air is rising). Air therefore rushes from the sea onto the land as an onshore breeze.
5. Weather describes short-term atmospheric conditions occurring over a small area (for example the events last over a weekend in a town). Climate describes long-term events over a larger area (for example ice ages).
6. **(a)** Air is rising, strong winds, high rainfall.
   (b) Calm, no clouds, no precipitation.

CHAPTER 3

1. **(a)** Plant roots will force joints apart, allowing water to enter into the rock. The scouring action of the water will erode the rock and/or the water may then be used in freeze–thaw. It should also be noted that the water will induce chemical weathering.
   (b) Steepness of slope, vegetation cover, size of debris, soil moisture, rainfall – water acts as a lubricant.
2.

Type of plate boundary	Movement of plates	Named example
convergent	together	Himalayas
divergent	apart	Mid-Atlantic ridge
conservative	past one another	San Andreas fault

CHAPTER 4

1. Air, water, organic material, minerals, biota.
2.

Characteristic	Effect of ploughing in the remains of a harvested crop
soil mineral content	binds particles together and produces air spaces, increasing water retention capacity
soil organic matter	provides nutrients
soil organisms	provide mixing of nutrients and aeration
soil water-holding capacity	increased clay humus content, increasing water retention

3. Advantages: Allows a global classification based upon climatic characteristics. This allows us to ignore highly variable but localised factors such as changes in parent material or biota.
   Disadvantages: Provides only a generalised classification. Does not account for localised conditions, such as the specific parent materials, biota, relief and human activity. Therefore cannot be used to predict soil type on a small scale.

CHAPTER 5

1. **(a)** Latitude has a great effect on insolation/temperatures/rainfall, etc., which affects GPP and hence NPP.
   (b) (i) They are high because the temperatures and the moisture availability are high. Much of the biomass is photosynthetic. Recycling is very rapid.
   (ii) They are much lower because of long periods of low temperature and low light intensities, which restrict photosynthesis. Water availability may also be a common limiting factor, as the ground is frozen for long periods of time.
   (iii) They are low because of lack of available water, low temperatures, harsh growing conditions and low light intensity in winter.
   (iv) They are high because of the mixing/upwelling which brings nutrients to phytoplankton in surface waters.
   (v) They are high because of continual input of nutrients/sediment from both the sea and rivers. The latter may be high in nitrates/phosphates if flowing through agricultural areas.
2. **(a)** Light provides energy for the production of ATP and the splitting of water, which yields hydrogen ions.
   (b) All enzymic reactions must occur in solution, but water also provides a source of hydrogen ions and gives leaf

cells turgidity, increasing their surface area and allowing for greater absorption of light.
   (c) Chlorophyll absorbs light of various wavelengths, which is transformed into chemical energy.
3. It would enable our crop plants to use water more effectively and to take advantage of the low levels of CO_2 which presently restrict the rate of photosynthesis in arid areas.

CHAPTER 6

1. (a) Rabbits eat soft faecal pellets in a second attempt to digest cellulose.
   (b) Cows regurgitate food from the rumen back into the mouth (this is called chewing the cud) in a second attempt to digest cellulose.
2. Respiration: the release of energy from food.
   Homeostasis: the maintenance of a constant internal environment.
3. (a) They use energy from the sun (photoautotrophs) or from inorganic chemicals (chemoautotrophs) to fix carbon, that is, to convert carbon dioxide into organic compounds such as carbohydrates.
   (b) (i) The total amount of energy fixed by a green plant minus that lost in respiration.
   (ii) A: Faeces; B: Respiration.
   (iii) To release energy from food.
   (iv) Food digested: 1100 kJ; Growth: 80 kJ; Process B: 1020 kJ.
   (v) 1100/3000 = 36.6 %
   (c) (i) The food quantity and quality is controlled so that all of the essential nutrients are supplied in the correct quantities and proportions.
   (ii) By restricting animals' movement, respiration losses are again reduced.
4. (a) Food supply, temperature, movement of the animal.
   (b) This is so that the birds do not use energy trying to keep warm. Like mammals, birds are homeotherms, that is, they try to maintain a constant body temperature. This means that the birds can use more of their food energy to build tissue (meat); by restricting movement, respiration losses are again reduced.
   (c) Calves; pigs.
5. (a) K is the most efficient, M is the least efficient.
   (b) Cabbages; potatoes.
   (c) Pig.

CHAPTER 7

1. (a) A community in which energy and matter are transferred, in interactions between the biotic and abiotic environment.
   (b) An area in which a species/community lives.
   (c) The interaction between populations of different species.
   (d) A number of organisms of the same species within a given area.
   (e) A mutually beneficial relationship between two species.
2. (a) (i) Salt spray, shifting sand, little organic matter (humus), and sand is very well drained so there is little fresh water.
   (ii) Good soil formation, including good organic content, a neutral pH, and low salinity due to leaching of salt.
   (b) Some plant species (for example oak) begin to dominate and out-compete other species (for example by blocking out sunlight), dry soil, remove nutrients.
   (c) Trampling destroys vegetation, decreasing diversity, reducing organic matter, lowering soil fertility and increasing soil erosion. Soil erosion may also increase due to the increased disturbance and compaction of the soil which decreases the pore spaces, so decreasing infiltration, increasing runoff and hence increasing erosion. This decrease in the quality of soil will prevent succession from occurring and trample-resistant species may dominate.
3. (a) (i) Tadpoles/insect larvae/water fleas.
   (ii) Pike/perch.
   (b) The number of water fleas, insect larvae and tadpoles would increase. The number of frogs and roach would increase as their food supply increased (that is, insect larvae and water fleas). There may be fewer perch due to a lack of water beetles, although perch also eat frogs and roach of which there will be more.
   (c) Respiration and egestion at each trophic level. Not all primary productivity will be consumed.

CHAPTER 8

1. The rabbit dies and is broken down by decomposers such as bacteria. During this process, the carbon in the eye of the rabbit is oxidised to CO_2 and released into the air. The CO_2 molecule is then fixed and assimilated into a protein molecule in a bean plant. You consume the bean and digest it, which releases amino acids which will then be assimilated into the tissues of your eye. The carbon atom in the amino acid molecule is the very same one that was in the eye of the rabbit.
2. (a) It increases CO_2 in the atmosphere, reducing the long-term terrestrial storage. Higher CO_2 concentrations increase the rate of photosynthesis which, in turn, work to reduce atmospheric concentrations (negative feedback).
   (b) Higher tropospheric temperatures increase the rate of photosynthesis, which lowers atmospheric CO_2 concentrations and increases the amount of carbon locked in plant biomass. However, higher temperatures also increase the rate of respiration, which releases CO_2. Higher temperatures and decomposition also lead to a warming of oceans, which decreases CO_2 solubility, lowering the amount of carbon which can be locked up in this 'sink'.
3. (a) (i) Biomass is the greatest in tropical rainforest, lowest in coniferous forest, and intermediate in deciduous forest.
   (ii) Tropical rainforests have greatest primary productivity per biomass, followed by coniferous forests, followed by deciduous forests. This is because tropical rainforests have the greatest photosynthetic biomass and the highest annual insolation, temperature and water availability. Coniferous forests have greater primary productivity than deciduous forests because conifers retain their leaves throughout the year, and so have greater potential for photosynthesis.
   (b) Animals will consume some of the new plant material so that it does not become plant litter. Substances such as pollen may blow away. Plants are growing, so there must be a net increase in material per year.
   (c) Every year there is significant production of new herbivore biomass. New plant material per year is high, 3.3 kg m^{-2}, and only 2.5 kg m^{-2} becomes plant litter. This illustrates that significant consumption of the new plant material occurs.
   (d) Approximate ratios, litter : humus content of soil – coniferous 9:1; deciduous 2:1; tropical rainforest 1:12.5. Therefore, tropical rainforests rapidly remove nutrients from the leaf litter, so the amount of humus is low. Conversely, the recycling of nutrients is slow in coniferous forests, with large amounts of humus accumulating. Deciduous woodland is intermediate.
   (e) Tropical rainforest has high levels of growth for replacement, whereas conifers and deciduous forests have a low replacement rate. However, tropical rainforest is very reliant on recycling of nutrients, so the removal of biomass will disrupt this ecosystem more than it will a coniferous or deciduous ecosystem.

CHAPTER 9

1. (a) An appropriate time-scale for the population to 'mingle'. Marking does not affect the species' chance of survival. Each individual in the population has an equal probability of being caught.
   (b) 94
   (c) (i) 10
   (ii) The figures given below 10 are highly variable and are very different from those above 10 marked, which also give a narrow ranging population size.
2. (a) (i) 2.1 years
   (ii) 2.3 years
   (b) Predator-prey relationship. The number of each species is significantly determined by the number of the other species. As the lemming population rises, more food is made available for the owls and more owls can be supported by the environment.An increasing number of owls increases the predation pressure on lemmings, reducing their numbers. Less food is available for the increased number of owls, so their number also declines (OWTTE).
   (c) (i) Density-dependent factors are those which increasingly limit population growth as the population in an area gets higher, whereas density-independent factors limit population growth irrespective of population density.
   (ii) Density-dependent.
   (d) Availability of suitable habitat (competition for space). Competition for food and territory.
   (e) The graph should show significant fluctuations, with no clear carrying capacity. There are wide fluctuations, as the population numbers are significantly affected by density-independent factors, which are random, so there is little stabilising density-dependent influence.
3. Individuals aim to meet their own needs. In low-technology, labour-intensive societies children provide a potential source of income and security in old age. Many countries also have religious limitations on birth control.
4. (a) (i) 6.4% + 5.4% = 11.8%
   (ii) 3.6% + 3.8% = 7.4%
   (b) 50 000 000 × 7.4/100 = 3 700 000
   (c) (i) In 1871 there is a steep continuous decline as age increases in this range, whereas in 1971 there is actually a small increase at first, followed by a much less significant decline. A much smaller percentage of the population are in this age range in 1971 than in 1871.
   (ii) In 1871 the numbers show a constant decline increasing through this age range, whereas in 1971, although the numbers decline from age 20 to age 34, they level off and indeed slightly increase thereafter.
   (iii) There are much smaller numbers in the top bands of this age range (60+) in 1871 than in 1971, and smaller numbers in the range 45–49. In 1871 we see the same steady decline throughout this period, whereas in 1971 numbers stay approximately constant from 40–59, only going into decline at 60+. The pyramid is more symmetrical for 1871, whereas in 1971 the significantly larger proportion of females surviving to ages above 70 is apparent.
   (d) (i) The dependency rate tells us how many non-working people (i.e. children and the retired) are, on average, dependent on each member of the population who is of working age (this would be higher if we considered only those members of that age group who were actually working!).
   (ii) A: (10.7 + 8.9 + 15.6 + 14.2) / (26.6 + 24) = 0.976
   B: (4.9 + 4.5 + 26.2 +23.5) / (20.1 + 20.8) = 1.445
   (iii) A: Not all individuals between 20 and 59 work – some are disabled, some unemployed. This would tend to increase the rate. However, some individuals deemed dependent – particularly 16–19 year-olds and 61–65 year-olds – do actually work. This would tend to decrease the rate. In the UK, the former will outweigh the latter – there is an increasing tendency for people to start work late and retire early – and the true dependency rate will be higher.
   B: In a low-technology, labour-intensive society such as India, individuals tend to start work much younger – often as children – and not stop until they are physically incapable of working. The effect of this is to make the true dependency ratio lower than that calculated.
   (iv) In a society with welfare provision in place, an almost inevitable high tax burden on the working population to finance health care for the young and old as well as costs of pensions, child benefit, education, grants and unemployment benefits. In a society with no such provision, the burden will fall upon members of the extended family, with financial pressure to support not only children but aged parents and grandparents and incapacitated relatives.
   (e) India's population pyramid shows a steady decline in numbers as age increases, with the decline accelerating towards age 30. This shape suggests a high birth rate, and the decline could be accounted for either by lower birth rates in the past or – as is the case for India – a high death rate. The labour-intensive economy with little welfare provision is an inducement to produce many children – they both supplement the workforce and act as pension provision for their parents' old age. The death rate is high due to health care being relatively primitive in many village. In the UK, economic conditions offer no advantage to those producing many children – indeed, financially they are disadvantaged since they are obliged to support them and can anticipate no financial return from them. The death rate is low – access to medical care is good. Accordingly, a much smaller proportion of the population are young, and a much higher proportion survive to old age.

CHAPTER 10

1. (a) Exhaust emissions lead to acid rain which is deposited over a Swedish lake. Acidity has a directly harmful effect on fish but may also lead to release of aluminium from the catchment which damages fish metabolism.
   (b) Acid rain has a wide range of harmful effects on the foliage and the root activity of trees. Weakened trees may succumb to disease which begins to kill the tree and foliage dies.
2. Urban pollutants are less likely to be dispersed in anticyclonic conditions – hot, sunny, wind-free days.
3. (a) (i) 40–60
   (ii) The cost carries on increasing for decreasing and finally no benefit. The cost may exceed the value of benefits.
   (b) Costs: Increased CO_2 leads to an increase in the enhanced greenhouse effect.
   Benefits: There are fewer health problems and reduced acid rain.

CHAPTER 11

1. (a) Westerly winds have crossed the Atlantic and have therefore picked up much moisture/orographic rainfall because of the upland nature of, for example Wales or the Lake District, and much moisture has been lost before the air mass reaches, for example, East Anglia.
   (b) Storage in reservoirs and then transfer using rivers/aquifers from higher rainfall areas. The use of groundwater, for example sandstone. Water treatment and sewage treatment works along the system.
2. (a) It is heated to make steam to drive the turbines. It is used as a coolant and is essential in the cooling process.

(b) Advantage: They reduce the demand on water supplies (OWTTE) and offer the potential to lower water bills. Disadvantage: They are expensive to install, are impractical with shared pipes/tower blocks/flats, and are unfair on large families.

(c) (i) $(83.3 \times 10^6)\ (57 \times 10^6) = 1.46\ m^3$/day/person.

(ii) There is an increased ownership of high water demand appliances/dishwashers/washing machines, and there is improved personal hygiene with the increased use of showers and baths.

(d) (i) The population of the south-east of England is rising, therefore there is increased demand. Aquifers/groundwater supplies are being depleted. There is an increasing standard of living, with increasing water demand (showers, dishwashers, washing machines, etc.).

(ii) Inter-basin/inter-region transfer schemes, desalination, aquifer recharge, metering to reduce demand, reduce leakage in transport, effluent reuse.

CHAPTER 12

2. Fit draught-proofing, insulate the loft, install cavity-wall insulation, fit a condensing boiler, fit double-glazing, fit energy-saving light bulbs.

CHAPTER 13

1. 1919: Forests were a strategic reserve, to supply construction timber and for pit props to allow for the mining of fossil fuels which could then be used to generate electricity for manufacturing of, for example, weapons. 2000: Forests will provide habitat, be a recreational resource, provide catchment protection, reclamation, play an important role in paper, board and timber production, etc.

2. Forestry provides the basis for hundreds of other industries which can range from small-scale, local activities such as fence-making, to major export industries such as plantation-grown hardwoods. See Table 13.1.

3. (a) (i) April/May 1966

(ii) There is an increased loss of Ca^{2+}, K^+ and NO_3^- because of increased leaching. Removal of the trees results in reduced interception, and a greater volume of rainfall enters the soil. There is a reduced loss of SO_4^{2-} because of the lack of foliage to filter out wind-blown sulphate particles.

(iii) It would have much less effect since the ground would have been frozen for long periods of time, which effectively prevents leaching.

(b) Nitrogen distribution reflects the speed of recycling. In the taiga, low temperatures and acid soils slow down decomposition and the root absorption of ions, leaving a lot of nitrogen in the soil. At the other extreme, high temperatures and moisture levels mean that microbial activity is very rapid, and fallen leaf litter, twigs and fruits, etc. are quickly broken down and reabsorbed. Hence a large proportion of the nitrogen is in the above-ground biomass.

4. (a) This restricts branching, which weakens future timber, preventing its use in construction. Canopy closure is achieved quickly, blocking out light to the forest floor, inhibiting weed growth which would otherwise compete for nutrients.

(b) This gives the strongest trees more light and reduces competition for water and nutrients.

(c) This reduces the accumulation of acidic leaf litter near streams, reducing acid pollution.

(d) Different species grow at different rates, so a varied canopy height will result. This increases the visual attractiveness. Different species attract different animal species (for example insects and birds). Also, monocultures are more prone to pest outbreaks.

CHAPTER 14

1. (b) The total increased from 631 million tonnes in 1950 to 1780 million tonnes in 1990, representing a 60% increase. After 1990 there has been a slight decrease to 1745 million tonnes. Per capita has increased overall, although there has been a decrease since 1985 from 343 kg to 318 kg.

(c) $\frac{318 - 247}{247} \times 100 = 28.7\%$

(d) Malnutrition is caused by a lack of a specific dietary component and may occur if a very narrow range of foodstuffs is available, as is often the case in developing countries. Agriculture is vulnerable (for example due to climatic variations and pests). There is overpopulation.

2. (a) Growth hormones, improvements in veterinary science, the development of controlled farming techniques, improvements in foodstock, genetic improvement.

(b) (i) Controlled pig and poultry production, e.g. battery farming is easier and cheaper than the controlled production of sheep and cows.

(ii) Fear of the effects of hormone injection in beef and there is less fat in chicken and pork.

3. (a) Increasing meat consumption stimulates greater production, which leads to increased emissions of methane and, as a result of conversion of forest to pasture, less absorption of carbon dioxide.

(b) Overgrazing leads to the destruction of soil vegetation, which in turn reduces humus input, hence weakening the soil structure and increasing erodibility.

4. Initiatives such as ESAs and the Countryside Stewardship Scheme reduce production by encouraging traditional rather than the most intensive farming techniques. Set-aside reduces the area of land which is used to grow crops. Some critics have argued that farmers have tended to put their least productive land under Set-aside, so that the actual reduction in production has not been as great as was first hoped. Support for organic agriculture will reduce production, since artificial fertilisers and pesticides are not used.

5. To meet the nutrient needs of plants through the utilisation of natural mineral recycling processes based on the decomposition of organic materials which also adds humus to the soil, thereby maintaining soil structure.

6. (a) B will be most fertile. The clay content is not too high and the soil is therefore not waterlogged but has some water retention. There is some humus present to form the clay-humus complex, but the soil is not too well drained, as may be the case with A.

(b) Organic fertilisers – add mulches, which release nutrients slowly and form humus. Bury crop residues – instead of removing nutrients by harvesting the whole plant, it is possible to burn the crop to release the nutrients back into the soil. Leave the land to fallow – allow natural vegetation to develop, to improve the nutrient status over time.

(c) (i) Irrigation which may lower the water table, removal of natural wind breaks, removal of vegetation cover, compaction by heavy machinery, ploughing which may expose the soil, harvesting which leaves the soil bare, exposing it to the wind and rain.

(ii) Siltation down the river, increased erosion in the river, the loss of fertile topsoil.

(d) (i) Tillage breaks up the soil surface, loosening and exposing particles to the effects of wind and water erosion.

(ii) Ploughing directly down a slope has the same effects, but also creates 'gullies' to aid runoff. By ploughing along the slope, 'ridges' are made to slow down runoff.

CHAPTER 15

1. Less mining therefore less habitat and landscape destruction. Stimulation of recycling and general raising of awareness of its importance.
2. (a) (i) Generally, greater abudance equals greater consumption, but this relationship is not perfect. A decline in abundance is not directly represented by a decline in consumption.
   (ii) There are a wide number of applications in society. Copper is of high economic value.
   (iii) It allows for the representation of a large range of values on one scale.
   (b) (i) (6×10^8) (9×10^6) = 67:1 ratio
   (ii) World reserves of copper will be exhausted in 67 years at the current rate of consumption. (OWTTE).
   (c) (i) If energy becomes cheaper or if economic demand increases, the cut-off grade becomes lower. If energy becomes more expensive or if the demand decreases, the cut-off grade increases.
   (ii) A lower cut-off grade equals an increased use of energy, habitat destruction, dereliction of land, production mining waste. A higher cut-off point or a higher cost of virgin copper raw material equals an incentive for recycling or reuse (i.e. the reverse of the above).

CHAPTER 16

1. Extraction of raw materials: some components of overburden and spoil; mining equipment waste (metals and plastics).
   Generation of energy: nuclear waste (low, intermediate and high); coal ash; used pollution-control equipment (for example FGD); used generating equipment (metals and plastics).
2. 'What to do with waste' is an environmental problem due to the availability of disposal sites and the growing waste volume with an increasing impact. It is also an economic problem when we must compare costs of different disposal techniques.
   'Cost of waste disposal' is an economic problem due significantly to the need for disposal to be environmentally friendly.
   'Environmental impact of disposal' is an environmental problem of waste stream. It also has economic implications through the dereliciton of land and the cost of remedial treatment.
   'Hazardous and toxic materials' is an environmental problem due to a significant impact on ecosystems. It is an economic problem due to the higher costs associated with the disposal of such materials.
   'High levels of raw material and energy consumption' is an environmental problem and is an economic problem due to the cost of extraction, processing and generating.
3. Industrial activities: The sale of waste material to other industries (for example tree bark waste from saw mills sold as a fertiliser mulch).
   Agriculture: The recycling of waste materials (for example organic farm waste used as a fertiliser).
   Mining activities: The exploitation of the highest-grade ore therefore minimising the amount of spoil produced.
4. Increased population, greater use of packaging, substitution of materials such as wood and leather for non-biodegradable materials such as plastics, increased use of disposable and convenience goods, built-in obsolescence.
5. See Section 16.3.
6. Problems: A lack of space in the home to store materials and in urban areas to site recycling points, the transport of materials to the recycling centre, and the separation of different components of the waste stream for recycling.
   Widespread benefits: Conservation of virgin raw materials, a reduction in mining (and associated environmental impacts), a lower demand for energy in extraction/processing, a reduction in waste disposal requirements, and a decline in pollution from processing industries.
7. Manufacturing: Recycle and reuse, increase designed lifespan and ease of repair, substitute non-biodegradable for biodegradable materials, reduce the use of packaging.
   General public: Recycle and reuse, the composting of organic waste, the use of reusable containers, product selection for recyclable/biodegradable/long-life/repairable/low levels of packaging.

CHAPTER 17

1. There are push and pull factors. There are few employment opportunities in rural areas (push), whereas the city offers the promise of jobs, homes, better amenities and a higher standard of living (pull).
2. (a) Collection of water takes a long time, so activities such as washing may be reduced, increasing the chances of infection. Water may be stored around the home in unhygienic conditions, further increasing the chances that it will become infected.
   (b) Sewage may be discharged onto the street, where it will form a direct health hazard and may also contaminate water supplies.
   (c) Refuse may contain dangerous materials (for example glass), but it will also contain organic material which will attract rodents and insects, both of which may carry disease.
3. Carbon monoxide, nitrogen dioxide, hydrocarbons, ozone. Bright sunlight catalyses the reaction between nitrogen oxides and hydrocarbons, which produces low-level ozone. Temperature inversions – when cold air is trapped beneath a layer of warmer air – mean that upward dispersion of pollutants is prevented, thus low-level concentrations of pollutants remain high.
4. (a) Impermeable surfaces such as roads, paths, car parks, etc. mean that runoff is extremely high.
   (b) Deforestation results from the need for fuelwood and from urban growth. The loss of trees accelerates soil erosion, as root binding decreases and soil erodibility increases.
5. (a) Co-ordinated monitoring of seismic waves, reduction of maximum building height, incorporation of safety measures in building design.
   (b) Effective drainage networks, increased proportion of green areas to increase infiltration, reduction of gradient of impermeable surfaces.
   (c) Creation of satellite reservoirs to reduce dependence on groundwater, pumping water back into the aquifer to prevent further subsidence, demolition or underpinning of existing buildings.

CHAPTER 18

1. (a) Destruction of forests, poaching and hunting, intensive use of pesticides.
   (b) (i) People have a responsibility to conserve wildlife. We are the guardians of the environment. We have a responsibility to future generations to maintain species diversity.
   (ii) Livings things, animals and plants, provide a source of pleasure for humans.
   (iii) Wildlife conservation is important because animals and plants help to maintain stability of a complex system. For example, photosynthesis (plants) and respiration (plants and animals) help to regulate atmospheric compaction.
2. For both parts, credit is given for reference to the moral and aesthetic arguments.
   (a) Tigers are the top predators in ecosystems, therefore, through predation, they regulate the numbers of other animals, for example herbivores. If these herbivores were unchecked, they would put severe grazing pressure on vegetation.
   (b) Mosquitoes are near the base of the food web, and therefore they provide a food source for many organisms. In addition, by carrying diseases such as malaria, it can be said that they restrict numbers of other organisms.

Index

Page numbers for figures and tables are in *italics*.

A after page number = Analysis
CS after page number = Case Study
Q after page number = Question
SA after page number = Summary Assignment